Introduction to Surface Engineering and Functionally Engineered Materials

Scrivener Publishing
3 Winter Street, Suite 3
Salem, MA 01970

Scrivener Publishing Collections Editors

James E. R. Couper	Ken Dragoon
Richard Erdlac	Rafiq Islam
Pradip Khaladkar	Vitthal Kulkarni
Norman Lieberman	Peter Martin
W. Kent Muhlbauer	Andrew Y. C. Nee
S. A. Sherif	James G. Speight

Publishers at Scrivener
Martin Scrivener (martin@scrivenerpublishing.com)
Phillip Carmical (pcarmical@scrivenerpublishing.com)

Introduction to Surface Engineering and Functionally Engineered Materials

Peter M. Martin

Scrivener

Co-published by John Wiley & Sons, Inc. Hoboken, New Jersey, and Scrivener Publishing LLC, Salem, Massachusetts.
Published simultaneously in Canada.

For general information on our other products and services or for technical support, please contact our Customer Care Department within the United States at (800) 762-2974, outside the United States at (317) 572-3993 or fax (317) 572-4002.

Wiley also publishes its books in a variety of electronic formats. Some content that appears in print may not be available in electronic formats. For more information about Wiley products, visit our web site at www.wiley.com.

For more information about Scrivener products please visit www.scrivenerpublishing.com.

Cover design by Russell Richardson

Library of Congress Cataloging-in-Publication Data:

ISBN 978-0-470-63927-6

Printed in the United States of America

10 9 8 7 6 5 4 3 2 1

To Ludmila, Katherine and Erin

Contents

Preface

Surface engineering has become an indispensible technology for improving virtually all the properties of solid surfaces. Almost all types of materials, including metals, ceramics, polymers, and composites can be coated with thin films or surface structures of similar or dissimilar materials. It is also possible to form coatings of newer materials (e.g., met glass. beta-C_3N_4), graded deposits, nanocomposites, and multi-component deposits etc. Functional surface engineering has provided advancements such as extending the life of optoelectronic devices, cutting tools, engine parts, medical implants, hardware and plumbing fixtures; improved corrosion resistance of ferrous materials; wear resistant decorative coatings for jewelry and architectural glass; improved reflectivity and hardening of laser and telescope mirrors, improved efficiency and manufacturing of photovoltaic cells; new energy efficiency glazings (low-e windows); thin film batteries; self cleaning surfaces and much more.

Surface engineering can be traced back as far as the mid-1900's to first efforts to modify the properties of solid surfaces to reduce wear, reduce friction and improve appearance. In the second half of the last century, however, surface engineering involved primarily application of thin film materials and plasma treatment to modify and enhance surface properties such as wear resistance, lubricity and corrosion resistance. It has evolved to cover the full range of surface properties, such as optical, electrical, magnetic, electro-optical properties, permeation barriers, and functionally engineered materials. Each of these properties can be subdivided into dozens of subtopics. It has further evolved to encompass artificial structures and surfaces such as low dimensional structures (super-lattices, quantum wires and quantum dots), nanotubes, sculpted thin films, nanocomposites, energy band engineering, and even biological structures. A number of technical conferences are solely or partially dedicated to surface engineering and functional materials engineering. Excellent resources are the Society of Vacuum Coaters (SVC), AVS and Materials Research Society (MRS) and their associated publications.

The scope of this book is an introduction to a wide variety of aspects of surface engineering and functional materials engineering. There are a number of excellent books on the market that cover the topics such as hard coatings, deposition technologies for thin films, physical vapor deposition, nanocomposites, and low dimensional materials. No one book, however, incorporates all aspects of surface engineering and engineered materials. To this end, this book is intended to present a wide variety of surface engineering and functional materials engineering aspects in less detail than specialized handbooks, but as a standalone resource.

This book is intended to serve as an introduction to a multitude of surface engineering and functional materials engineering topics and should viewed as such. In most cases, only the basics are addressed. Because of this, thin films with more than three components have generally been omitted. Advanced engineered materials such as carbon and titania nanotubes, nanocomposites, metamaterials, sculpted thin film, photonic crystals and low dimensional structures show promise of enhanced structural, optical and electrical properties. Only basic mechanical, optical and electrical properties are presented. As much math as logically possible has been omitted without damaging basic concepts. It was unavoidable, however, to go into more detail with regard to thin film nucleation, energy band engineering and nanoelectronics.

Engineered materials are now being developed for and used in advanced photovoltaic devices, dye sensitized solar cells, quantum cascade lasers, advanced electronics, drug delivery, medical devices, metamaterials, optical photonic bandgap devices, negative refractive index devices, superlenses, artificial magnetism, cloaking devices, thermoelectric power generation and much more. Structures and properties not possible in naturally occurring materials are synthesized by a number of lithographic, etching, plasma and deposition processes.

The structure and properties of thin films are almost entirely dependent on deposition process. Many properties are directly related to the energy of atoms and molecules incident on the substrate surface. It is essential to understand how each deposition process synthesizes thin film structure and composition and the adatom energetic of each process. To this end, a significant amount of text, namely chapters 2 and 3, is dedicated to deposition processes and structure of thin films. In the ensuing chapters, it will become evident how the properties of each type of thin film (tribological, optical, electrical, etc.) depend on deposition conditions, structure and bonding. Tables are presented in each chapter that summarize these relationships. It was impossible to survey all literature on each deposition process and thin film material, and as a result, many tables are incomplete due to lack of available information.

The surveys of low dimensional structures, metamaterials and nanotubes may seem out of place but these technologies are being increasingly used to improve the tribological, optical, electrical and optoelectronic performance of thin film structures and surface devices. Sculpted thin films synthesized by glancing angle deposition (GLAD) are now used to achieve properties not possible with conventional thin film materials. Metamaterials are one of the most significant technical developments of this decade and are being developed with optical properties (e.g., negative refractive index) not possible with solid thin films. Applications include cloaking devices and superlenses than can resolve below the diffraction limit.

It is hoped that this book will give the readers enough background information to begin to solve critical surface engineering and materials problems, or provide enough information and resources to spring board them to generate new solutions and materials.

Peter M. Martin
June, 2011
Kennewick, WA

1

Properties of Solid Surfaces

1.1 Introduction

Wear and corrosion of structural materials are ubiquitous reliability and lifetime problems that have existed since the inception of mechanical devices and structures. Additionally, the optical, electrical, and electro-optical properties of solid surfaces were determined by crystalline, compositional, and electrical properties of the bulk solid. Until the advent of surface engineering, these properties belonged to the surface of the bulk materials being used and could be modified to only a limited degree by various metallurgical and plasma surface treatments. Surfaces of bulk materials could be hardened and wear corrosion resistance increased by a number of external treatments, including plasma bombardment, ion implantation, anodization, heat treatment, plasma nitriding, carburizing and boronizing, pack cementation, and ion implantation. They could also be polished or etched to modify optical properties and electrical properties to a limited degree.

Surface Engineering provides additional functionality to solid surfaces, involves structures and compositions not found naturally in solids, is used to modify the surface properties of solids, and involves application of thin film coatings, surface functionalization and activation, and plasma treatment. It can also be defined as the design and modification of the surface and substrate of an engineering material together as a system, to give cost effective performance of which neither is capable alone.

Surface engineering techniques are being used in the automotive, aerospace, missile, power, electronic, biomedical, textile, petroleum, petrochemical, chemical, steel, power, cement, machine tools, and construction industries. Surface engineering techniques can be used to develop a wide range of functional properties, including physical, chemical, electrical, electronic, magnetic, mechanical, wear-resistant and corrosion-resistant properties at the required substrate surfaces. Almost all types of materials, including metals, ceramics, polymers, and composites can be coated on similar or dissimilar materials. It is also possible to form coatings of newer materials (e.g., met glass. b-C_3N_4), graded deposits, multi-component deposits, etc.

In 1995, surface engineering was a £10 billion market in the United Kingdom. Coatings, to make surface life resistant to wear and corrosion, was approximately half the market.

In recent years, there has been a paradigm shift in surface engineering from age-old electroplating to processes such as vapor phase deposition, diffusion, thermal spray and welding using advanced heat sources like plasma, laser, ion, electron, microwave, solar beams, pulsed arc, pulsed combustion, spark, friction and induction. Biological materials for self-healing, self-cleaning and artificial photosynthesis are now becoming involved.

It is estimated that loss due to wear and corrosion in the U.S. is approximately $500 billion. In the U.S., there are around 9524 establishments (including automotive, aircraft, power, and construction industries) who depend on engineered surfaces with support from 23,466 industries.

There are around 65 academic institutions world-wide engaged in surface engineering research and education.

Surface engineering can be traced as far back as Thomas Edison in 1900 with the plating of gold films [1]. In 1938, Berghaus was among the first to develop plasma and ion modification of surfaces to improve surface properties and properties of vacuum deposited coatings [2]. The ion plating process, developed in the early 1960's, was a significant step forward in plasma-assisted coating deposition [3, 4, 5]. Ion plating was the first true industrial surface engineering process. Because conventional dc-diode sputtering used for ion plating did not provide sufficient levels of ionization to permit deposition of dense ceramic coatings with adequate mechanical properties, post deposition processes such as peening were often required to densify the coating. After the early 1970's, the history of surface engineering is intimately connected to the development of thin film deposition and plasma processes and closely parallels the history of physical vapor deposition (PVD) coatings and processes (magnetron sputtering, ion assisted deposition), plasma processing, and chemical vapor deposition (CVD) processes in particular.

The majority of surface engineering technology has focused on enhancement of tribological properties (hardness, wear resistance, friction, elastic moduli) and corrosion resistance. The purist might think that surface engineering encompasses only tribological and wear resistant treatments, as initiated by Ron Bunshah as far back as 1961 [6]. Many engineering components need wear or corrosion resistant surfaces as well as tough, impact-resistant substrates. These requirements can be best met by using treatments that alter surface properties without significantly modifying those of the core, or bulk, material. If these principles are applied correctly, surface engineering brings many benefits, including:

- Lower manufacturing costs
- Reduced life cycle costs
- Extended maintenance intervals
- Enhanced recyclability of materials
- Reduced environmental impact

There are, however, many more properties of a solid surface that can be enhanced by application of thin films, plasma treatment, patterning and nanoscale structures. This is reflected in the programs of a number of technical conferences dedicated solely to surface engineering (International Conference on Metallurgical Coatings and Thin Films, for example), starting as early as 1974 [6]. The first conferences focused on modification of the surface of a component to enhance its the overall performance. This area, however, has grown much broader than just this technology, as demonstrated by the symposia presented at the 2010 International Conference on Metallurgical Coatings and Thin Films (ICMCTF), sponsored by the Advanced Surface Engineering Division of AVS. Conference symposia include:

A.   Coatings for Use at High Temperature
B.   Hard Coatings and Vapor Deposition Technology
C.   Fundamentals and Technology of Multifunctional Thin Films:
D.   Carbon and Nitride Materials: Synthesis Structure-Property Relationships: Towards Optoelectronic Device Applications
E.   Tribology and Mechanical Behavior of Coatings and Thin Films
F.   Characterization: Linking Synthesis Properties and Microstructure
G.   Applications, Manufacturing, and Equipment
H.   New Horizons in Coatings and Thin Films
TS1.   Experimental and Computational Studies of Molecular Materials and Thin Films
TS2.   Coatings for Fuel Cells and Batteries
TS3.   Bioactive Coatings and Surface Biofunctionalization
TS4.   Surface Engineering for Thermal Transport, Storage, and Harvesting

Thus the optical, electrical, magnetic, thermal and even biological properties of a solid surface can also be modified using surface engineering techniques. Glass surfaces can be transformed into highly reflective or high selective reflector, transmitters or emitters. Thin film coatings are applied to glass to reflect heat, transmit heat and create heat. Decorative coatings can change color with viewing angle. The color of a thin film, whether in reflection or transmission, is critical in many applications, including low-e windows, antireflection coatings, hardware, plumbing fixtures, high reflector coatings, jewelry, automotive parts (including paints) and architectural glass. The "color" of a thin film results from its optical properties: transmittance, reflectance and absorption. Thin film coatings can be both wear resistant and colorful, and often must function in extreme environments (such as salt spray, missile domes, underground, windshields). Hard and durable gold, silver, and bronze colored coatings are applied to hardware, jewelry, plumbing fixtures and even auto bodies.

The options for surface engineering are limitless. For example, in addition to wear resistant coatings the following types of coatings are being developed:

- Decorative coatings
- Photocatalytic thin films, such as TiO_2, can transform a glass surface into a self cleaning surface
- Piezoelectric thin films can transform a glass or insulating surface into a high frequency transducer
- Polymer/dielectric multilayer films can decrease the water and gas permeation of a plastic surface by six orders of magnitude
- Transparent conductive oxides can make a glass or insulating surface almost as conductive as a metal while still preserving high optical transmission
- Semiconductor thin films can transform a glass, plastic or metal surface into a photovoltaic device
- Oxygen and water permeation barriers for sensitive electronics, plastics, and food packaging
- Organic thin films can transform a glass or plastic surface into a light emitting device

Often an application requires a thin film coating to be multifunctional. Common examples of this are

- Transition metal nitrides (TiN, ZrN, TaN, HfN) applied to hardware, jewelry, and plumbing fixtures for wear resistance and color
- Antireflection, heater coatings
- Wear resistant optical coatings
- Deicing optical filters
- Conductive, wear resistant coatings
- Decorative coatings

This book addresses the fundamentals of modifying and enhancing the tribological, optical, electrical, photo-electric, mechanical, and corrosion resistance of solid surfaces and adding functionality to solids by engineering their surface, structure, and electronic, magnetic, and optical structure. Note that *thin film*, *film*, and *coating* are used interchangeably in this book and all describe structures with thicknesses ranging from the nanometer (nm) scale to several microns (μm). While thin film applications will be emphasized, the increasing use of low dimensional structures and nanocomposites will integral to this discussion. Adhesion of thin films and elastic properties will not be directly addressed, although they cannot be omitted in some discussions. These topics have been addressed in detail in other books and the reader should refer to these books for in depth discussions [7, 8, 9, 10, 11]. It should be emphasized that, while this book will cover a broad range of surface engineering and engineered materials topics, it is strictly an introduction. Advanced topics and detailed analysis are beyond the scope of this book. Advanced topics, for example, are covered in the Third Edition of the *Handbook of Deposition Technologies for Films and Coatings* (P M Martin, Ed.) and *Handbook of Nanostructured Thin Films and Coatings* (Sam Zang, Ed.).

Virtually every advance in surface engineering and engineered materials has resulted from thin film and related technologies. Engineered materials are the future of thin film technology. Properties can now be engineered into thin films that achieve performance not possible a decade ago. Engineered structures such as superlattices, nanolaminates, nanotubes, nanocomposites, smart materials, photonic bandgap materials, metamaterials, molecularly doped polymers, and structured materials all have the capacity to expand and increase the functionality of thin films and coatings used in a variety of applications and provide new applications. New advanced deposition processes and hybrid processes are now being used and developed to deposit advanced thin film materials and structures not possible with conventional techniques a decade ago.

Engineered materials are now being developed for and used in advanced photovoltaic devices, dye sensitized solar cells, quantum cascade lasers, advanced electronics, drug delivery, medical devices, metamaterials, optical photonic bandgap devices, negative refractive index devices, superlenses, artificial magnetism, cloaking devices, thermoelectric power generation, functional biological materials, and much more. Structures and properties not possible in naturally occurring materials are synthesized by a number of lithographic, etching, plasma, and thin film deposition processes.

In order to fully understand how surface engineering technologies improve performance of solid surfaces, it will be useful to review the properties of solid surfaces that can be enhanced. We will address the following surface properties:

- Wear resistance
- Hardness
- Lubricity
- Corrosion and chemical resistance
- Optical (transmittance, reflectance, emittance)
- Electrical (conductivity)
- Electro-optical (photoconductivity, stimulated emission)
- Photocatalysis
- Surface energy (hygroscopicity and hydrofobicity)
- Temperature stability

1.2   Tribological Properties of Solid Surfaces

Wear and corrosion are ubiquitous problems that affect virtually every type of surface, especially those that come in contact with other surfaces. Tribological coatings are used to mitigate these problems. The word tribology is derived from the Greek word tribos, meaning "rubbing". Tribology is defined as the science and engineering of interacting surfaces in relative motion, including the study and application of the principles of friction, lubrication, and wear. These materials are concerned with friction, wear, corrosion, hardness, adhesion, lubrication, and related phenomena. They essentially combine the best of hard, wear resistant, and low surface energy coatings. Degradation of a material, tool, or components involves interactions occurring at their surfaces. Loss of material from a surface and subsequent damage may result from tribological interactions of the exposed face of the solid with interfacing materials and environment. Mass loss results from chemical driving forces and wear results from interfacial mechanical forces. Tribological coatings are applied to bearings, engine components, valves, computer disc heads and discs, biological joints, motor bushings, cutting tools, shaving heads, surgical tools, pump components, gears, molds, and the list goes on. Tribological thin film coatings are discussed in detail in chapter 4.

Tribological coatings are applied to surfaces to

- Increase wear resistance and hardness,
- Increase chemical and corrosion resistance
- Improve environmental stability
- Reduce coefficient of friction
- Change color of the surface
- Simulate metals such as gold, brass, copper, bronze surfaces

Three categories of tribological wear behavior can be defined:

- Both friction and wear are low (usually in bearings, gears, cams and slideways)
- Friction is high but wear is low (usually in power transmission devices such as clutches, belt drives, and tires)
- Friction is low and wear of one component is high (machine cutting, drilling, and grinding).

1.2.1   Wear

Before we can address surface engineering techniques used to improve tribological properties, it will be instructive to define and understand each property and associated test procedures in depth. Only then can we understand the significance of improvements in performance.

Wear is defined as the erosion of material from a solid surface by the action of another surface. It is related to surface interactions and, more specifically, the removal of material from a surface as a result of mechanical action. It is important to note that mechanical wear requires some form of contact resulting from relative motion, opposed to other processes that can produce similar results. This definition does not include dimensional changes due to mechanical deformation (no removal of material). Impact wear, where there is no sliding motion, cavitation, where the opposing material is a fluid, and corrosion, where the damage is due to chemical rather than mechanical action are also not included [12, 13].

Wear can also be defined as a process in which interaction of the surfaces or bounding faces of a solid with its working environment results in dimensional loss of the solid, with or without loss of material. Working environment here includes loads (such as unidirectional sliding, reciprocating, rolling, and impact loads - pressure), speed, temperature, type of opposing material (solid, liquid, or gas), and type of contact (single phase or multiphase, in which the phases involved can be liquid plus solid particles plus gas bubbles).

Seven mechanical wear mechanisms are listed in Table 1.1 [14]. These results, however, from only three types of surface to surface interactions: sliding (one surface sliding relative to another over long distances), fretting (one surface oscillates over minute distances relative to the other) and erosion (solid particles impinging on a single surface from an external source). The actual wear mechanism for dry sliding wear depends on a number of variables including, surface finish, surface geometry, orientation, sliding speed, relative hardness (of one surface relative to the other or relative to the abrasive particles between the surfaces), material microstructure, and

Table 1.1 Wear classification and mechanisms [3].

Classification	Wear Mechanism	Wear Coefficient K (range)
Wear dominated by mechanical behavior of materials	Asperity deformation and removal	10^{-4}
	Wear caused by plowing	10^{-4}
	Delamination wear	10^{-4}
	Adhesive wear	10^{-4}
	Abrasive wear	10^{-2}–10^{-1}
	Fretting wear	10^{-6}–10^{-4}
	Wear by solid particle impingement	
Wear dominated by chemical behavior of materials	Solution wear	
	Oxidation wear	
	Diffusion wear	
	Wear by melting of surface layer	
	Adhesive wear at high temperatures	

more. From these variables, it can be seen that wear rate is not a pure material property and does not always occur uniformly.

The tribological material or coating must address the issues connected with each mechanism. *Adhesive friction* encompasses *cohesive adhesive forces* and *adhesive wear*. Cohesive adhesive forces hold two surfaces together. The atoms and molecules of surfaces in contact actually never touch each other but are separated by atom-atom forces or cluster interactions, i.e., cohesive adhesive forces (thus, adhesion of two contacting surfaces arises from the attractive forces that exist between the surface atoms of the two materials). *Adhesive wear* occurs when surface features, such as microscopic roughness, are brought into contact under a load. It can be described as plastic deformation of very small fragments within the surface layer when two surfaces slide against each other: asperities (i.e., microscopic high points) found on the interacting surfaces will penetrate the opposing surface and develop a plastic zone around the penetrating asperity. Obviously, adhesive wear increases with increased surface roughness. During initial contact, fragments of one surface are pulled off and adhere to the other, due to the strong adhesive forces between atoms [12]. However, energy absorbed in plastic deformation and movement is the main cause for material transfer and wear. Stress builds up at the microscopic peaks on the surfaces and they deform and tend to weld together. These microscopic welds will eventually shear the crystallites from one surface and transfer them to the other surface, usually from the softer material to the harder material. If further rubbing occurs, particles of the softer material will form and smearing, galling, or seizure of the surfaces will occur. In the case of adhesive wear, the volume of the wear material is proportional to the distance over which the sliding occurs and to the applied load, and is inversely proportional to the hardness or yield stress of the softer material. This can be summarized by the Archard equation [15, 16]:

$$V = \frac{KFL}{3H},$$ (1.1)

where V is the volume of wear material, K is the dimensionless wear coefficient, F is the applied force (~pressure), L is the length of slide, and H is the yield stress of the soft material. Table 1.1 lists several wear mechanisms and associated coefficient K [14]. Note that K can be decreased by many orders of magnitude by lubricating the surfaces or applying a tribological coating.

It is generally not possible to deform a solid material using direct contact without applying a high pressure (or force), and at some time during this interaction, the process must accelerate and decelerate. It is necessary that high pressure be applied to all sides of the deformed material. Also, flowing material will immediately exhibit energy loss and reduced ability to flow if ejected from high pressure into low pressure zones. Thus, once the wear process is initiated, it has the capacity to continue because energy stored when the system is under high pressure will facilitate lower pressure needed to continue sliding.

Abrasive wear occurs when a hard rough surface slides across a softer surface [12]. The American Society for Testing and Materials (ASTM) defines it as the loss of material due to hard particles or hard protuberances that are forced against and move along a solid surface [17].

It is also classified according to the type of contact and the contact environment [18]. The type of contact determines the mode of abrasive wear. The two modes are defined as two-body and three-body abrasive wear. Two-body wear occurs when the grits, or hard particles, are rigidly mounted or adhere to a surface, when they remove the material from the surface. Sandpaper is the typical example of two-body wear. Obviously, the smoother the surfaces, the less abrasive wear will occur. In three-body wear, particles are not constrained, and are free to roll and slide down a surface. The contact environment determines whether the wear is classified as open or closed. An open contact environment occurs when the surfaces are sufficiently displaced to be independent of one another.

Three factors that influence abrasive wear and hence the manner of material removal are:

- Plowing
- Cutting
- Fragmentation

Plowing, shown in Figure 1.1, involves the formation of grooves that form when material is displaced to the side, away from the wear particles, and does not involve direct material removal. The displaced material forms ridges adjacent to grooves, which may be removed by subsequent passage of abrasive particles, i.e., smoothing. Cutting involves separation of material from a surface in the form of primary debris, or microchips, with little or no material displaced to the sides of the grooves, and closely resembles conventional machining. Fragmentation, shown in Figure 1.2, occurs when material is separated from a surface by a cutting process and the indenting abrasive causes localized fracture of the wear material. These cracks then freely propagate locally around the wear groove, resulting in additional material removal by spalling [18].

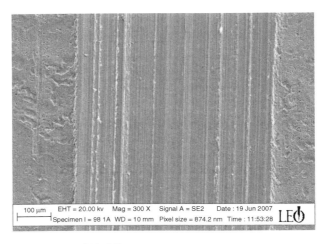

Figure 1.1 Plowing wear [19, 20].

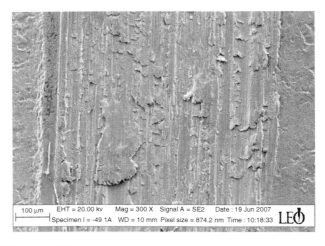

Figure 1.2 Fragmentation wear.

Wind and water erosion are familiar examples *erosive wear*. Sand and bead blasting are common industrial processes that involve this type of wear. Filtration and sealing are the two main methods used to minimize this type of wear. Erosive wear is shown in Figure 1.3. Also, as demonstrated by wind and sand erosion, gases can be a very powerful method of mass removal [21]. The rate of erosion is very dependent on incidence angle. Particles impinging on the surface of a brittle material remove material faster at higher angles of incidence, as shown in Figure 1.4. The inverse occurs for ductile materials. Components that are strongly affected by erosion are mining equipment, gas turbines, and electrical contacts in motors and generators.

Note that a Taber Abrasion Test is commonly used to assess abrasive wear. Fretting is the cyclical rubbing of two surfaces and *fretting wear*, shown in Figure 1.5, is the removal of material from one or both surfaces as a result of fretting. This type of wear can occur during the contact of round, cylindrical, or spherical surfaces, such as roller bearings. Fretting fatigue results from fretting wear. Cracks in one or both of the surfaces in contact form as a result of fretting fatigue. This process is accelerated if particulates are ejected by the contact. These particulates can oxidize and further abrade the surfaces. Fretting corrosion will occur if there are vibrations in the system. Pits or grooves and oxide debris can are typical of this type of damage, which is typically found in machinery, bolted assemblies, and ball or roller bearings. Contact surfaces

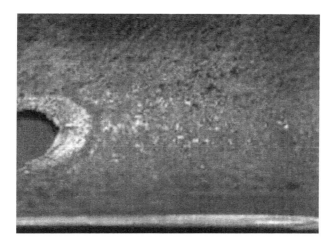

Figure 1.3 Erosive wear.

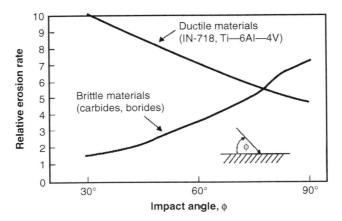

Figure 1.4 Relative erosion rates in brittle and ductile materials as a function of particle impact angle [22].

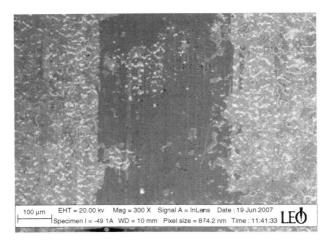

Figure 1.5 Fretting wear.

exposed to vibration during transportation are exposed to the risk of fretting corrosion.

1.2.2   Coefficient of Friction: Lubricity

The coefficient of friction (COF) between two surfaces is a major factor that affects wear rate. We are interested here in kinetic friction (as opposed to static friction), which generates a force directed opposite to the velocity (v) of one surface over the other. This force is defined as

$$F_k = -\mu_k N v$$

where μ_k is the coefficient of kinetic friction. Note that μ_k is independent of v (for small v) and that F_k is independent of contact area. COF is characteristic of the physical and chemical states of the two surfaces and is also affected by temperature, humidity, and environment. Kinetic friction involves dissipation of mechanical energy into heat and eventually wearing away of the softer surface. Frictional wear involves breaking of interfacial bonds on one part of the surface and creation of new bonds on another part of the surface. Obviously, if a bond cannot be broken, friction will be less. The work needed for bond destruction is supplied by the relative motion of the two surfaces. Creation of new bonds releases energy in the form of solid excitations (phonons = heat). The energy created into heat can be expressed as

$$E_h = \int \mu_k \, F_k dx$$

1.2.3   Hardness

Hardness is vaguely defined as "Resistance of metal to plastic deformation, usually by indentation" [23], that is, as the measure of how resistant a solid surface is to various kinds of permanent shape change when a force is applied. Hardness is related to the minimum stress needed to produce irreversible plastic deformation to the surface of a solid [24]. Hardness and wear resistance are intimately connected, although there is some evidence that elastic modulus also plays a role. Hardness, however, is not a basic property of a material, but rather a composite with contributions from the yield strength, work hardening, true tensile strength, elastic modulus, and others factors. In many cases, hardness is defined by the type of test used to measure it.

The greater the hardness of the surface, the greater resistance it has to deformation. We are interested primarily in micro and nano-hardness of surfaces, which is generally characterized by strong intermolecular bonds. Hardness is characterized in various forms, including scratch hardness, indentation hardness, and rebound hardness. The values of these types of hardness are reflected in specific types of tests. Hardness is dependent on ductility, elasticity, plasticity, strain, strength, toughness, viscoelasticity, and viscosity.

A number of tests have been developed to quantify hardness of a solid and compare the hardness of various solids:

- Mohs
- Brinell
- Rockwell hardness
- Rockwell superficial hardness
- Vickers
- Knoop
- Scleroscope and rebound hardness
- Durometer
- Barcol
- Microindenter
- Nanoindenter

Note that in most cases there is no one-to-one correspondence between the above tests, and hardness measurements must be compared using the same test methodology. Also note that virtually every hardness measurement made on thin films uses a micro or nanoindenter technique.

Scratch hardness defines the resistance of a surface to fracture or plastic (permanent) deformation due to friction from a sharp object. The scratching material is obviously harder than the surface under test. The most common scratch hardness test is Mohs scale.

Indentation hardness measures the resistance of a surface to permanent plastic deformation due to a constant compression load from a sharp object, usually a stylus or indenter tip. Hardness is deduced from critical dimensions of an indentation left by a specifically dimensioned and loaded indenter. Common indentation hardness scales are Rockwell, Vickers, and Brinnel. Most modern techniques give hardness in gigapascals (GPa).

Rebound hardness, or dynamic hardness, is related to elasticity and measures the height of the "bounce" of a diamond-tipped hammer dropped from a fixed height onto a material. Two scales that measure rebound hardness are the Leeb rebound hardness test and Bennett hardness scale.

Hardness and tensile strength (σ_{TS}) are proportional to each other, but ultimately hardness is related to bonding and microstructure of the solid [25].

The Mohs hardness test is typically used only to identify a mineral and is arguably one of the oldest tests [26]. This test involves observing whether a material's surface is scratched by a substance of known or defined hardness. The Mohs scale involves 10 minerals, shown in Figure 1.6, and is used to provide a numerical "hardness" value. The hardness of each of the 10 minerals is given an arbitrary value. Minerals are ranked along the Mohs scale, which is composed of 10 minerals that have been given arbitrary hardness values. This test, therefore, is not suitable for accurately gauging the hardness of industrial materials such as steel or ceramics.

The Brinell hardness test, developed in 1900, uses a desktop machine to apply a specified load to a hardened sphere of a specified diameter. The Brinell hardness number (Brinell), is obtained by calculating the ratio of the load used (in kg) and measured surface area of the surface indentation (mm²), in square millimeters. Figure 1.7 shows the test geometry. This test is often used to determine the hardness metal forgings and castings that have a large grain structures, and provides a measurement over a fairly large area that is less affected by the coarse grain structure of these materials than micro and nanoindentation tests. A typical test uses a 10 mm diameter steel ball as an indenter with a 3,000 kfg (29 kN) force. For softer materials, a smaller force is used; for harder materials, a TiC ball is substituted for the steel ball. The indentation is measured and hardness calculated as:

$$\mathrm{BHN} = \frac{2\mathrm{P}}{2\pi} \left[\mathrm{D} - (\mathrm{D}^2 - \mathrm{d}^2)^{\frac{1}{2}} \right] \qquad (1.2)$$

Where P = applied load (kg), D = diameter of indentor, d – diameter of indention. A number of standards have been developed for this test [27, 28].

All subsequent tests described here are variations of the indentation test. The Rockwell Hardness Test employs an instrument to apply a specific load and then measure the depth of the resulting indentation. The indenter is either a steel ball of a specified diameter or a spherical diamond-tipped cone of 120° angle and 0.2 mm tip radius, called a brale. A minor load of 10 kg is first applied, which causes a small initial penetration to seat the indenter and remove the effects of any surface irregularities. Next, the load measurement is zeroed and a major load is applied. Upon removal

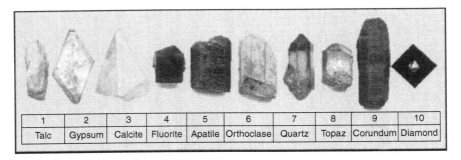

1	2	3	4	5	6	7	8	9	10
Talc	Gypsum	Calcite	Fluorite	Apatile	Orthoclase	Quartz	Topaz	Corundum	Diamond

Figure 1.6 Minerals used in Mohs hardness test.

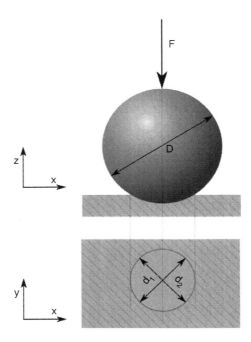

Figure 1.7 Brinell hardness test geometry.

of the major load, the depth reading is taken while the minor load is still on. The hardness number may then be read directly from the scale. The indenter and the test load used determine the hardness scale that is used (A, B, C, etc). In order to get a reliable reading the thickness of the test-piece should be at least 10 times the depth of the indentation and standards [29, 30, 31].

This test has the following scales:

A -Cemented carbides, thin steel and shallow case hardened steel

B -Copper alloys, soft steels, aluminum alloys, malleable iron, etc.

C -Steel, hard cast irons, pearlitic malleable iron, titanium, deep case hardened steel and other materials harder than B 100

D -Thin steel and medium case hardened steel and pearlitic malleable iron

E -Cast iron, aluminum and magnesium alloys, bearing metals

F -Annealed copper alloys, thin soft sheet metals

G -Phosphor bronze, beryllium copper, malleable irons

H -Aluminum, zinc, lead

K, L, M, P, R, S, V -Bearing metals and other very soft or thin materials,including plastics.

For example, 50 HRB indicates that the material has a hardness reading of 50 on the B scale.

The Rockwell Superficial Hardness Tester is used to test thin materials, lightly carburized steel surfaces, or parts that might bend or crush under the conditions of the regular test. All indenters and test geometry are the same except that loads are reduced. A lighter minor 3 kg load is used and the major load is either 15 or 45 kg, depending on the indenter used. Additionally, a 0.0625" diameter steel ball indenter is used. "T" is added (= thin sheet testing) to identify the superficial hardness designation. For example, a superficial Rockwell hardness of 23 HR15T, indicates the superficial hardness as 23, with a load of 15 kg using a steel ball.

Vickers and Knoop hardness are macrohardness and microhardness tests developed specifically to measure the hardness of thin films and surfaces [32, 33]. The indenter used in these tests is a small diamond pyramid, shown in Figure 1.8, which is pressed into the sample under loads significantly less than those used in the Brinell test. A typical indentation is also shown in the figure. Figures 1.9 and 1.10 show the geometry of these two tests. Basically, the only

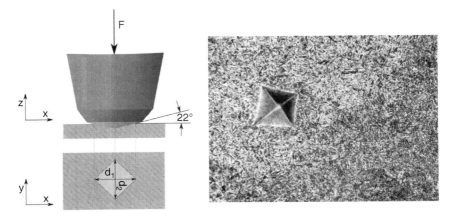

Figure 1.8 Indenter used in Vickers hardness tests and typical indentation.

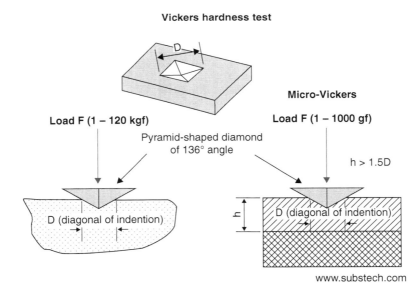

Figure 1.9 Geometry of Vickers hardness test.

difference between these tests is the shape of the diamond pyramid indenter. The Vickers test uses a square pyramidal indenter which is prone to crack brittle materials while the Knoop test uses a rhombic-based (diagonal ratio 7.114:1) pyramidal indenter. For

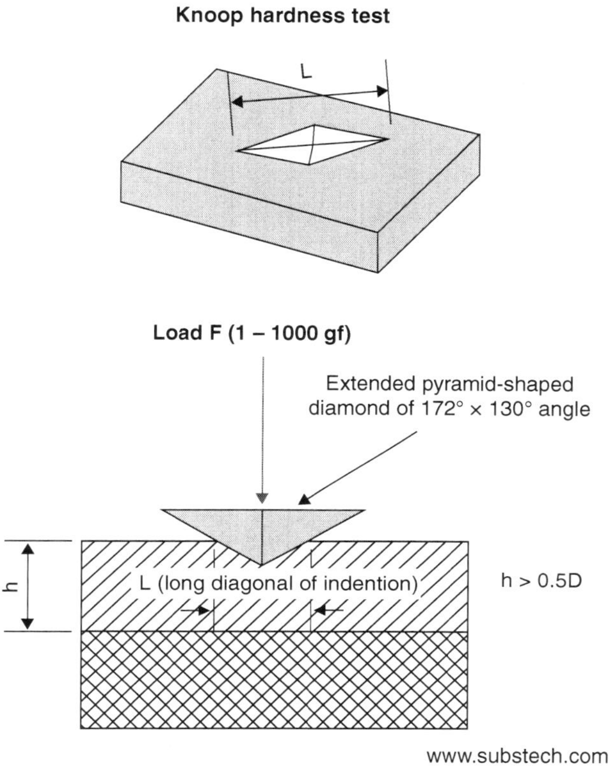

Knoop hardness test

Figure 1.10 Geometry of Knoop hardness test.

equal loads, Knoop indentations are about 2.8 times longer than Vickers indentations. The yield strength of a material can be approximated by

$$\sigma_y \sim \frac{H_V}{c} \tag{1.3}$$

c is constant ranging between 2 and 4.

Vickers hardness can be calculated using [34]

$$HV = \frac{1.8554F}{d^2}(kg/mm^2) \tag{1.4}$$

Here F is the indenter load and d is the width of the pyramidal indentation.

Knoop hardness can be found using the following [35]:

$$HK = \frac{F}{A} = \frac{F}{C_p L^2} \qquad (1.5)$$

With A = area of indentation, L = length of indentation and C_p = correction factor for shape of indenter (~0.070729).

The Scleroscope test involves dropping a diamond tipped hammer, which free falls inside a glass tube from a fixed height onto the test surface [36]. The height of the rebound travel of the hammer is measured on an arbitrarily chosen 100-unit graduated scale, called Shore units. One hundred Shore units represents the average rebound from pure hardened high-carbon steel. Because there are harder materials than the calibration material, the scale is continued higher than 100. The Shore Scleroscope measures hardness in terms of the elasticity of the material; the hardness number is related to the hammer's rebound height. The harder the material, the higher the rebound.

The Rebound Hardness Test Method employs the Shore Scleroscope for testing relatively large samples [37]. Typically, a spring is used to accelerate a spherical, tungsten carbide tipped mass towards the surface of the test surface, producing an indentation on the surface which takes some of this energy from the impact body. The harder the surface, the smaller the indentation: the indenter will lose more energy and its rebound velocity will be less when testing a softer material. The velocities of the impact body before and after impact are measured and the loss of velocity is related to Brinell, Rockwell, or other common hardness value.

Hardness testing of plastics, polymers, elastomers, and resins can be tricky. The Durometer and Barcol tests are used specifically for these elastic materials. A durometer uses a calibrated spring to apply a specific pressure to an indenter, which can be either cone or sphere shaped. Depth of indentation is measured. A durometer employs several scales, depending on the type of material:

- ASTM D2240 type A and type D scales (A - softer plastics, D - harder plastics)
- ASTM D2240-00 consists of 12 scales, depending on the intended use; types A, B, C, D, DO, E, M, O, OO, OOO, OOO-S, and R. Higher values correspond to a harder material.

The Barcol hardness test obtains a hardness value by measuring the penetration of a sharp steel point under a spring load. The sample is placed under the indenter of the Barcol hardness tester and a uniform pressure is applied until the dial indication reaches a maximum. The governing standard for the Barcol hardness test is ASTM D 2583.

Nanoindentation testing is used to measure hardness and elastic constants of a small area and volumes and thin films [24, 38]. This measurement generally employs an atomic force microscope (AFM) or a scanning electron microscope (SEM) to image the indentation. This test improves on macro and micro indentation tests discussed above by indenting on the nanoscale with a very precise tip shape, high spatial resolutions to place the indents, and by providing real-time load-displacement (into the surface) data while the indentation is in progress.

Small loads and tip sizes are used in this test, and as a result, the indentation area may only be a few μm^2 or nm^2. The problem here is that the indentation is not easy to find. A typical indentation is shown in Figure 1.11. Instead of trying to image the indentation, an indenter with a geometry known to high precision (usually

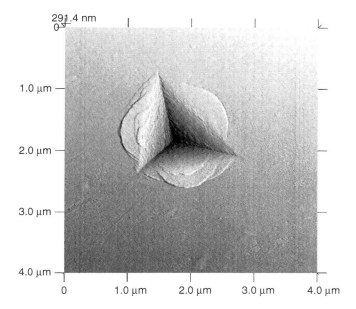

Figure 1.11 Typical nanoindentation.

a Berkovich tip, which has a three-sided pyramid geometry) is employed. Instead, the depth of penetration is recorded, and then the area of the indent is determined using the known geometry of the indentation tip. The response of the film to indentation, shown in Figure 1.12, is obtained for loading and unloading of the tip; load and depth of penetration are measured. Figure 1.12 shows a load-displacement curve, which is used to extract mechanical properties of the material [38].

This test can determine two types of hardness:

- Single hardness value
- Hardness as a function of depth in the material

Again, the above relation for hardness ($H = P_{max}/A_r$) is also valid for micro and nanoindenters. As mentioned above, some nanoindenters use an area function based on the geometry of the tip, compensating for elastic load during the test. Use of this area function provides a method of gaining real-time nanohardness values from a load-displacement graph. Figure 1.13 demonstrates this type of measurement, plotting hardness of a nanolaminate against load [39]. Software has been specifically designed to incorporate all test parameters and analyze the displacement vs. load curve [40, 41]. Nanoindenters are often incorporated into AFMs.

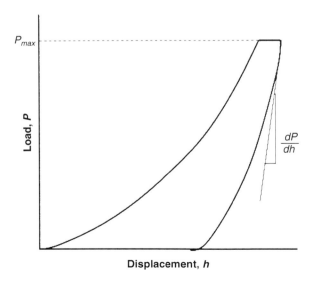

Figure 1.12 Load displacement curve for nanoindentor [38].

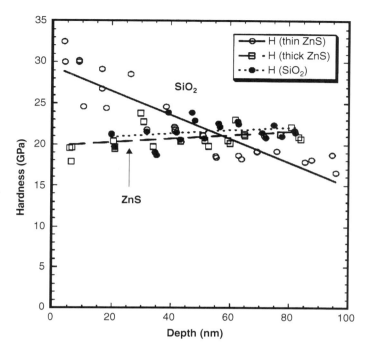

Figure 1.13 Nanohardness of a nanolaminate plotted against load for ZnS and SiO$_2$ substrates [39].

A number of tests have been developed to measure hardness of solid surfaces and thin films. Indentation tests work best for thin films. Most modern methods use nanoindentation coupled with some form of computer and AFM analysis. Because of the types of indenter used, it is difficult to cross-compare hardness values obtained by the various indentation techniques (Vickers, Knoop, nanoindentation). Most modern tests give hardness values in GPa, while a few decades ago, this unit was not used. It is, therefore, judicious to compare hardness values obtained by the same test method.

1.3   Optical Properties of Solid Surfaces

Unlike many other properties, modification of the optical properties of a surface can totally change the optical properties of the substrate. However, in addition to modifying the optical properties, in many cases they serve two or more surface engineering functions, such as improving abrasion and wear resistance, electromagnetic shielding, and as gas, chemical and water permeation barriers. Surface engineering of optical properties is presented in more detail in Chapter 5. Optical properties of a surface can be modified using:

- Polishing
- Thin films
- Patterned structures
- Etching
- Low dimensional structures
- Photonic crystals
- GLAD coatings

Electromagnetic radiation (which includes light) when incident on a solid surface can be transmitted, reflected, absorbed, or scattered. Light can also be diffracted and refracted. The film/substrate interface is important for many optical coatings since it contributes to the overall optical properties of the entire optical system and will be addressed in Chapter 5. The optical properties of a surface depend on a number of factors:

- Surface quality (roughness, pits, digs, scratches)
- Optical constants (refractive index and extinction coefficient)
- Absorptance and emittance

Transmittance (T), reflectance (R), and absorptance (A) are directly related to the optical constants of the material's real and imaginary parts of the refractive index (n & κ) . Transmittance, reflectance, and absorptance (see Figure 1.14) at an air interface (n_{air} = 1) are related to optical constants by

$$n = n + i\kappa \text{ (complex refractive index)} \qquad (1.6)$$

$$T = 1 - R - A \qquad (1.7)$$

$$R = \frac{[(n-1)^2 + \kappa^2]}{[(n+1)^2 + \kappa^2]} \text{ (normal incidence)} \qquad (1.8)$$

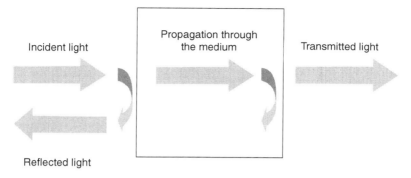

Figure 1.14 Relationship between T, R, and A.

$$A = -\log_{10} T \tag{1.9}$$

$$\alpha = \frac{4\pi\kappa}{\lambda} \text{ (absorption coefficient)} \tag{1.10}$$

Here λ is the wavelength of light.

All surfaces scatter incident electromagnetic radiation at some level. Scattering causes attenuation of a light beam similar to absorption. The intensity of light due to scattering cross section σ_S is

$$I(z) = I_0 \exp(-\sigma_S N_z) \tag{1.11}$$

where N is the number of scattering centers and z is the propagation direction through the solid. Note that scattering only gets worse at shorter wavelengths

$$\sigma_S \sim \frac{1}{\lambda^4} \tag{1.12}$$

Except for reducing scattering (by polishing), there is very little that can be done to modify the optical properties of a bulk surface without some external process (such as application of thin films, smoothing layers, etching, and patterning). Because the optical constants are ultimately properties of the lattice and bonding in materials, stress and strain can also affect their behavior.

Methods to modify and enhance the optical properties of solid surfaces will be discussed in more detail in Chapter 5. Treatments listed above can

- Change a simple transparent ceramic window into a reflector, EMI shield, heat mirror, cold mirror, emissive display, laser shield, or switchable optical device
- Increase transmission of ceramic, semiconductor and semi-opaque substrates
- Modify the optical band structure of a surface
- Increase or decrease optical scattering of a surface
- Enable read-write capability on a substrate

Color is another important optical property of the solid surface and can be modified over wide ranges by surface engineering techniques. Color is critical in decorative coating applications. One of the primary applications of thin films is to change the color or reduce the color of a surface, window, optic, or object. Color is an intrinsic property related to transmission, reflectance, and emissivity of a surface. Color is defined as the visual perceptual property corresponding in humans to the categories called red, green, blue, and others. We all "see" the color of an object differently. Color results from the visible light spectrum interacting in the eye with the spectral sensitivities of the light receptors located in the retina. Color categories and physical specifications of color associated with objects, materials, light sources, etc., are based on optical properties such as visible light absorption, transmission, reflection, or emission spectra. Several attempts have been made to quantify color by identifying numerically their coordinates in a color space model [40, 41, 42, 43, 44, 45, 46, 47, 48, 49, 50].

A number of models have been presented to quantify color perception, which will be discussed in Chapter 5. These models relate color coordinates to the actual color experience by the human eye. Figure 1.15 summarizes color vision characteristics and associated models [46, 50]. The figure diagrams color perception factors and how they are mapped into a chromaticity model, discussed in Chapter 5. The entire picture ends up in the bottom chromaticity diagram and includes:

- Vision characteristics
- Color perception

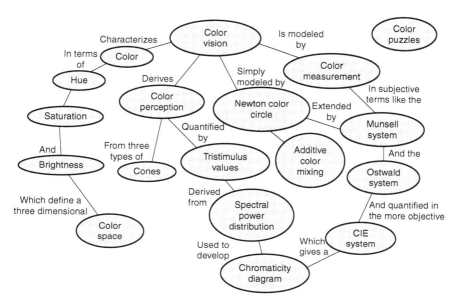

Figure 1.15 Mapping of human color perception [46].

- Color mixing
- Color measurement
- Spectral power distribution
- Chromaticity model

Chapter 5 summarizes several chromaticity models and relates these models to colors and change in color of reflective and transmissive surfaces.

1.4   Electric and Opto-electronic Properties of Solid Surfaces

The electrical and opto-electronic properties of a solid are determined by the transport and excitation of electrons and holes in the solid and can be attributed to

- Energy band structure
- Lattice structure:
    ○ Short range order or lack of short range order
    ○ Crystalline phase composition

- Wide variations in density
- Grain boundary effects
- Surface and interface effects
- Quantum effects

Many of these factors are interdependent, e.g., energy band structure is derived from composition, lattice structure, and defects. In this introduction, we will address classical and quantum mechanical models for electrical conduction in solids and relate these to electrical conduction in thin films and low dimensional structures (superlattices, nanowires, quantum dots, nanotubes). We briefly review the classical theory of electrical conduction. The basic relationship between current density $\mathbf{J}(\mathbf{r})$ and electric field $\mathbf{E}(\mathbf{r})$ is given by [25]

$$\mathbf{J}(\mathbf{r}) = \sigma\mathbf{E}(\mathbf{r}) \qquad (1.13)$$

Here $\mathbf{J}$, $\mathbf{E}$ and $\mathbf{r}$ are vectors and σ is the electrical conductivity tensor. If the medium is isotropic, σ is a scalar quantity and $\mathbf{J}$ and $\mathbf{E}$ are parallel. Drude theory defines conductivity in terms of an electron's charge (e), free electron density (n), mass (m) and relaxation time between collisions (τ):

$$\sigma = ne^2\frac{\tau}{m}\text{ or in terms of the mobility }(\mu) \qquad (1.14)$$

$$\sigma = ne\mu \qquad (1.15)$$

Table 1.2 displays conductivity for some common metals. It will be useful to refer back to this table when addressing conductivity of thin films. *We will also see that all the quantities shown above become dependent on energy band structure in the quantum world and are functions of the shrinking dimensions of thin films and low dimensional structures and defect density; i.e., properties of smaller structures are more severely affected by factors that would not significantly affect bulk materials.*

Without electron (or other charge carrier) collisions σ and τ would be infinite and the material would be a perfect conductor. We will see in Chapter 6 that collisions also have dramatic effects on the conductivity of thin films and low dimensional structures.

Table 1.2 Conductivity parameters for some common metals [25].

Metal	Conductivity (10^5S/cm)
Ag	6.28
Al	3.65
Au	4.92
Cu	5.96
Ni	1.43
Fe	1.0
Ti	2.38
C	1.43
In	1.14
K	1.39
Li	1.07
Mg	2.23

Optoelectronic properties result from the interaction of photons with electrons in solid surfaces, as is the basis for photovoltaics, lasers, photoconductivity, photon detectors, and photodiodes. Solids can absorb and emit photons, both of which involve movement of an electron between atomic or molecular energy levels. An electron is excited into a higher energy level when absorbed by atom, and relaxes to a lower energy level as a result of photon emission. The band structure of solids will be discussed in more detail in Chapter 7. Figure 1.16 shows energy transitions of an electron when a photon is either absorbed or emitted [51]. When a photon is absorbed, the electron is excited to a higher energy ($E_1 \rightarrow E_2$), essentially absorbing the photon's energy (= Plank's constant X frequency = hν). A photon is emitted when the electron decays from $E_2 \rightarrow E_1$. As shown in Figure 1.17, assuming the energy of the photon is greater than the band gap (E_g) of a semiconductor, electrons can also be excited from the valence band (E_V) into the conduction band (E_C) to create an electric current, i.e., photoconductivity.

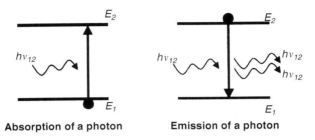

Absorption of a photon   **Emission of a photon**

Figure 1.16 Electronic transitions due to the absorption or emission of a photon [51].

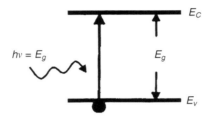

Figure 1.17 Electron excitation from the valence band into the conduction band of a semiconductor [51].

Light emitting diodes (LED) and organic light emitting diodes (OLED) function on the emission of a photon when electron and electron-hole pairs are excited in semiconductors and conductive polymers. In this case, charge carriers are excited to higher energy levels by application of a voltage (energy) and decay by giving off photons [51, 52]. Electrons can be excited to higher energy levels thermally. Thermal energy can excite electrons into higher energy levels and photons emitted when they decay to lower energy levels.

The interaction of photons with electrons is also responsible for optical properties such as the plasma frequency, surface plasmons, evanescent waves and, of course, photosynthesis. The plasma frequency is particularly important in metals and transparent conductive oxides in that it determines the frequency dependent electrical conductivity, and at the wavelength of light that high reflectivity initiates. The reflectivity of metals is totally dependent on the plasma frequency [53].

The frequency dependent conductivity can be expressed as [54]

$$\sigma(\omega) = \left(ne^2 \frac{\tau}{m} \right) \left(\frac{1 + i\omega\tau}{(1 + \omega^2\tau^2)} \right) \qquad (1.16)$$

where ω is the frequency of the electromagnetic wave (light in this case), τ is the relaxation time between electron collisions, and m is the mass of the electron. Thus, as ω decreases, $Re[\sigma(\omega)]$ increases and we get essentially the dc conductivity $ne^2\tau/m$. If the frequency dependent permittivity is

$$\varepsilon(\omega) = \varepsilon_0 + \frac{i\sigma(\omega)}{\sigma(\omega)} \qquad (1.17)$$

The frequency dependent dielectric function is given by

$$\varepsilon(\omega) = \frac{1 - \omega_p^2}{\omega\left(\omega + \dfrac{i}{\tau}\right)} \text{ with the } \textit{plasma frequency } \omega_p^2 = \frac{ne^2}{m\varepsilon_0}.$$

$$(1.18)$$

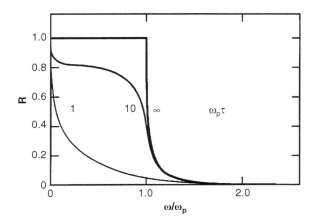

Figure 1.18 Dependence of the electrical conductivity of a metal with frequency, showing the plasma frequency.

The plasma frequency is very significant in that it is the natural frequency of the electron gas in the conduction band. The plasma frequency in metals generally occurs at ultraviolet and short visible wavelengths, and separates the frequency domains that are either absorbing ($\omega \sim \omega_p$), reflecting ($\omega < \omega_p$), or transparent ($\omega >> \omega_p$). Optical reflectivity of metals is thus derived from this relation. Figure 1.18 shows the behavior of the electrical conductivity with frequency. The imaginary part peaks at the plasma frequency.

We can thus relate electrical conductivity and other physical properties to energy transitions of electrons in a solid due to a number of energy sources (light, heat, electrical). This is complicated in thin films by the fact that defects, lack of long range order, and compositional variations can affect and degrade their band structure [54]. These factors can totally smear out band structure in thin films, not to mention size effects. Quantum theory also predicts magnetic properties, magnetoresistance, thermal conductivity, thermoelectric effects, and optical phenomena.

Thin films are also applied to non-piezoelectric surfaces to provide a number of piezoelectric and ferroelectric properties. Materials such as AlN, ZnO, Si_3N_4 $BaTiO_3$, $PbTiO_3$, $LiNbO_3$ and $KNbTiO_3$ possess excellent piezoelectric performance, if the microstructure is correct [55]. Such materials can be used for high frequency transducers, sensors, actuators and motors.

1.5 Corrosion of Solid Surfaces

In addition to wear, corrosion causes significant damage and economic losses and reliability problems. Corrosion is the degradation and removal of an engineered material's surface into its constituent atoms due to chemical reactions (oxidation, sulfidation, chlorides/salts) with its environment [10]. It is essentially the electrochemical oxidation of metals in reaction with an oxidant such as oxygen or sulfur. The severity of corrosion is determined by chemical kinetics, which can have a strong dependence on temperature. Formation of Fe_2O_3, or rust, on an iron surface is due to oxidation of the iron atoms in solid solution and is a common example of electrochemical corrosion. Corrosion can also occur in other materials than metals, such as ceramics and polymers. Corrosion can also be self-limiting when a protective scale forms on the surface of the metal.

Corrosion can be classified as dry, wet, or stress enhanced corrosion [10]. Dry corrosion is a chemical process that involves interaction of a gaseous environment with a solid surface, for example, oxidation in air. Gaseous sulfides and halides can also be strongly corrosive to metal surfaces. Wet corrosion involves exposing the surface to liquids, which may or may not function as electrolytes (wet galvanic or nongalvanic). Ionic charge transport in an electrochemical redox process occurs in galvanic corrosion, which involves an aqueous medium with dissolved salts. If a deposit does not form, corrosion can proceed until the entire object or component is consumed. In wet nongalvanic corrosion, the surface is dissolved by the aqueous medium in a non-redox reaction. Again, it is possible for the entire solid structure to be dissolved.

Stress enhanced corrosion is characterized by corrosion concentrated locally to form a pit or crack, or it can extend across a wide area more or less uniformly corroding the surface. Because corrosion is a diffusion controlled process, it occurs on exposed surfaces. Passivation and chromate conversion can increase the corrosion resistance of a surface. Applied stress or fatigue can enhance localized corrosion, even in otherwise benign conditions. Stress-corrosion cracking occurs when a surface is simultaneously exposed to a corrosive environment and applied stress. Intergranular regions and intragranular regions are extremely susceptible to this type of corrosion. Corrosion fatigue occurs when loads and stresses are variable.

A number of surface treatments are possible to mitigate the above types of corrosion, including:

- Corrosion resistant thin film coatings
- Reactive coatings
- Galvanization
- Anodizing
- Cathodic protection

Corrosion resistant thin films have the following general characteristics:

- Inert to the environment
- Wide range of thicknesses
- High density
- Continuous over the entire area (prevent permeation)

- Low permeability
- Very low porosity
- Low stress

This completes our brief overview of many of the properties that a solid surface can display. All these properties can be significantly modified by application of thin films. low dimensional structures, and other surface engineering techniques.

Each chapter of this book is dedicated to a specific area of surface engineering or materials engineering. Thin film deposition processes are described in chapter 2. Structure and microstructure of the thin film depend almost entirely on the deposition process, and are addressed in chapter 3. chapters 4 – 6 address tribological, optical, electric, and opto-electronic aspects of surface engineering. Films and coatings are often multifunctional, and to this end, chapter 7 is an introduction to how materials and physical properties are engineered onto surfaces and reflect the most up-to-date aspects of surface engineering science. Hybrid surface engineering applications are described in chapter 8. We look ahead to the future in chapter 9 with emphasis on functional biological materials.

References

1. R K Waits, *J Vac Sci Technol* A 19 (2001)1666.
2. B Berghaus, UK Patent No. 510,993 (1938).
3. D M Mattox, *Electrochem Technol* 2 (1964) 95.
4. D M Mattox, US Patent No. 3,329,601 (1967).
5. D. M. Mattox, *J Vac Sci Technol* 10 (1973) 47.
6. William D Sproul, *J Vac Sci Technol* 21(5) (2003) S222.
7. P M Martin, Ed., *Handbook of Deposition Technologies for Films and Coatings*, 3rd Ed., Elsevier (2009).
8. M Ohring, *Materials Science of Thin Films*, Elsevier (2002).
9. Donald Mattox, *Handbook of Physical Vapor Deposition* (PVD), 2nd Ed., Elsevier (2010).
10. Rointan Bunshah, Ed., *Handbook of Hard Coatings*, William Andrew (2001).
11. Sam Zhang, Ed., *Handbook of Nanostructured Thin Films and Coatings*, CRC Press (2010).
12. E Rabinowicz, *Friction and Wear of Materials*, John Wiley and Sons (1995).

13. J A Williams, *Tribology International* 38(10) (2005) 863.
14. John M Thompson and Mary Kathryn Thompson, A Proposal for the Calculation of Wear, ansys Paper -2 (2006).
15. J F Archard, *J. Appl. Phys*, 24 (1953) 981.
16. J F Archard and W Hirst, *Proc, Royal Soc.* A-236 (1958-06-23) 71.
17. Standard Terminology Relating to Wear and Erosion, *Annual Book of Standards*, Vol 03.02, ASTM (1987) 243.
18. ASM Handbook Committee *ASM Handbook. Friction, Lubrication and Wear Technology*. U.S.A., ASM International, Vol 18 (2002).
19. D Sarker *et al., J. Europ Cer Soc*, 26 (2006) 2441.
20. I Serre *et al., Appl Mech & Matls*, 13 – 14 (2008) 163.
21. P M Martin, Ed., *Handbook of Deposition Technologies for Films and Coatings*, 3rd Ed., Elsevier (2009).
22. P M Martin, *Process Technology Tutorials, Vacuum Technology & Coating 2007–2009.*
23. *Metals Handbook*, 2nd Ed., J R Davis, Ed, ASM International (1998).
24. M Ohring, *Materials Science of Thin Films*, Elsevier (2002).
25. Joel I Gersten and Frederick W Smith, *The Physics and Chemistry of Materials*, Wiley.
26. American Federation of Mineralogical Societies. "Mohs Scale of Mineral Hardness" (2001).
27. International (ISO) and European (CEN) Standard
    a. EN ISO 6506-1:2005: Metallic materials - Brinell hardness test - Part 1: test method.
    b. EN ISO 6506-2:2005: Metallic materials - Brinell hardness test - Part 2: verification and calibration of testing machine.
    c. EN ISO 6506-3:2005: Metallic materials - Brinell hardness test - Part 3: calibration of reference blocks.
    d. EN ISO 6506-4:2005: Metallic materials - Brinell hardness test - Part 4: Table of hardness values.
28. US standard(ASTM International).
    a. ASTM E10-08: Standard method for Brinell hardness of metallic materials.
29. International (ISO) ISO 6508-1: Metallic materials -- Rockwell hardness test -- Part 1: Test method (scales A, B, C, D, E, F, G, H, K, N, T).
30. ISO 2039-2: Plastics -- Determination of hardness -- Part 2: Rockwell hardness.
31. US standard (ASTM International) ASTM E18 : Standard methods for Rockwell hardness and Rockwell superficial hardness of metallic materials.
32. R L Smith & G E Sandland, *Proc Inst of Mech Eng*, I (1922) 623.
33. F Knoop *et al., Journal of Research of the National Bureau of Standards*, V23 #1, July 1939, Research Paper RP1220, 39.
34. ISO 6507-1:2005.

35. ASTM D1474.
36. ASTM A956 "Standard Test Method for Leeb Hardness Testing of Steel Products".
37. R T Mennicke, ICASI 2008 & CCATM 2008 Congress Proceedings (2008).
38. W C Oliver and G M Pharr, *J Mater Res*, 19 (2004) 3 (review article).
39. P M Martin *et al.*, *Proceedings of the 50th SVC Annual Technical Conference* (2007) 653.
40. David Shuman, *Microscopy-Analysis* (May 2005) 21.
41. A C Fischer-Cripps, *Nanoindentation*, Springer (2004).
42. Richard Sewall Hunter, *JOSA*, 38 (7) (1948) 661.
43. Richard Sewall Hunter, *JOSA*, 38 (12) (1948) 1094.
44. CIE Commission internationale de l'Eclairage proceedings, 1931. Cambridge University Press (1932).
45. Thomas Smith and John Guild, *Transactions of the Optical Society*, 33 (3) (1931-32). 73.
46. H Albert Munsell, *The American Journal of Psychology*, 23 (2), University of Illinois Press (1912) 236.
47. W Ostwald, *Die Farbenfibel*, Leipzig (1916).
48. W Ostwald, *Der Farbatlas*, Leipzig (1917).
49. F Birren, *The Principles of Color*, New York (1969).
50. See HyperPhysics: Color Vision.
51. Manijeh Razeghi, *Fundamentals of Solid State Engineering*, Kluwer Academic Publishers (2002).
52. See Hyperphysics.com.
53. S M Sze, *Physics of Semiconductors*, 2nd Ed., Wiley Interscience (1981).
54. Charles Kittel, *Introduction to Solid State Physics* (Eighth Ed.), Wiley (2005).
55. P M Martin *et al.*, *Thin Solid Films*, 379 (2000) 253–258.

2

Thin Film Deposition Processes

Surface engineering (SE) requires that thin film and low dimensional structures be deposited on solid surfaces. The electrical, optical, mechanical, and tribological properties as well as the structure and microstructure of thin film materials can vary over wide ranges, and are highly dependent on the deposition process used. To fully apply and utilize SE technology, one must be familiar with thin film deposition technology and materials. There are a number of excellent books that describe the various deposition processes [1, 2, 3, 4] and thousands of publications that describe the properties of thin films. Each process has its strengths and weaknesses and deposition conditions, and their effect on film properties, are quite different for all processes involved. One distinct advantage of thin film processes, because they are nonequilibrium in nature, is that they are able to synthesize compositions not possible with bulk processes. While electrochemical deposition (plating) processes were first used to deposit protective films such as chromium (Cr), and are still used to some extent, physical vapor deposition processes (PVD) have taken over the lion's share of deposition of surface engineering materials. Chemical vapor deposition (CVD), plasma enhanced chemical vapor deposition (PECVD), filtered cathodic arc deposition, hollow cathode deposition (FCAD), and atmospheric pressure plasma processes are also used extensively. Anodizing and plasma treatments such as nitriding are still used extensively in some applications. A brief description of these processes will be given here. A more detailed treatment of deposition technologies for films and coatings can be found in the *Handbook of Deposition Technologies for Films and Coatings*, 3rd Ed, Peter M Martin, Ed., Elsevier (2009).

The following deposition processes will be briefly described:

1. Thermal evaporation (TE) - including ion assist
2. Electron beam evaporation (including ion assist)
3. Ion plating
4. Sputtering
   a. Planar magnetron
   b. Cylindrical magnetron
   c. High power impulse magnetron sputtering (HIPIMS)
   d. Unbalanced magnetron
   e. Closed field
   f. Ion beam
5. Chemical vapor deposition (CVD)
6. Plasma enhanced chemical vapor deposition (PECVD)
7. Atomic layer deposition (ALD)
8. Filtered cathodic arc deposition (FCAD)
9. Hollow cathode deposition
10. Vacuum polymer deposition
11. Plasma processes (anodization, nitriding, etc.)

For simplicity, we will address three broad categories: physical vapor deposition (PVD), chemical vapor deposition (CVD), and plasma processes. Hybrid deposition processes include combinations of these technologies.

2.1   Physical Vapor Deposition

Physical vapor deposition processes encompass a wide range of vapor phase technologies, and is a general term used to describe any of a variety of methods to deposit thin solid films by the condensation of a vaporized form of the solid material onto various surfaces. PVD involves physical ejection of material as atoms or molecules and condensation and nucleation of these atoms onto a substrate. The vapor phase material can consist of ions or plasma and is often chemically reacted with gases introduced into the vapor, called reactive deposition, to form new compounds. PVD processes include

- Thermal evaporation
- Electron beam (e-beam) evaporation and reactive electron beam evaporation
- Sputtering (planar magnetron, cylindrical magnetron, dual magnetron, high power pulsed magnetron, unbalanced magnetron, closed field magnetron, ion beam sputtering, diode, triode) and reactive sputtering
- Filtered and unfiltered cathodic arc deposition (nonreactive and reactive)
- Ion plating
- Pulsed laser deposition

Variants on these processes are

- Bias sputtering
- Ion assisted deposition
- Glancing angle deposition (GLAD)
- Hybrid processes

Hybrid processes combine the best attributes of each PVD and/or CVD process. Among the combinations are

- Magnetron sputtering and e-beam evaporation
- Magnetron sputtering and filtered cathodic arc deposition
- E-beam evaporation and filtered cathodic arc deposition
- Vacuum polymer deposition, polymer flash evaporation + magnetron sputtering/evaporation

The basic PVD processes are evaporation, sputtering and ion plating. Materials are physically created in the vapor phase by energetic bombardment of a source (for example, sputtering target), and subsequent ejection of material. A number of specialized PVD processes have been derived from these processes and extensively used, including reactive ion plating, reactive sputtering, unbalanced magnetron sputtering, high power pulsed magnetron sputtering, filtered cathodic arc deposition, etc. There is also the possibility of confusion since many of these processes can be covered by more than one name. For example, if the *activated reactive evaporation* (ARE) process is used with a negative bias on the substrate, it is very often called *reactive ion plating*. Simple evaporation using an RF heated crucible is often called *gasless ion plating*. There is even more confusion over the ion plating process where the material is converted from a solid phase to the vapor phase using a number of processes involving thermal energy (evaporation), momentum transfer (sputtering), electrical energy (cathodic arc), or supplied as vapor (similar to CVD processes). Logically, one could define all PVD processes as ion plating, but this ignores the most important aspect of ion plating: modification of the microstructure and composition of the deposit caused by ion bombardment of the deposit resulting from bias applied to the substrate.

2.1.1   Thermal and Electron Beam Evaporation

Thermal evaporation (TE) is a physical vapor deposition (PVD) technique that employs a resistive element to heat a source to its vaporization temperature. For an in depth review of TE, please refer to Ron Bunshah's and Don Mattox's handbooks [1, 2, 3]. Over the past century it has been used to deposit metals (aluminum and silver in particular) for plating of glass and telescope mirrors, metalizing capacitors, gas and water diffusion barrier coatings, metalizing plastic wrappings and reflectors, small molecule materials and polymers for organic light emitting devices OLEDs, semiconductors, transparent conductive coatings, and even high temperature superconductors.

Figure 2.1 shows a simplified diagram of a TE system. Our attention here will be on chamber process details that directly affect the properties and microstructure of the deposited films. Virtually all PVD processes, and evaporation is no exception, are line-of-sight processes. This means that evaporated atoms/molecules move directly from the source to the substrate with very little scattering. Material is deposited over a wide range of angles to the substrate with a specific flux distribution, which directly affects

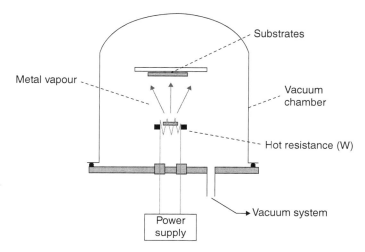

Figure 2.1 Basic thermal evaporation chamber geometry.

microstructure and several film properties (mechanical stress for example). There are advantages and disadvantages for line-of-sight deposition:

- Masking and patterning of the substrate is straight-forward
- Poor step coverage
- Off normal incidence can cause shadowing and degrade coating properties
- Off normal incidence can be used to advantage to deposit sculpted coatings [5]

Films are deposited at pressures less than 10^{-4} Torr and usually in the range of 10^{-6} Torr, but sometimes as low as 10^{-9} Torr, which reduces scattering by gas atoms in the chamber and thermalization of the evaporated atoms. Table 2.1 shows the relationship between mean free path as a function of chamber pressure [6]. As with all deposition processes, mean free path is important in determining the energy of species incident on the substrate surface.

A large number of materials can be thermally evaporated, including metals and refractory metals, alloys, semiconductors, and insulators [3]. Films can also be co-evaporated using multiple sources. The purity of the deposited film is directly related to the purity of

Table 2.1 Pressure dependence of gas mean free path.

Vacuum Range	Pressure (Pa)	MFP
Ambient	1013	68 nm
Low vacuum	300–1	0.1–100 nm
Medium vacuum	1–10^{-3}	0.1–100 mm
High vacuum	10^{-3}–10^{-7}	10 cm–1 km
Ultra high vacuum	10^{-7}–10^{-12}	1 km–105 km
Extremely high vacuum	$<10^{-12}$	$>10^5$ km

the evaporation source. Since the chamber pressure is low, very little gas is incorporated, or trapped, into the film. Thus, very pure thin film materials can be deposited.

There is virtually no energetic or plasma bombardment of the growing film during deposition. Atoms arriving at the substrate are essentially thermalized. As a result, the film's surface is not damaged by this process and no material is ejected from the surface. This is both a blessing and a curse. Many thin film materials, fluorides, and some semiconductors, for example, can be easily damaged by back reflected neutral atoms and plasma bombardment during the sputtering process. However, high energy particle bombardment has been shown to increase the density and reduce voids in thin films. Also as a result of the heating of the source, the substrate may experience high radiant heat loads and possible x-ray heating.

The energy of the evaporated atoms/molecules and their mobility on the surface of the substrate is critical in determining physical, optical, and mechanical properties and microstructure. The energy of an incident evaporated particle is in the range 0.10–0.50 eV. At these low incident energies, adhesion to the substrate is generally poor and the microstructure depends on substrate temperature, as shown by the Structure Zone Models described in the next chapter. The e-beam process employs, as the description states, a high energy beam of electrons to heat a crucible or boat that contains the material/s to be deposited. Figure 2.2 shows a diagram of e-beam process geometry. Deposition pressures are much the same as for thermal evaporation. The material to be evaporated

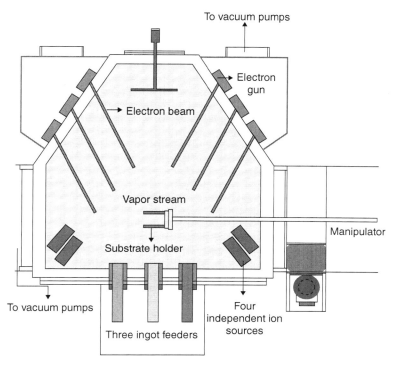

Figure 2.2 Diagram of e-beam evaporation geometry.

is typically held in crucibles or boats and as many as six electron guns, each having a power from a few tens to hundreds of kW are used. Materials usage can be poor to high. We often find significant amounts of unused material in the crucible. The kinetic energy of the electrons is converted into thermal energy as the beam bombards the surface evaporant. A number of materials can be used to hold the evaporated material, but must be electrically conductive. Tungsten, molybdenum, tantalum, and tantalum-zirconium alloys are common materials used in crucibles. Electrically conducting ceramics, such as graphite, BN-TiB_2, and some cermets can also be used [2]. As with thermal evaporation, deposition rates can be very high and large substrate areas can be covered. E-beam process is particularly well suited for refractory materials, such as TiC, TiB_2, ZrB_2, B_4C, Ti, W, and Ta.

Certain refractory oxides and carbides have a problem in that they undergo fragmentation during their evaporation by the electron beam, resulting in a stoichiometry that is different from the initial material. To remedy this situation, reactive and activated reactive evaporation are employed [2, 3]. Both e-beam and thermal evaporation processes are used for reactive deposition. Because the energy of the evaporated material and the resulting heat of fusion are relatively low, reactive deposition requires a little help. Without adding a little energy to the reaction, most evaporated oxides and carbides are substoichiometric, and nitrides are a real problem. Without activation, deposition of metal oxides from metal and semiconducting sources results in substoichiometric films. For example, alumina (Al_2O_3), when evaporated by electron beam, dissociates into aluminum, AlO_3 and Al_2O. Active oxygen must be introduced into the chamber to obtain stoichiometric Al_2O_3.

Some refractory carbides like silicon carbide and tungsten carbide decompose upon heating and the dissociated elements have different volatilities. These compounds must be deposited either by reactive evaporation or by co-evaporation. In the reactive evaporation process, the reactive gas (oxygen, nitrogen, hydrocarbon) is introduced into the chamber while the metal is evaporated. When the thermodynamic conditions are met, the metal atoms react with the gas in the vicinity of the substrate to form films. Plasmas are also used to activate the reactive gas (plasma assisted PVD).

If we do not consider ion assist or ion plating, substrate temperature is the most important deposition condition that affects all properties of the films, if plasma activation is not used. At low substrate temperature (see Zone 1 in Chapter 3), the microstructure is a porous columnar structure with small grain size, but grain size increases with increased substrate temperature. Recrystallized grain growth occurs at the highest substrate temperatures. Note that adatom mobility also increases with increased substrate temperature. For deposition of many metals, the substrate is actually cooled to reduce grain size and keep the microstructure as dense as possible.

Table 2.2 shows materials that are typically evaporated. Thermal evaporation and subsequently electron beam evaporation is used extensively for metallization of flexible plastics, telescope mirrors, plastic and glass reflectors. A wide range of metals are deposited, including Ag, Al, Cr, Au, Cu, Ni, W, Fe, Zn. The primary advantages of this process are the very high deposition rates, large area coverage, and low cost. Metals must be deposited at the highest possible rate and at lowest possible temperature to reduce surface roughness and impurities. Most evaporated metals need a bonding

Table 2.2 Commonly evaporated materials.

Material Family	Materials
Carbides	TiC, HfC, ZrC, VC, W_2C, TaC
Carbonitrides	Ti(C,N)
Nitrides	TiN, HfN, ZrN
Oxides	TiO_2, ZrO_2, SiO, Al_2O_3, SiO_2
Sulfides	TiS_2, MoS_2, MoS_3
Superconductors	Nb_3Ge, $CuMoS_8$, $YBa_2CU_3O_{7-8}$
Optical	CaF_2, MgF_2, HfF_4, ThF_4
Optoelectronic	ITO, ZnO
Photovoltaic	a-Si:H, $CuInS_2$

or "glue" layer when deposited on glass or plastics. Cr and Ti are commonly used. It has been found that the affinity of the metal for oxygen plays an important role in determining the adhesion, and the results appear to confirm the theory that the formation of an intermediate oxide layer at the metal/glass interface is necessary for good adhesion [7].

Until magnetron sputtering matured as a deposition process, virtually all optical coatings were deposited by thermal and electron beam evaporation. Even today, it is the preferred process for deposition for fluoride thin films (CaF_2, MgF_2, HfF_4, ThF_4). Before ion assist was used, the main problem with evaporated optical thin films was that their reduced density (due to a porous microstructure) caused water pick up and poor mechanical properties. Many coatings are easily scratched. Physisorbed water causes the refractive index to decrease and as a result, the optical properties (transmittance, reflectance) are not stable. This is a huge problem for precision multilayer optical coatings [8]. The refractive index of the film can be expressed as

$$n_f = pn_s + (1-p)n_v \qquad (2.1)$$

Packing density is p, n_s is the refractive index of the solid part of the film, and n_v is the refractive index of the voids [9]. The refractive index of air is 1 and the index of water is ~1.3. Thus we see that for virtually all optical materials, water incorporated into voids and the porous microstructure decreases the refractive index. Baking out water does not help in this case since water is replaced by a lower index material, air. The real challenge is to keep p as large as possible, which is difficult according to the structure zone model.

This makes environmental stability a real issue for evaporated thin films [10]. Water is initially chemisorbed from residual gas in the deposition chamber. This water is bonded in the lattice of the film. However, more water is actually physisorbed when the film is exposed to air. This shifts the spectrum of the film to longer wavelengths. The amount of wavelength shift depends on relative humidity and temperature. Figure 2.3 shows the fractional wavelength shift of an evaporated Ta_2O_5 film with increased temperature (ignore the IADL curve for now) [10]. Environmental stability tests for optical films are documented in Mil-Specs. This post deposition shift, however, is reversible. Ion assist is needed to stabilize the optical and electrical properties of evaporated films.

Evaporated films have been used extensively in tribolological applications [1, 2, 3, 4]. Important properties here are hardness, wear resistance, mechanical stress, and elastic constants. We will focus here on single layer films. Multilayer, nanolaminate, and superlattice structures are not considered in this comparative review. The hardness and ductility of Ni and Ti metals depends to a certain

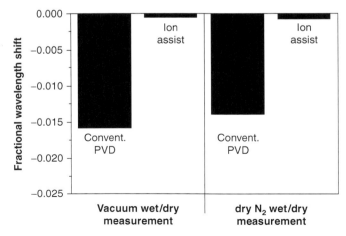

Figure 2.3 Fractional wavelength shift of an evaporated film with increased temperature [10].

degree on the structural zone in which they are deposited [2]. Zone 2 appears to yield films with highest hardness but low ductility. Refer to Bunshah for a detailed treatment [2] Figure 2.4 shows the variation of microhardness with substrate temperature [4]. As expected, carbides and TiC in particular have highest hardness values ($\sim$4000 kg/mm^2 $\sim$ 39 GPa) [3]. The yield strength, ductility, and hardness of evaporated metals and alloys are essentially the same as their counterparts produced by casting, mechanical working, and recrystallization.

The deposition rate for evaporation can be as low as 1 nm per minute to as high as a few micrometers per minute. The material utilization efficiency can be high compared to other methods, however, the process offers limited structural and morphological control compared to other processes. Sputtering processes are catching up in materials usage. Due to the very high deposition rate, this process has potential industrial application for wear resistant and thermal barrier coatings in aerospace industries, hard coatings for cutting and tool industries, and electronic and optical films for semiconductor industries. Evaporation processes are the most economical of all PVD processes and their large area coverage is attractive. Thin films evaporated without ion assist must often be heat treated after deposition to increase their density and adhesion.

Evaporation processes are still widely used for metallization; however, bonding layers must usually be used to achieve good adhesion. One of the most important applications of e-beam evaporation is aluminization of PET film packaging in roll-to-roll coaters. Because of the high deposition rates, large areas and lengths of flexible plastic, usually PET, can be coated economically [11, 12, 13].

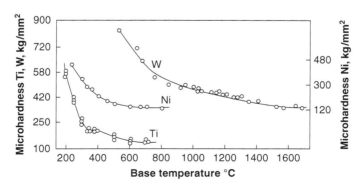

Figure 2.4 Variation of microhardness with substrate temperature [3].

Evaporated aluminum films also have extensive applications in high energy density capacitors [14]. While e-beam evaporation can be used for tribological and wear coatings, it is not the process of choice (we will discuss this in more detail in future chapters).

Another advantage of the evaporation processes is that it can be precisely controlled and film thickness easily monitored. This is extremely important for deposition of multilayer optical coatings and semiconductor films. Crystal rate monitors and optical monitoring are extremely effective in controlling film thickness.

2.1.2   Ion Treatments in Thin Film Deposition: Ion Assisted Deposition

Plasma and ion treatments can be used during physical vapor and chemical vapor deposition processes to modify the microstructure and properties of thin films. They are used to activate and clean substrate surfaces, enhance the mobility of atoms deposited on the substrate, etch materials on the substrate, and enhance chemical reactions occurring in the deposition chamber. Ion assisted deposition (IAD) can be used to controllably modify thin film properties such as adhesion, hardness, index of refraction, packing density, mechanical stress, stoichiometry, and microstructure. The degree of modification depends on the energy of the bombarding ions and deposition parameters. Ion energies in the 100–200 eV range are usually required to achieve property and microstructural modifications. Energies higher than 300 eV usually damage the film or implant gas atoms. Etching is achieved at higher energies.

Energetic ions react with matter (i.e., the film growing on the substrate) by imparting their energy to the surface atoms through elastic and inelastic collisions. The impinging ion loses energy due to collisions with atoms on the substrate and to electronic excitations. The ion can also be stopped in the film or substrate and trapped interstitially, or it can chemically react with the surface atoms, forming a new compound. The energy and mass of the bombarding ion, as well as the structure and bonding in the thin film, determine what type of reaction will occur at the surface.

It is sufficient to know that the ion has several barriers to overcome to have any effect on the atoms in the film and this interaction is complicated by interatomic potentials and the energy of the ions. To complete the picture of how an ion interacts with the growing film we must consider what happens to the kinetic energy of the ion after a collision with an atom in the film, the resultant energy imparted to the atom, the range of the ion in the film, and how the ion energy is transferred to the atoms in the film. When an ion penetrates the growing film, it collides with several atoms and electrons in the film. As a result, the farther it penetrates, the more energy it loses. Thus, it loses energy to the atoms in the film and it also is implanted in the film. The rate at which an ion loses energy is called the stopping power. Two things can happen: collisions with nuclei result in atom displacement (and possible ejection) and collisions with electrons result in electronic excitations (higher energy states). The distance it takes for the ion to do its thing in the solid until it comes to a stop is called the ion range. Figure 2.5 shows the path

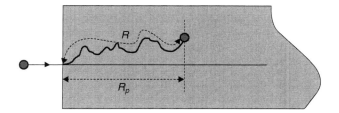

Figure 2.5 Path of an ion penetrating a solid [15].

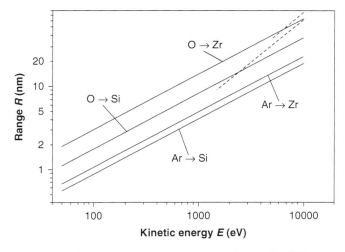

Figure 2.6 Ion range of various types of gas ions and energies [15].

of an ion penetrating a solid and Figure 2.6 shows the ion range for various types of ions and their energies [15].

Thus, during deposition of a thin film, an energetic ion can

- Increase the energy of deposited atoms (increased surface mobility),
- Displace atoms (damage),
- Eject atoms (sputtering, etching),
- Excite electronic states, and
- Become implanted.

There is a specific energy range for these actions to take place. Ion assist occurs when the bombarding ion imparts sufficient energy to increase the mobility of the deposited atom on the surface and modify a specific property of the film, but not to remove or eject the atom from the surface. Ion bombardment causes many of the same effects as increased substrate temperature (but without excessive substrate heating). For etching or sputtering to occur, the kinetic energy of the ion must be great enough to overcome the interatomic potential and break the bond between an atom and its nearest neighbors (actually other atoms are also involved but we're looking at the simplest case here).

Factors that primarily affect film growth with IAD are [15]

- Type of ions (single atoms or molecules)
- Kinetic energy distribution of the ions
- Ion current density
- Ion flux
- Angle of incidence
- Ion to atom ratio
- Surface temperature
- Ion energy and momentum transfer to the atoms

Figure 2.7 shows the energy range required for property modifications [16]. Ion energies greater than ~10 eV/atom can cause etching and sputtering of the growing film.

IAD can be used to modify a number of properties of thin films by influencing nucleation and growth, surface roughness, interface mixing, densification, defect generation, and gas trapping. Mechanical properties, including stress, can also be controlled. Film adhesion to the substrate is usually improved with ion assist and

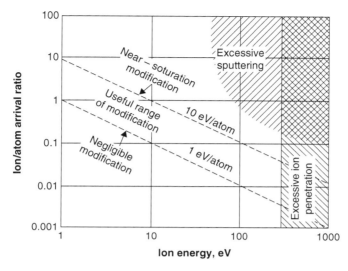

Figure 2.7 Ion energy ranges for property modification [16].

an ion cleaning prior to deposition. Kaufman determined that an ion energy dose of 1–10 eV/deposited atom is needed for measurable property modification [16]. Property modification, however, is mainly due to increased mobility of the deposited atoms due to ion bombardment. This results in a denser film with few voids (due to the increased mobility of the atoms on the surface).

As expected, densification of the growing film due to IAD also changes the elastic and mechanical properties. In particular, the hardness of the film is increased with ion assist. There is, however, a limit to which we can apply ion assist. Films deposited by IAD usually have a high compressive stress, which limits the useful thickness of the film. When the thickness gets large enough, the stress in the film will cause delamination and loss of adhesion. Compressive stress also increases with hardness [17]. So above a certain thickness, the film will lose adhesion and will not be useful, no matter how hard it is [17].

The three broad beam ion sources used most often for ion assist during deposition of a thin film are the Kaufmann, end Hall, and gridless types. The Kaufmann design was developed in the late 1980's and consists of a cathode and anode located in a discharge chamber, and a magnet to confine and focus the ion beam. This type of ion source produces positively charged ions and usually needs a filament to keep the net charge of the beam neutral. The ion beam is usually well collimated and has a very narrow energy spread so that the energy of the ions is well defined. A wide range of ion energies is possible from this source. The working gas is usually Ar or another inert gas. This type of ion source has several grids and is complicated to operate. The end-Hall type ion source has a much simpler design, consisting of a filament that acts as a cathode, an anode at positive potential, and a magnetic field to generate, confine, and focus the ions. The alternating current applied to the cathode initiates ionization of the working gas atoms. Further ionization occurs in the chamber in the magnetic field and as a result of electron-ion interactions in the chamber, ions are discharged past the cathode. The ions are not well collimated and the beam has considerable divergence. An external filament is not needed to neutralize the beam. The beam current and voltage (energy) depend on the gas pressure, and the ions have a wide energy distribution. This type of ion source generates ions with low energy (~100 eV) but large beam currents. Note that both these sources work best with an inert working gas. Reactive gases added to the mixture will rapidly degrade their filaments. The filamentless ion source, just as it name implies, employs no filament that can degrade and cause problems during a coating run. The hollow cathode ion source is the most widely used of this type of instrument. A plasma discharge is sustained within the source's cavity by electrons produced in a cold cathode system, and ions are extracted at the anode by applying a potential across the cathode and anode. Much like a magnetron sputtering cathode, a magnetic field is applied across the plasma to focus and confine the ions. This type of ion source can employ grids to narrow the energy distribution of the ions emerging at the anode. Very precise, and often monoenergtic, energy distributions are possible. These sources often employ RF or ECR power to the cathode. Table 2.3 summarizes the salient properties of these ion sources.

Use of IAD is as much an art as it is a science. The effects of ion treatment are different for each material, system, and process with which it is associated. As much as the suppliers would like you believe, it is not a turn-key process. To achieve the optimum benefit from IAD, one must understand the mechanisms and physical principles involved and be able to apply them to each situation.

2.1.3 Ion Plating

One of the most important goals of any deposition process is to improve the energetics of atoms or particles being deposited onto

Table 2.3 Summary of typical ion source performance.

Type	Ion Energy (eV)	Ion Current Density (mA/cm²)	Ion Energy distribution	Working Gas	Working Pressure (mTorr)
Kaufmann	0–2000	1–5	Narrow	Inert	~10^{-4}
End-Hall	0–200	~1	Broad	Inert	~10^{-4}
Cold cathode	0–1000	0.5–2	Narrow-Broad	Inert, reactive	~10^{-4}

the substrate. As we have seen in all previous Process Technology Tutorials, adatom energetics is responsible for improving the density, microstructure, adhesion, physical and tribological properties. Ion plating was originally developed by Don Mattox as early as 1968 to increase the energy of the atoms incident on the substrate [18]. He best describes this process as

"Ion plating is an atomistic vacuum coating process in which the depositing film is continuously or periodically bombarded by energetic atomic-sized inert or reactive particles that can affect the growth and properties of the film. The source of depositing atoms can be from vacuum evaporation, sputtering, arc vaporization or from a chemical vapor precursor" [19]. For a detailed description of this process please refer to Don's chapter in the Third Edition *of Handbook of Deposition Technologies for Films and Coatings* [20] or the second edition *of Handbook of Physical Vapor Deposition (PVD) Processing* [21]. We see here that virtually any PVD or CVD process can possibly benefit from the ion plating technology, and processes such as cathodic arc deposition are a form of ion plating. Processes that are referred to as ion plating are [19]

- Bias sputtering
- Bias sputter deposition
- Ion vapor deposition
- Ion beam enhanced deposition (IBED)
- Accelerated ion deposition
- Plasma enhanced vapor deposition
- Ion assisted deposition (IAD)
- Biased activated reactive deposition (BARE)
- Plasma surface alloying

More recently, ion plating is often referred to as "energetic condensation" [22]. Numerous applications include tribological coatings, optical coatings, transparent conductive coatings, adherent metal coatings, and metal coatings.

The process was originally developed to enhance the energy of thermally evaporated, sputtered and CVD species reaching the substrate. In one configuration, a plasma is generated at the substrate which is negatively biased. Bombarding species are generally either ions accelerated from a plasma in the deposition chamber (plasma-based ion plating) or ions from an 'ion source (vacuum-based' ion plating). The stages of ion plating can be separated into surface preparation, nucleation and interface formation, and film growth. Ion plating was first described in the early 1960s and was initially used to enhance film adhesion and improve surface coverage. Later it was shown that controlled bombardment could be used to modify film properties such as density, morphology, index of refraction, and residual film stress. More recently the bombardment has been used to enhance chemical reactions in reactive and quasireactive deposition processes. Presently, ionization and acceleration of the depositing film atoms (film ions) is being used for directed deposition to improve filling of surface features in semiconductor processing.

Figure 2.8 shows the basic ion plating configuration for thermal evaporation [22]. Recall from the above discussion that evaporation processes are characterized by very high deposition rates, but, without some help, evaporated thin films generally have poor adhesion and mechanical properties, a porous columnar structure, and reduced density compared to bulk materials. Many of these poor properties can also be remedied by ionizing a part of the depositing material, i.e., an ionized plasma, just before it impinges on the substrate. In the ion plating process ions in partially ionized plasma are attracted to the substrate by a negative bias applied to the substrate. This ion bombardment affects coating properties in much the same way as ion assist, but without the ion source.

In ion plating the energy flux and mass of the bombarding species along with the ratio of bombarding particles to depositing particles are important processing variables. There are a number of methods used to achieve partial ionization, including thermionic arcs (cathodic arc for example), hollow cathode discharge, DC diode and RF discharge, microwave discharge, electron emitter

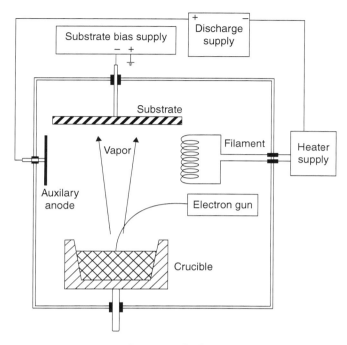

Figure 2.8 Basic ion plating configuration [22].

discharge, magnetron discharge, and plasma enhancement [2, 5]. Anders [22] defines the degree of ionization as

$$\alpha = \frac{n_i}{\left(n_i + n_o\right)} \text{ where} \tag{2.2}$$

n_i is the ion density and n_o is the density of neutral atoms.

Ion densities differ according to deposition process. Ion densities for several processes are [23]

- Electron beam evaporation: 0.1–0.50
- Sputtering: <0.10
- Cathodic arc: 0.50–0.80

Since the 1960s, ion plating has been used in hundreds of deposition processes to improve virtually every type of coating property. The basic technology developed by Don Mattox has given birth to a wide range of ion plating techniques with hundreds of related publications. The main advantages of ion plating are

- Compatible with virtually any PVD and many CVD processes
- High density films possible
- Films have excellent adhesion to substrate
- Process is scalable to industrial needs

Disadvantages are

- Added complexity to the deposition process
- Unwanted substrate heating possible
- Defects and gas incorporation introduced by plasma bombardment

2.1.4 Planar Diode and Planar Magnetron Sputtering

Sputter deposition processes have become the most widely used method of thin film and surface engineering treatments. While the diode sputtering process was the first to be developed (back in 1852), the many forms of magnetron sputtering are now the most widely used deposition methods for high quality thin films.

Figure 2.9 diagrams the general sputtering process, in which, atoms or molecules of a solid material (a target or sputtering source) are ejected into a gas form or plasma due to bombardment of the material by energetic gas ions and deposited on a substrate

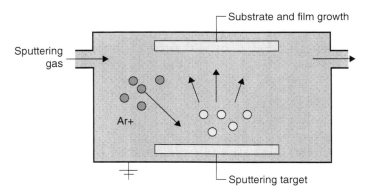

Figure 2.9 Diagram of the sputtering process.

above or to the side of the target. A high vacuum is required to initiate a plasma whose ions bombard the target. The sputtering process is essentially a momentum exchange [24], between the gas ions; the more intense and concentrated the plasma in the region of the target, the higher the atom removal rate (or deposition rate). There will be negligible sputtering with light atomic weight gases such as hydrogen and helium. Reactive gases typically used are oxygen (O_2), nitrogen, fluorine (F_2), hydrogen (H_2), and hydrocarbons (methane, butane, etc.). One of the major advantages of this process is that sputter-ejected atoms have kinetic energies significantly larger than evaporated materials and sputtered films receive many of the benefits described for IAD in section 2.3. The growing film is subjected to a number of energetic species from the plasma. Figure 2.10 shows a comparison of the energetics of selected PVD processes [25].

An important advantage and difference between sputtering and evaporation processes is that the sputtered atoms can have relatively high kinetic energies. For example, the average ejection energy of Ge atoms sputtered with 1.2 keV Ar is ~15 eV, while the average energy of an evaporated Ge atom is ~0.1 eV [24]. The average ejection energy of metal atoms is in the range 15–20 eV, with highest energies near 35 eV reported for W, U, Ta and Pt. Note that heavier elements generally have higher average energies.

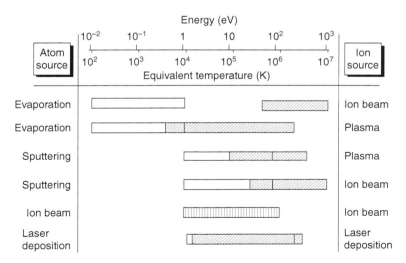

Figure 2.10 Comparison of the energies and equivalent temperatures of selected PVD processes [25].

The diode, or more specifically the planar diode system, is the simplest sputtering configuration. It consists of a cathode and anode separated by 5–20 cm spacing, and can be operated in DC and RF power modes. In DC sputtering, the cathode is both the sputtering target and the electrode for sustaining the glow discharge. Much of the original work in development of sputtered semiconductors and optical coatings was accomplished using diode sputtering [26, 27]. The glow, or plasma, is confined between the anode and cathode. Pressures needed to sustain a glow discharge are higher than for other sputtering processes, typically between 10 and 50 mTorr. The dark space at the cathode becomes too large at lower pressures and the glow is extinguished. Ground shields are used to obtain a uniform erosion rate of the target. This geometry can be used in the sputter up or sputter down mode, although substrate heating and holding is much easier in the sputter down geometry. This configuration also generates more substrate heating than other sputtering techniques because the plasma is in intimate contact with the substrate.

As the name implies, planar magnetron sputtering utilizes a flat sputtering target in the cathode enclosure shown in Figure 2.11. Magnetron targets can be as small as 1 inch or as large as several meters. The anatomy of a basic planar magnetron cathode consists of magnets (called magnetics), which are placed under the target in

Figure 2.11 Planar magnetron cathode mounted in large planetary batch coater.

various configurations to confine the plasma or spread the plasma above the region of the target. The magnetic lines of force focus the charged gas atoms (ions). The stronger the magnetic field, the more confined the plasma will be (consequences are discussed below). Stronger magnets are required for sputtering ferromagnetic materials such as Fe, permalloy (Fe_xNi_y) and Ni. The magnetics focus the plasma at the surface of the target and an erosion pattern, commonly called a "racetrack," is formed as sputtering proceeds. Magnet configuration is different for planar circular, planar rectangular, external and internal cylindrical magnetrons.

One of the major advantages of magnetron sputtering over diode sputtering is operation at pressures in the range 0.5 mTorr to 10 mTorr. As a result, there is less scattering and subsequent thermalization of the sputtered atoms by gas ions and atoms, which can reduce their energy. This is important for large throwing distances to the substrate. The mean free path of a sputtered atom is ~5 cm at 1 mTorr pressure. It is important to work at low pressures to achieve fully dense films by the sputtering process, again comparable to IAD processes. Deposition rates, however, are lower than evaporation processes.

Both direct current (DC) and radio frequency (RF) power sources can be used with most magnetron cathodes. Pulsed DC and low and mid frequency power supplies are also used in magnetron sputtering. New concepts such as DC with RF overlay are also being developed for a number of coating applications (transparent conductive oxides for example).

Magnetron sputtering is particularly well suited for reactive deposition. It should be noted that there are a number informative papers on the physics, chemistry, and kinetics of reactive sputtering [28] and the Society of Vacuum Coaters (SVC) and AVS offer several excellent short courses on sputtering. Virtually any oxide, nitride, and carbide thin film material can be deposited. Fluorides, however, present a problem due to reflected neutrals off the target. The operating reactive gas partial pressure depends on deposition rate (target power), target material, chamber pressure, and substrate target spacing [24, 29]. Figure 2.12 shows a typical hysteresis plot for an oxide coating, which plots target voltage against oxygen partial pressure. Target voltage remains constant or increases slightly with increased oxygen, and then plummets at a transition concentration. The surface of the target is oxidized for percentages higher than the transition voltage and in the "metal" mode below the transition.

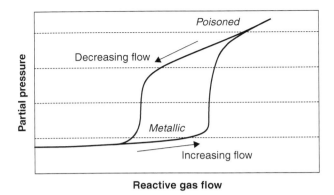

Figure 2.12 Hysteresis plot of target voltage versus percentage of oxygen.

Metallic and substoichiometric coatings result on the left side of the transition. If the percentage oxygen is too high, the target will become "poisoned" and an insulating crust will form on its surface. The deposition rate will decrease significantly and arcing may occur. The resulting coating may be less dense and of poor quality. The operating point is just on the cusp of the transition. The coating is fully stoichiometric and the deposition rate is optimum. Note that if the oxygen is decreased, there will be a hysteresis effect due to cleaning up of the surface oxide. The other technique actually determines the partial pressure of reactive gas. As shown in Figure 2.13. Reactive gas pressure is plotted against reactive gas flow. The operating point occurs just before the partial pressure increases with flow. The operating point is the flow at which the reactive gas is no longer "reacting" with the target material. There is also a hysteresis effect for this technique. Whatever method one uses to determine the operating reactive gas pressure, it should be monitored as the target ages. Generally, less reactive gas will be needed with increased erosion of the target. It should be noted that magnetron power supplies are available that incorporate reactive gas flow control.

Magnetron sputtering is also particularly well suited for deposition of multilayer thin film structures, including optical designs, electrochromic coatings, nanolaminates, superlattices, tribological coatings and barrier coatings. High quality thin film optical materials have been deposited since the 1970s, and a wide range of optical thin film materials has been developed. The list includes most oxides, nitrides, carbides, transparent conductive materials, semiconductors, and many polymers. Fluorides are best left to evaporation processes and hybrid processes. The main advantage is that

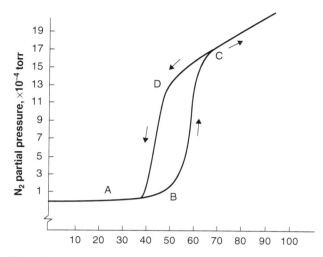

Figure 2.13 Plot of reactive gas partial pressure against gas flow. The operating point is the flow at which the reactive gas is no longer "reacting" with the target material [24].

these materials can be deposited using both nonreactive and reactive processes with excellent control of composition, layer thickness, thickness uniformity, and mechanical properties. Additionally, deposition of multilayer structures, including combinations of the materials listed above, is straightforward using multiple cathodes and reactive gases. A small sampling of the optical materials deposited by planar magnetron sputtering is listed in Table 2.4. Deposition of alloys, oxinitrides, alloy oxinitrides, and carbon nitrides is straightforward. Use of multiple cathodes and reactive gases also makes deposition of multilayer structures possible. It is possible to deposit virtually any type of optical design, from the simplest antireflection coating to complex beam splitters and multiple cavity filters. Also, since sputtering is a nonequilibrium process, alloys and compounds not found in binary and ternary phase diagrams can be deposited.

2.1.5   Unbalanced and Closed Field Magnetron Sputtering Processes

The unbalanced magnetron, along with the closed field magnetron, is the next step in the evolution of this deposition technology. The placement and strength of the magnets in the magnetron cathode make it possible to manipulate the magnetic field, and thus the

Table 2.4 Examples of optical thin film materials deposited by planar magnetron sputtering.

Material	n @ 550 nm
SiO_2	1.45–1.48
Si_3N_4	1.95–2.05
Al_2O_3	1.60–1.67
AlN	2.05
SiAlON	1.48–2.05
AlON	1.60–2.05
SiON	1.48–2.05
Ta_2O_5	2.10
Nb_2O_5	2.19
ITO	~1.80
SnO_2	~1.80
ZnO	~1.90
Y_2O_3	1.95
H_fO_2	~1.9
CeO_2	1.7–2.3
TiO_2	2.2–2.5
WO_3	~2.4
ZrO_2	2.18
In_2O_3	~1.80

plasma, in the area of the cathode and substrate. The magnet assembly consists of north and south pole configurations. It is important to note that the magnetic field acts as an electron trap that directly determines the trajectories of secondary electrons ejected from the target. The electrons are forced into a cycloidal path that greatly increases the probability of ionizing the sputtering gas. The performance of the unbalanced and closed field magnetrons relies on the design of the magnets and the placement of the magnet assemblies. An unbalanced magnetron is the design where the magnetic field strength of one pole is significantly stronger than the other pole, as shown simply in Figure 2.14. As a result of the magnetic field lines extending further from the target surface, electrons can ionize inert sputtering gas atoms near the substrate surface, which increases substrate bombardment, and acts effectively like an ion source for IAD. The degree of balance (or unbalance) depends on the relative strength of the magnet poles. There is, however, a price to pay for the added substrate ion bombardment in poorer target utilization and easier poisoning during reactive deposition. Target utilization can be increased if the poles are only slightly unbalanced, but substrate bombardment is significantly reduced. IAD is also often used in this case to augment the magnetron plasma.

The closed field magnetron, shown in Figures 2.15a and 2.15b, employs a magnet geometry that confines the plasma to within the space of the sputtering targets and keeps it away from chamber walls where it cannot be used [30]. This configuration uses a series (usually no more than six) of unbalanced magnetrons to confine the plasma [31]. Substrates are placed above the center of the cathode arrangement. The closed field geometry can increase the ion current density at the substrate by as much as an order of magnitude [31]. Using this technique, fully dense coatings with excellent adhesion can be deposited.

Unbalanced and closed field magnetrons have significantly advanced tribological coating development during the last decade [31, 32]. Table 2.5 summarizes some of the hardness results. Improved single layer, multilayer, superlattice, nanolaminate, and

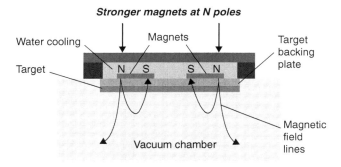

Figure 2.14 Placement of stronger north magnetic poles than south magnetic poles in an unbalanced magnetron.

(a)

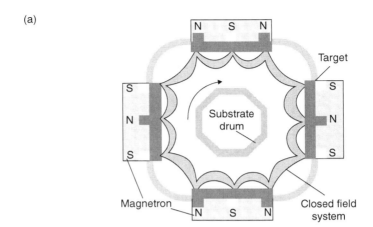

Figure 2.15a Closed field magnetron configuration using four cathodes.

(b)

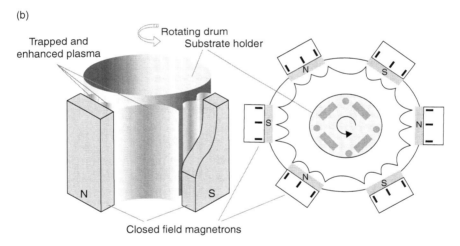

Figure 2.15b Closed field magnetron configuration using six cathodes [30].

nanocomposite coatings have been demonstrated [33–43]. Single layers of TiN, TiAlN, TiCN, VN, DLC, MoS_2 and CrN first demonstrated hardnesses in the 15–25 GPa range [35, 39, 40, 41, 43, 44]. Ion current and bias voltage are now added as a variable deposition condition, which also makes substrate placement important. Substrate placement in the chamber is always important, but has a stronger influence on coating properties in the presence of ion bombardment and particularly asymmetric ion bombardment [45]. Hardness of

Table 2.5 Summary of hardness data for unbalanced and closed field magnetron sputtered films.

Material/SL	Structure	Hardness (GPa)
TiAlN	Single Layer	25 [39]
	TiAlN/VN*	39 [40]
	TiAlNYN-VN*	78 [40]
	TiAlN-CrN*	55 [40]
TiN	Single Layer	15 [40], 20 [34]
	TiCN/TiN*	25 [39]
	CN/TiCN/TiN*	35 [39]
	TiN/TiC$_x$N$_y$/ Si$_3$N$_4$/ SiCN*	55
TiAlN	Single Layer	25 [39]
AlN	Single Layer	18 [33]
	AlN/Si$_3$N$_4$*	35 [33]
	nc-TiN/aSi$_3$N$_4$/ nc-&TiSi$_z$*	>100 [48]
	nc-TiN/aSi$_3$N$_4$*	>50 [48]
	C/Cr/CrN	[45]
	NbN/CrN	[45]
	CNx/TiN	35 [49]
	CNx/HfN	40 [49]
W-DLC	nanocomposite	~14 [50]

*Superlatice or nanolaminate.

TiC films was found to increase from 3 GPa to 20 GPa simply by increasing substrate distance from the target, and elastic modulus also increased from 8 GPa to near 24 GPa [45]. Coefficient of friction also decreased, but additionally depended on target power. The improvement in tribological and wear properties was attributed to increased coating density resulting from ion bombardment. Tribological coatings are addressed in more detail in Chapter 4.

Unbalanced and closed field magnetron sputtering is very amenable to large area and industrial applications [39, 46]. CrN/NbN films are deposited onto stainless steel hydraulic spools to improve corrosion resistance [47]. TiN, ZrN, TiZrN and TiCN decorative hard coatings are deposited onto various metals to create a range of new colors and improve wear resistance [48].

Unbalanced and closed field magnetron processes have all the advantages presented last month for planar magnetrons. Additional advantages and disadvantages are:

- Additional ion bombardment resulting in dense, adherent films
- Ion assist and substrate cleaning possible
- Improved tribological, wear and corrosion resistant films
- Amenable to multilayer, superlattice, nanolaminant, and nanocomposite films
- Poorer materials usage
- More complex cathode configuration/expense

Unbalanced and closed field magnetron sputtering revolutionized the properties achievable by magnetron sputtering processes. As successful with improved thin film tribological, corrosion resistance, and optical properties as this technology was, improvements in magnetron technology had just begun.

2.1.6 Cylindrical and Rotating Magnetron Sputtering

Planar, unbalanced, and closed field magnetron sputtering processes work well with planar and mildly nonplanar substrates held above the magnetron cathodes. Uniformly coating a wire, cylinder, sphere, or three-dimensional surfaces, however, can be challenging for these processes. In addition to other applications, the cylindrical and rotating cylindrical magnetron were developed to meet these challenges. There are three basic cylindrical magnetron configurations:

- Rotatable cylindrical magnetron (interior magnets)
- Cylindrical post magnetron (exterior magnets)
- Inverted cylindrical magnetron (exterior magnets)

The rotatable cylindrical magnetron (RCM) goes back two decades, and was developed to improve target materials usage [51]. The RCM can replace planar magnetrons in many applications. A cross section shown in Figures 2.16a and 2.16b. The magnetics for this type of cylindrical magnetron are located *inside* a cylindrical tube target, and the target is rotated around and over the magnets. This exposes all target surfaces to the magnetic field and results in uniform erosion of the entire target surface, and as a result, improves target usage efficiency. Also there is reduced chance of debris and insulator build up on the surface of the target, which makes for smoother spark free operation and fewer particulates at the substrate. This type of cathode is also used extensively in vacuum roll coating processes.

(a)

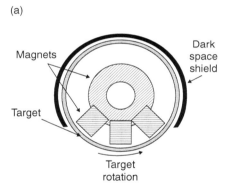

Figure 2.16a Schematic of rotatable cylindrical magnetron.

(b)

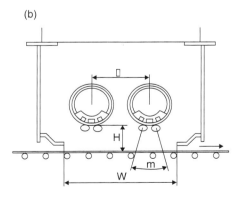

Figure 2.16b Cross section of rotatable cylindrical magnetron [51].

As shown in Figure 2.17, the basic post cylindrical magnetron cathode consists of a center cylindrical cathode and concentric tube anode. The magnets (magnetic coils are often used) are located outside the anode and arranged so the magnetic field is aligned along the axis of the central post. Variations are

- Cylindrical post magnetron (target in center)
- Cylindrical hollow magnetron (cylindrical shell target)
- Cylindrical post magnetron with reflecting surface
- Cylindrical hollow magnetron with reflecting surface

One of the problems with this geometry is that the magnetic field and thus the deposition rate falls off at the ends of the tube and can actually accumulate material sputtered from the interior section. This is rectified by using electrostatic and magnetic end confinement.

The cylindrical magnetron can deposit virtually any material that other magnetrons are capable of. Reactive sputtering, RF sputtering, and mid frequency sputtering are readily accomplished using standard techniques and power sources. One disadvantage is that the sputtering targets are more challenging to fabricate, and this may add cost (but high materials usage should compensate for this added cost). Many semiconductor materials are very difficult to form in cylindrical and shell-like shapes. Because of the high materials

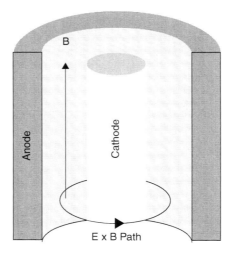

Figure 2.17 Basic cylindrical post magnetron configuration and associated magnetic field [3].

usage, the capability to coat tubes, wires, three-dimensional and very wide substrates, and wide range of materials, the cylindrical magnetron is perfectly positioned for industrial applications.

A wide variety of thin film materials are deposited by reactive sputtering using cylindrical magnetrons [51, 52]. This process works particularly well for tribological coatings since many parts are three dimensional. High quality TiN and ZrN have been reactively deposited. Table 2.6 compares the costs for conventional magnetron and cylindrical magnetron processes [51]. We see that for comparable target masses, significantly more parts can be coated using the cylindrical magnetron and the cost per part is ~1/5 that of planar magnetrons. Hardness of the TiN and ZrN films was comparable to the best films deposited by unbalanced and closed field magnetron sputtering processes.

Deposition of multilayer coatings or nanocomposites requires at least two different target materials in the cathode. The dual inverted cylindrical magnetron, shown in Figure 2.18, consists of two coaxial targets in a common magnetic field, and has the advantage of providing a larger target area and the capability of cosputtering and deposition of different layer materials.

Below is a partial list of materials that have been successfully deposited by cylindrical magnetrons:

- Metal oxides
- Metal nitrides
- Carbides
- Metals and metal alloys
- Semiconductors
- Multilayers
- Graded compositions
- Ceramics
- Borides
- Transparent conductive oxides and nitrides
- Alloys
- Nanocomposites
- Coating combinations
- Nanocomposites: figure of sectioned ICM and magnetic field
- Variations
- New models
- Typical substrates and components

Table 2.6 Comparison of planar and cylindrical magnetron processes for TiN and ZrN coatings [51].

	Planar Magnetron	Cylindrical Magnetron
Target mass (kg)	4.5	4.5
Efficiency	50%	75%
Utilization	30%	70%
Parts per target	11,700	41,100
Target cost	$1,500	$1,000
Material cost/part	$0.128	$0.024

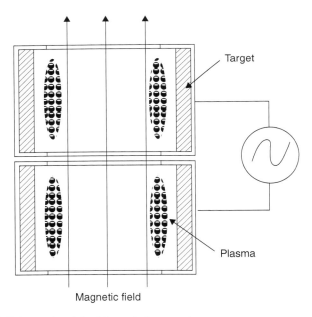

Figure 2.18 Schematic of dual inverted magnetron.

The dual inverted cylindrical magnetron is also used to deposit nanocomposites, such as Mo-Ti-N [53]. Mo and Ti targets were placed in the basic dual inverted configuration. The target materials were reactively sputtered in N_2/Ar mixtures for forming the nanocomposite. Here particles TiN were embedded in a Mo-N matrix. Hardness depended on the particle size as shown in Figure 2.19. Impressive hardness values ~ 32 GPa were obtained for particles sized ~ 2.2 nm.

The rotatable cylindrical magnetron is used in large area vacuum web coating applications [54, 55]. Glass panels as wide as 3.2 m have been coated. The ITO target is fabricated in a single piece; no tiling is used. Regions between tiles can cause arcing, nodule formation on the substrate, particulates, and leaking of target bonding material. Production advantages of the rotatable cylindrical magnetron are:

- Threefold increase in power density to the target
- Improved target utilization of 2.5 times better than planar magnetrons
- Reduced process times and costs
- Deposition on plastic, ceramic, and magnetic substrates

2.1.7 High Power Pulsed Magnetron Sputtering (HPPMS)

The latest development in magnetron sputtering technology is high power pulsed magnetron sputtering (HPPMS), also known as high power impulse magnetron sputtering (HIPIMS) [56]. HPPMS

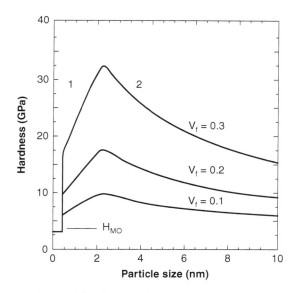

Figure 2.19 Dependence of hardness on TiN particle size for Ti-Mo-N nanocomposites [53].

also involves advanced power supply development. The requirements for the magnetron power supply have increased over the last several decades. At a minimum, the power supply must be stable during long coating runs (often as long as several days), be stable during arcing of the target, and be able to handle voltage spikes. The type of power output ranges from radio frequency (RF), direct current (DC), bipolar pulsed DC, and mid frequency. Power can be controlled by constant voltage, constant current, or constant power modes. RF power is generally more complicated and is used for nonconducting target materials (insulators, ceramics, some semiconductors). The sputtering target must be electrically conducting for use with DC and mid frequency supplies. This, however, does not preclude the use of RF supplies with conducting target materials. Output powers range from 0.5 kW to well over 30 kW.

Table 2.7 lists the various types of magnetron power supplies.

Arc suppression has always been a necessity for reactive sputtering and a challenge for the power supply. Early DC supplies used mercury vapor rectifiers [57]. Arc suppression units evolved from silicon controlled rectifiers to switch mode technology. While arc suppression is not as critical a problem as with DC and low frequency supplies, it is also available for RF supplies [58]. There is considerably less arcing with bipolar pulsed DC supplies.

HPPMS uses a large energy impulse supplied to the cathode in a very short time period, typically ~100 µs. This requires a very

Table 2.7 Magnetron power supplies.

Type	Power Range (kW)	Frequency Range	Arc Suppression
DC	Up to 200	<60 Hz	Necessary
RF	Up to 100 W RF	13.56 MHz	
Bipolar pulsed DC	Up to 200	0–5 kHz	Reduced arcing
Mid frequency	Up to 60	40–460 kHz	
Pulsed	3 MW peak/ 20 kW avg.	500 Hz	Yes

different type of power supply [59]. The HPPMS process delivers a large low energy flux of ions to the substrate. Peak powers up to 3 MW/pulse with pulse widths between 100 and 150 μs must be generated by the power supply. Average powers are ~20 kW with frequencies up to 500 Hz. In addition to supplying pulsed power, arc suppression is also necessary. This process takes advantage of enhanced ionization resulting from the high energy pulse. Power densities applied to the target are in the neighborhood of 1–3 kW/cm^2 (compare this to traditional magnetron sputtering with power densities ~1–10 W/cm^2) [60].

The high power pulse is the core of this process. Figure 2.20 shows a typical waveform of the target voltage, target current, and ion current of one of the first HPPMS systems sputtering from a $Ti_{0.5}Al_{0.5}$ target [61]. The voltage pulse is between 1.3–1.5 kV and total pulse duration is ~120 μs. Note that the target current increases and peaks at ~200 A as the voltage pulse decays. The power density at this point is ~600 W/cm^2. Ion current has a broad peak at ~60 mA at 50 μs into the voltage pulse.

Another manifestation of HPPMS uses another method of generating a high density plasma based on formation of one long modulated pulse that creates weakly and strongly ionized plasmas [62]. Pulse durations for the modulated power plasma method are 1–2 ms. Figure 2.21 shows the breakdown of the waveform from a

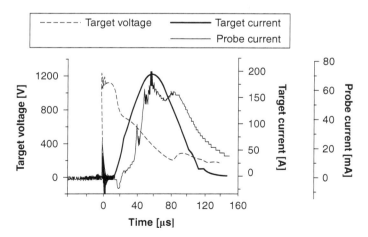

Figure 2.20 Typical waveform of the target voltage, target current, and ion current during the HPPMS process [61].

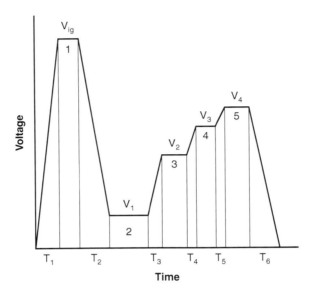

Figure 2.21 Stages of a high voltage pulse from a modulated power plasma generator [63].

modulated power plasma generator using Ti and Al targets [63]. Stage 1, between times T_1 and T_2 correspond to plasma ignition (ignition voltage V_{ig} is 1500 V). Different voltage levels are observed for the four subsequent stages. Power levels of 10–200 kW per pulse are required to sustain a discharge, and pulse repetition can vary from 4 to 1000 Hz.

Even though very high current densities are used, the resultant deposition rate for HPPMS can be lower than conventional magnetron sputtering; as much s 25–30 % lower [64, 65]. The modulated power plasma technique claims to have overcome this disadvantage [62, 63]. Deposition rates comparable to or higher than DC sputtering are attributed to the longer pulse length (1–3 ms compared to 150 µs for pulsed supplies). Most of the sputtering occurs during the strongly ionized plasma stage (T_3–T_6 in Figure 2.21).

As a result of the highly ionized plasma and very high power density, films deposited by HPPMS can have very different microstructures and properties. One distinct advantage of HPPMS over DC magnetron sputtering is that the films are generally denser and smoother, and can have lower mechanical stress. Figure 2.22 shows SEM images of two Ta films, one deposited by DC sputtering (2.22a) and the other deposited by HPPMS (2.22b). The columnar structure is much more pronounced in the DC sputtered film [63].

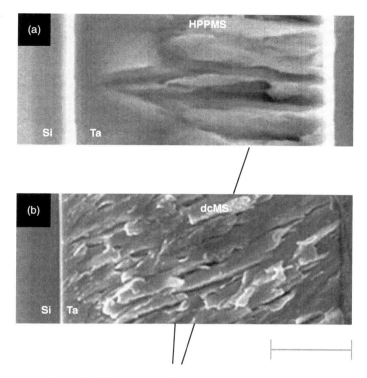

Figure 2.22a and2. 22b SEM images of the microstructure of DC magnetron sputtered and HPPMS Ta films [63].

As a result, HPPMS films should then have improved tribological, optical, electrical, and environmental properties. The improved microstructure improves the optical properties as well [66, 67, 68, 69, 70, 71, 72]. It is well known that the refractive index depends on the density of the optical thin film. The refractive index of less dense films is generally lower than films with higher density. TiO_2 films are a perfect example of the dependence on density. The refractive index of this material can vary from 2.2–2.5 depending on density. These films also have a density of 3.83 g/cm^3 and very low surface roughness of 0.5 nm, compared to a density of 3.71 g/cm^3 and surface roughness of 1.3 nm for DC films. Very high refractive indices ~2.72 are also reported for TiO_2 films, if the optimum duty cycle for the power supply is used [68]. Improved optical properties for SiO_2, ZnO, Al_2O_3, Ta_2O_5, and ZrO_2 films have also been reported [67, 70, 71]. These results are impressive, but have to be taken with certain caveats. HPPMS films do not always have improved performance compared to films deposited using other power supplies [73]. Optical constants always depend to a large extent on deposition conditions, and the optimum set of conditions must be determined for each material and deposition system. Very high refractive indices are also reported for planar magnetron sputtered films. In fact, in some cases, mid frequency magnetron sputtering resulted in better performance than HPPMS [73].

Another advantage of the HPPMS process should be improved chemical and environmental stability. Silver thin films are notorious for lack of chemical and environmental stability, particularly very thin films. While there is still much testing to be performed, Ag films deposited by HPPMS appear to have improved stability and optical performance in multilayer structures [72].

With improved density and smoothness, this process has the potential to improve the performance of tribological coatings [74–78]. Films with dense microstructure and smooth surfaces are preferred in many applications due to the increased corrosion and wear resistance, and reduced friction. Hard materials such as TiN, CrN_x, Cr_xN_y, and Ti_3SiC_2 have all been deposited by HPPMS. TiN films with a very fine grain structure compared to those deposited by DC magnetron sputtering have been reported [1976]. CrN films also have a much finer microstructure than even UBM sputtered films. Hardness values near 25 GPa were reported, and the sliding wear coefficient decreased from 7 to 0.2 [78].

While HPPMS is a promising process to deposit dense smooth films with reduced mechanical stress, it must still compete with other magnetron and deposition processes. In my opinion, the jury is still out on whether or not it is a giant step forward or just useful in certain applications. The real proof might be how well it adapts to industrial applications.

2.1.8 Dual Magnetron and Mid Frequency Sputtering

Biomedical applications for sputtered thin films are increasing. An interesting example is that cylindrical magnetron sputtering is being used to deposit protective TA coatings on batches of medical stents. A picture of stents implanted in a patient is shown in Figure 2.23.

High deposition rate and target utilization are very important for industrial processes. While deposition rates of magnetron cathodes are relatively high, they are not as high as those of evaporation processes. The dual magnetron/pulsed magnetron configuration achieves both high deposition rates and improved

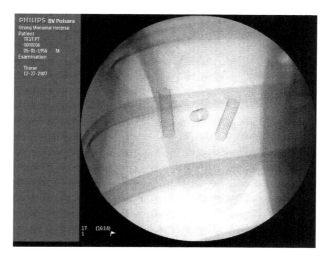

Figure 2.23 Ta-coated stents implanted in a heart patient (courtesy Isoflux).

materials utilization [79, 80, 81]. Dual magnetron sputtering uses a mid- frequency (~40 kHz – 300 kHz) pulsed power source and two magnetron cathodes. The dual magnetron configuration is shown in Figure 2.24, and a typical power pulse is shown in Figure 2.25. In its simplest form, this power source supplies a positive pulse to one magnetron cathode during the first half of the cycle while negatively biasing the other cathode and then supplying a positive pulse to the other magnetron cathode while negatively biasing the other cathode. In this manner, one cathode acts as an anode while the other is the sputtering cathode. Sputtering only occurs during negative bias. This process is very amenable to reactive sputtering [82, 83, 84].

Pulsed magnetron sputtering overcomes two disadvantages of magnetron sputtering:

- Loss (or hiding) of anode due to deposition of insulating layers
- Charging and arc formation during reactive sputtering

As a result of the pulsed power supply and pulse polarity reversal, charging of the target during reactive deposition is virtually eliminated.

Dual magnetron technology was initially developed to deposit insulating materials such as oxides, nitrides, and transparent

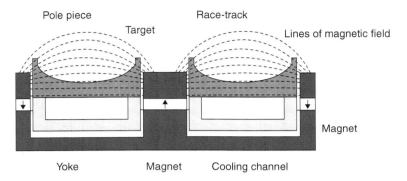

Figure 2.24 Dual magnetron cathode configuration [79].

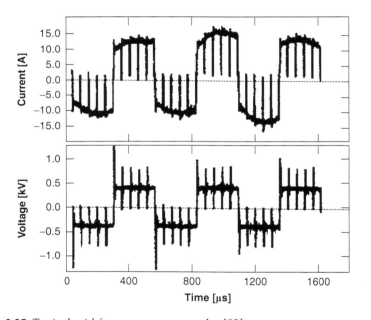

Figure 2.25 Typical mid frequency power pulse [82].

conducting oxides [82]. The planar magnetrons with identical targets are usually located next to each other but can be farther apart with different target materials in cosputtering configurations [83]. Arcing is reduced by secondary electrons ejected by the negatively biased target (cathode) being attracted to the positively charged target (anode) and essentially neutralizing positively charged surfaces that have built up where insulators have been deposited. Table 2.8 compares the deposition rate and optical properties of

Table 2.8 Comparison of the optical properties of DC and mid frequency sputtered films [79, 80].

Material	Deposition Rate (nm/min)	n @ 550 nm
TiO$_2$ (DC)	9	2.35
TiO$_2$ (MF)	50	2.65
Si$_3$N$_4$ (DC)	7	2.01
Si$_3$N$_4$ (MF)	40	2.05
SnO$_2$ (DC)	40	1.95
SnO$_2$ (MF)	90	2.01
ZnO (DC)	45	2
ZnO (MF)	120	2.05

four materials deposited by conventional DC magnetron sputtering and mid frequency sputtering [79, 80]. We see in all cases that the deposition rate can increase by as much as a factor of 5. The refractive index at 550 nm also increases in all cases, demonstrating increased density.

Other thin film materials that have benefited from mid frequency dual magnetron sputtering are SiO$_2$, Al$_2$O$_3$, MgO, CrN, ITO.

An advantage of cosputtering using dual magnetrons and a mid frequency power supply is that the amount of each component can be precisely controlled [83]. Obviously two magnetrons are needed for this process, but using a mid frequency supply allows reliable regulation of the power supplied to each target and independently regulating power to each target. This results in more control over the relative compositions.

2.1.9   Ion Beam Sputtering

Ion beam sputtering combines ion source and sputtering technologies, and has produced dense, defect-free thin films. As a result numerous applications, such as optical coatings, tribological coatings, semiconductors, heads for disk drives, and electronic materials can all be deposited by this process. Figure 2.26 shows the basic ion

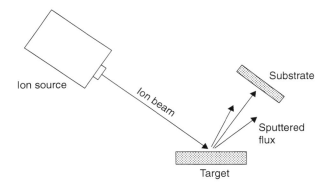

Figure 2.26 Ion beam sputtering configuration.

beam configuration. Ion beam sputtering (IBS) permits control over the energy and current density of the bombarding ions. Referring to Figure 2.1, energetic ions from an ion source are incident obliquely onto a sputtering target. Substrates are suitably placed to receive the sputtered material. Ion current and voltage can be independently controlled, and because an ion source is used, the deposition takes place at very low chamber pressure (0.1 mTorr). As we shall observe in several examples, IBS has the following advantages:

- Films have excellent adhesion and high density due to the high energy of the ion beam
- Independent control over ion kinetic energy and current density
- Directional control of the ion beam
- Low deposition pressures
- High deposition rates
- No sputtering gas trapping in films
- No substrate heating
- All types of materials (conductors, insulators, semiconductors, etc.) can be sputtered

The main disadvantages are

- An independent ion source is needed
- Substrate size is limited by the size of the ion beam
- Possible contamination from the ion source
- Ion source maintenance

The ion source is the heart of the IBS process The three major types, Kaufman-type, end-Hall, and cold cathode produce ions with very different energy distributions and energy ranges (see Table 2.3). Many ion sources can operate either with DC or RF power supplies.

The thickness distribution at the substrate is directly related to the profile of the ion beam [85, 86]. Masking of the ion beam to improve thickness uniformity is often involved [86]. Figure 2.27 shows the profile of Ti deposited through a mask in the configuration shown in the inset. The thickness distribution is somewhat narrower than a cosine distribution and thickness uniformity is ±5% over the region 15° from target normal.

Contaminants can be a problem in some cases. While inert gas trapping is minimal, reactive gases can be incorporated. Metal contaminants result from components of the system subjected to ion bombardment. Similar to ion assist, material sputtered from the filament of a Kaufman-type ion source can be a problem.

Deposition rate depends on ion current density and substrate-target configuration.

Reactive deposition is also possible with ion beam sputtering. Three methods that are generally used:

- Inert gas ion beam + reactive gas at the substrate
- Inert + reactive gas in the ion beam
- Dual ion beam using inert gas to sputter the target + reactive gas ion beam to bombard substrate

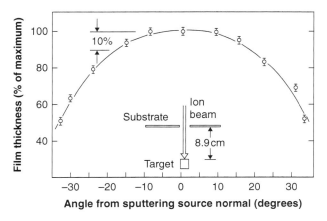

Figure 2.27 Thickness distribution of ion beam sputtered Ti film [86].

With IBS, as with reactive evaporation, it is usually necessary to add reactive gas to achieve a stoichiometric or desired composition at the substrate. This is accomplished by adding the reactive gas in the region of the substrate. Because the inert gas is used to sputter the target, deposition rates are highest in this case. Deposition rates decrease slightly when reactive gas is added to the inert gas, but the reactive gas is activated and more energetic than in the first case. In the third case, the sputtering beam and reactive gas can be varied independently giving more control over the composition of the film and deposition rate. As with reactive magnetron sputtering, there is hysteresis behavior and it is possible to poison the target and significantly reduce deposition rate [87].

The likelihood of target poisoning is reduced using the third technique. Stoichiometric films are deposited at reactive gas levels above a minimum value [88]. Films have similar properties as RF sputtered films.

Related processes include dual ion beam sputtering (DIBS) and ion beam enhanced deposition (IBED). The processes are essentially the same: DIBS involves two independent ion sources, one used for sputtering and one used for ion assist, while IBED combines ion bombardment with ion beam sputtering or evaporation. DIBS and IBED configurations are shown in Figure 2.28 [84, 89]. The system used in Figure 2.28 uses two gridded RF ion sources, a larger 16 cm source to sputter the target and a smaller 12 cm source for ion assist.

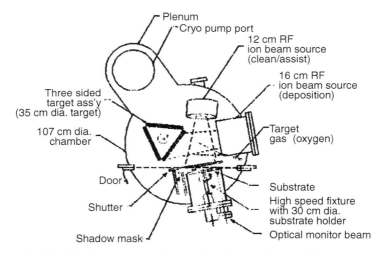

Figure 2.28 Dual ion beam sputtering and ion beam enhanced deposition configurations [84, 89].

Ion beam parameters are independently adjustable. A rotating substrate holder is used here. High deposition rates are not an advantage of this process; deposition rate is a low 2–4 Å/s. Several process improvements are possible in the dual ion source configuration:

- Elemental targets are used for reactive sputtering, thus increasing deposition rate
- Replacing Ar sputtering gas with Kr significantly increases sputtering yield and deposition rate (elastic properties can also be modified)
- Deposition rate increases at larger angles of incidence
- Deposition rate can be optimized by choice of ion source parameters (voltage and current)

As mentioned earlier, the advantages of IBS, DIBS, and IBED are very dense films with excellent adhesion. This is important in all aspects of thin film technology, but particularly useful in optical, tribological, and conductive coating materials [89, 90]. Even with the restrictions on substrate size, IBS optical coatings have demonstrated low scatter and absorption [84], low laser damage thresholds, near-bulk refractive indices, and improved mechanical properties [91]. This process is used to deposit high reflectors, partial reflectors, polarizing and nonpolarizing beam splitters, telecommunication filters, TCOs, metal films, and multilayer coatings. Scatter and optical absorption losses as low as 1 ppm and AR reflectance of < 0.01% have been reported [92]. Because of their low scatter and optical absorption, IBS coatings are particularly well suited for laser optics applications.

IBS can improve the mechanical properties of materials, such as fluorides, that typically demonstrate poor adhesion and tensile stress [13, 93, 94]. Stress in PVD YF_3 and LiF coatings tends to be unstable and becomes more compressive with aging, while stress is always compressive in IBS films (needed for good adhesion and stability) and remains ~ constant with aging for YF_3 films and decreases with aging for LiF films.

2.1.10   Filtered Cathodic Arc Deposition

Cathodic arc deposition (CAD) is a subset of the PVD family but deserves a full detailed description since it is now being used extensively in a number of applications and in hybrid processes.

The cathodic arc process has been around since the 1880s, and became an industrial process in the 1970s in the then Soviet Union and elsewhere in the 1980s. Cathodic arcs take a special place among PVD processes because of high ion densities; the "vapor" is ionized (up to 80%) at cathode source. Therefore, any bias applied to a substrate has a significant effect without the need for further ionization or plasma production (although ion assist can still be used). Other members of the "arc" family are cold cathode and anodic arc deposition. The first rudimentary system was invented by Thomas Edison in 1884 [95]. This vacuum process was originally used to deposit low grade hard coatings. Arc vaporization occurs when a high current, at low voltage, is passed through a source material placed between two electrodes, the anode and cathode. The surface of the electrode (cathode or anode), often heated to improve conductivity, is vaporized and forms a plasma of the ejected molten material. Cathodic arc processes thus result in a very high degree of ionization of the source material and create ion acceleration up to supersonic velocities (4700–23100 m/s) [96]. Ion plating is thus readily accomplished even in the absence of substrate bias. This process, however, produces particles with a wide range of sizes, and resulting coatings have large particulates (known as macroparticles) in them. Macroparticles, ions and neutrals can degrade coating performance and overall coating integrity.

In addition to the macroparticle issues, the major problems faced by the cathodic arc process are:

- Variable particle count and deposition rate
- Large area coverage not possible
- Poor thickness uniformity
- Poor film quality
- Poor arc control
- Poor repeatability and reliability for production processes
- High maintenance

The filtered cathodic arc (FCA) process was developed to address these problems and eliminate the macroparticles that degrade coating performance. Figure 2.29 shows a schematic diagram of a typical FCA deposition system. An ignition device is used to generate the arc. The arc source and cathode and anode are basically the

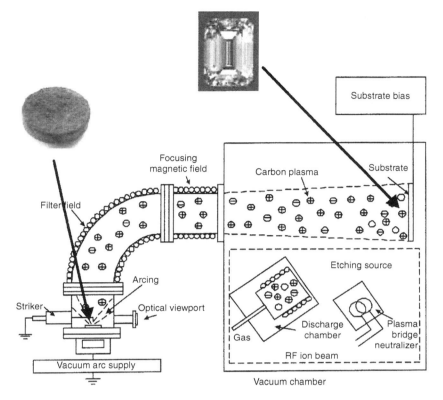

Figure 2.29 Schematic diagram of filtered cathodic arc deposition system.

same as the cathodic arc process, but the ejected particles are fed into a curved duct that has a focusing magnetic field and a steering field that separates the particles by mass, thus "filtering" them (much like a mass spectrometer). Unwanted macroparticles and neutral atoms are filtered out by mechanical filters, and coating species reaching the substrate are pure ions. The thickness distribution is nonuniform over large areas and, as a result, the beam of particles must be rastered over the substrate to coat large areas with acceptable uniformity.

Basic physics tells us that a moving charged particle (in this case an ion) is deflected by a magnetic field. Only ions are deflected by the magnetic field. The force on the ion is $F = q v X B$, where q is the charge on the ion, v is its velocity and B is the magnetic field strength, both are vector quantities. The cross product is $v X B$, which deflects the particle with a force perpendicular to the direction of flight, thus forming a curved path. The radius of the path of the deflected ion is proportional to the mass of the ion and its velocity [97]. The energy and path of the ions can thus be tuned to the mass of the ion by adjusting the curvature of the filter field and particle duct.

Historically, FCAD had the following drawbacks:

- Variable process control and stability
- Poor thickness uniformity
- Deposition of multilayer films difficult
- Codeposition requires a hybrid process
- Anode caking and poisoning a major problem
- Limited substrate rotation options

An ignition device is used to create the arc, and particles are vaporized at the cathode. All particles are passed through a filtering electric field and focusing magnetic field, and the ion-containing plasma exits from the curved duct and are deposited onto the substrate. The energy of the ions can be tuned by the electric field strength and can range from 20 to 3000 eV, similar to that of an ion source. Compare this to the 10–100 eV energy of sputtered atoms and we see that a much wider range of ion energies is possible with FCA. Typical chamber pressures during deposition are $\sim 10^{-6}$ Torr (much the same as evaporation) so incorporated gas impurities can be kept to a minimum. While deposition parameters are system dependent, the process has been improved over the last decade and state-of-the-art production systems now can claim the following performance results [98]:

- Substrate temperature <80 °C
- Medium substrate areas
- Solid source material
- Thickness uniformity ~4%
- Repeatability ~5%
- Good film quality
- Low operating costs

It should be noted that a conductive source is required, but reactive deposition can also be achieved by introducing a reactive gas into the plasma. Often the source must be heated to make it conductive. For example, semiconductors become very conductive when heated. Deposition rates can be very high ~ mm/hr.

Table 2.9 shows the materials that have been deposited by FCVA processes. Nanocomposites and diamond-like carbon (DLC) hard wear resistant films have been successfully deposited. High quality optical coatings are now being deposited [99].

FCAD processes have been very successful for deposition of tribological coatings, e.g., Ti-Si-N compositions with superhardness values in excess of 40 GPa. The enhanced properties of these coatings are attributed to the refinement of grain size resulting from the control of particle sizes and energies. One or more phases are present at the nano-scale to form a nanocomposite layer. Nanocomposite coatings consisting of TiN nanoparticles embedded in a Si_3N_4 amorphous matrix have been deposited by a hybrid combination of FCA and magnetron sputtering processes [100]. In general, the composite can be described by the composition nc-MeN/a-Si_3N_4 (where MeN is a transition metal nitride such as VN, TiN, NbN, ZrN, etc. in an amorphous silicon nitride matrix).

Table 2.9 Thin film materials deposited by FCVA processes.

Application	Material
Data storage	ta-C
	Al_2O_3
Precision engineering and tools	ta-C
	TiN, TiC, AlN, WC
Antiscratch	Al_2O_3,ZrN
Microelectronics: metallization	Cu, Ta, TaN
Metallization	Ti, Cr
Transparent electrode	ITO
Wear resistance	DLC
	Ti-C-N
	TiN/Si_3N_4 nanocomposite
Optical coating	TiO_2, SnO_2, WO_3, AlTiOx, $AlSiO_x$, ZnO, AlN, $AlSiN_x$, Si_3N_4, and Nb_2O_5

Nanocrystals are relatively free of dislocations, and when high stress is applied any dislocation movement that may be present is trapped at the grain boundaries by the amorphous matrix resulting in an increase in strength.

A number of high quality oxide and nitride films have been reported, including SnO_2, WO_3, $AlTiO_x$, $AlSiO_x$, ZnO, AlN, $AlSiN_x$, Si_3N_4, and Nb_2O_5. SnO_2 films with low resistivity $\sim 8 \times 10^{-4}$ Ω.cm have been reported [101, 102]. WO_3 films have been developed for electrochromic windows [103].

The drawbacks with FCAD for optical coatings are

- Control and reliability of the process and properties
- Uniformity over large areas
- Deposition of multilayer films

Multilayers can be deposited using two independent sources, but the precise control needed for optical thin films has not been fully achieved. Another problematic material is SiO_2.

Virtually every deposition process has strengths and weaknesses. The major advantages of FCA process are high deposition rate, tunable ion energies that can improve coating adhesion and density, only ions are deposited, low substrate temperatures, dense coatings and combinable into hybrid processes.

2.2   Chemical Vapor Deposition

Chemical vapor deposition (CVD) is parent to a family of processes whereby a solid material is deposited from a vapor by a chemical reaction occurring on or in the vicinity of a normally heated substrate surface. The resulting solid material is in the form of a thin film, powder, or single crystal. By varying experimental conditions, including substrate material, substrate temperature, and composition of the reaction gas mixture, total pressure gas flows, etc., materials with a wide range of physical, tribological, and chemical properties can be grown. A characteristic feature of the CVD technique is its excellent throwing power, enabling the production of coatings of uniform thickness and properties with a low porosity even on substrates of complicated shape. Another important feature is the capability of localized, or *selective*, deposition on patterned substrates.

CVD and related processes are employed in many thin film applications, including dielectrics, conductors, passivation layers, oxidation barriers, conductive oxides, tribological and corrosion resistant coatings, heat resistant coatings, and epitaxial layers for microelectronics. Other CVD applications are the preparation of high temperature materials (tungsten, ceramics, etc.) and the production of solar cells, of high temperature fiber composites, and of particles of well-defined sizes. Recently, highT_c superconductors, and more recently carbon nanotubes, have also been made by this technique [102]. Since oxygen activity in the vapor can be precisely controlled during the deposition, no annealing in oxygen is needed to achieve superconductivity.

There exist a multitude of CVD processes, listed in Table 2.10. In thermally activated CVD (TACVD), the deposition is initiated and maintained by heat. However, photons, electrons, and ions, as well as a combination of these (plasma activated CVD), may induce and maintain CVD reactions. In this chapter, the underlying principles of TACVD are introduced. In addition large-area deposition and selective CVD on patterned substrates are discussed.

Table 2.10 Summary of CVD process family.

Type	Pressure Range	Description
Atmospheric pressure CVD (APCVD)	High - atmospheric	Processes at atmospheric pressure
Low pressure CVD (LPCVD)	Low	Processes at subatmospheric pressures
Ultrahigh vacuum CVD (UHVCVD)	Typically below 10^{-6} Pa (~10^{-8} Torr)	Processes at a very low pressure
Aerosol assisted CVD (AACVD)		Precursors are transported to the substrate by means of a liquid/gas aerosol, which can be generated ultrasonicwally

Table 2.10 (cont.) Summary of CVD process family.

Direct liquid injection CVD (DLICVD)		Precursors are in liquid form (liquid or solid dissolved in a convenient solvent). Liquid solutions are injected in a vaporization chamber towards injectors (typically car injectors). Then the precursor's vapors are transported to the substrate as in classical CVD process.
Microwave plasma-assisted CVD (MPCVD)		
Remote plasma-enhanced CVD (RPECVD)		Utilizes a plasma to enhance chemical reaction rates of the precursors, and allows deposition at lower temperatures,
Atomic layer CVD (ALCVD) or ALD		Deposits successive layers of different substances to produce layered, crystalline films.
Hot wire CVD (HWCVD)		Also known as catalytic CVD (Cat-CVD) or hot filament CVD (HFCVD). Uses a hot filament to chemically decompose the source gases
Metallorganic chemical vapor deposition (MOCVD)		Based on metalorganic precursors

Table 2.10 (cont.) Summary of CVD process family.

Hybrid physical chemical vapor deposition (HPCVD)		Vapor deposition processes that involve both chemical decomposition of precursor gas and vaporization of solid a source.
Rapid thermal CVD (RTCVD)		Use heating lamps or other methods to rapidly heat the substrate.
Vapor phase epitaxy (VPE)		

2.2.1   Basic Chemical Vapor Deposition (CVD)

CVD processes are generally characterized by the decomposition of one or more feed gases (precursors) and reaction of these gases on a substrate to form a thin film. They are used extensively in the semiconductor industry and excel at coating nonplanar objects. Other applications include tribological coatings (wear and corrosion resistance), erosion protection, high temperature protection, integrated circuits, sensors and optoelectronic devices, structural parts (on a mandrel), optical coatings for telecommunications, composites (carbon-carbon, carbon-silicon-carbon, silicon carbide), powder production, catalysts, and micromachined devices.

Figure 2.30 outlines the basic CVD process and reactor. Referring to the figure, a typical CVD apparatus consists of several basic components:

- Gas delivery system for the supply of precursors to the reactor chamber, consisting of process gases, gas metering, and manifolds
- Reactor or deposition chamber with substrate handling, heating
- Substrate loading system
- Process control equipment, including gauges, controls, etc. to monitor process parameters such as pressure, temperature, and time. Safety devices would also be included
- Energy source to provide heat, electrical or magnetic energy for the reaction/decomposition to take place

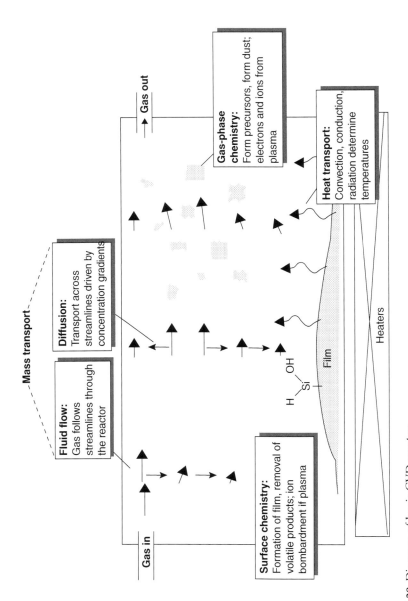

Figure 2.30 Diagram of basic CVD reactor.

- Exhaust system for removal of volatile by-products from the reaction chamber
- Exhaust treatment systems – exhaust gases may not be suitable for release into the atmosphere and may require treatment or conversion to safe/harmless compounds.

The basic chemical reaction is

Gaseous reactants → Solid material + Gaseous products [103]

Precursors used in CVD processes are usually toxic, corrosive, and expensive, and, above all, safe handling and delivery systems must be in place. Gas cabinets are used to store the gases safely. Precursor gases (often diluted in carrier gases) are delivered into the reaction chamber at approximately ambient temperatures. Precursors for CVD processes must be volatile, but at the same time stable enough to be able to be delivered to the reactor. As they pass over or come into contact with a heated substrate, they react or decompose and form a solid phase which is deposited onto the substrate as a thin film. The substrate temperature is critical and can influence what reactions will take place. The actual CVD process is much more complicated than this and I will present mainly simple examples of each type of reactor and material deposition. Energy is supplied to the reaction by resistive heating (tube furnace), radiant heating (halogen lamps), RF heating (microwave, induction), lasers, and high energy light sources.

A list of common precursors is shown in Table 2.11.

CVD processes involve a detailed understanding of gas chemistry, gas flow dynamics, thermodynamics, and physics. CVD reactions can be classified into a few important categories:

- Thermal decomposition reactions
- Reduction reactions
- Exchange reactions
- Disproportionation reactions
- Coupled reactions

The general *thermal decomposition* reaction is defined by

$$AX(g) \rightarrow A(s) + X(g)$$

Where AX is thermally broken up into solid (s) A and gaseous (g) reaction product X. Examples of this are [102]

$SiH_4 \rightarrow Si(s) + 2H_2(g)$ (silane decomposition: actually the Si has residual H bonded in an amorphous network)

$$Si(CH_3)Cl_3(g) \rightarrow SiC(s) + 3HCl(g)$$

Reduction reactions can be expressed as

$$2AX(g) + H_2(g) \rightarrow 2A(s) + 2HX(g).$$

Examples are

$$WF_6(g) + 3H_2(g) \rightarrow W(s) + 6HF(g) \text{ and}$$

$$SiCl_4(g) + 2H_2(g) \rightarrow Si(s) + 4HCl(g)$$

In an *exchange* reaction, an element replaces another:

$$AX(g) + E(g) \rightarrow AE(s) + X(g), \text{with examples}$$

$$Zn(g) + H_2(g) \rightarrow ZnS(s) + H_2(g)$$

Table 2.11 Common CVD precursors.

Type	Examples
Halides	$TiCl_4$, $TaCl_5$, WF_6
Hydrides	SiH_4, GeH_4, $AlH_3(NMe_3)_2$, NH_3
Metal organic compounds	Alkoxides such as dimethylami-noethoxide [$OCHaCH_2NMe_2$], or methoxyethoxide [OCH_2CH_2OMe]
Metal alkyls	$AlMe_3$, $Ti(CH_2tBu)_4$
Metal alkoxides	$Ti(OiPr)_4$
Metal dialylamides	$Ti(NMe_2)_4$
Metal diketonates	$Cu(acac)_2$
Metal carbonyls	$Ni(CO)_4$

$$SnCl_4(g) + O_2(g) \rightarrow SnO_2(g) + 2Cl_2(g)$$

Disproportionation reactions are seldom used in CVD processes. Here the oxidation number of an element both increases and decreases through the formation of two new species [102]. There are several types:

$$2AX(g) \rightarrow A(s) + AX_2(g)$$

$$3AX(g) \rightarrow 2A(s) + AX_3(g)$$

$$4AX(g) \rightarrow 3A(s) + AX_4(g)$$

Finally, coupled reactions involve two or more individual reactions. An exchange reaction can be coupled with a reduction reaction [103].

The main types of CVD reactors are

- Horizontal or vertical tube
- Showerhead
- High density plasma
- Linear injector
- Atmospheric

Each design employs different methods for performing the fundamental operations common to all CVD reactors: dispensing gases, controlling temperature, introducing a plasma if desired, and removing byproducts. Much like PVD processes, flow meters and mixing manifolds are needed to mix and dispense the correct gas mixture to control the reaction. A number of processes take place for thin film deposition to occur. First the gas must flow uniformly into the chamber. Gas can also diffuse through the chamber across streamlines created by concentration gradients. Gas phase chemistry is critical to form the required precursors, electrons, and plasma species in the reactor. Heat transport is by convection, conduction, and radiation and determines the temperature profile in the reactor and on the substrate. Finally, surface chemistry determines the final composition of the deposited thin film and is also important for the removal of volatile byproducts. Plasma bombardment will also affect surface chemistry as it does in PVD processes.

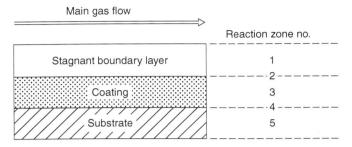

Figure 2.31 Reaction zones used in CVD [102].

As a result of gas flowing over the substrate and the temperatures used for the reaction, there are five reaction zones above the substrate, as shown in Figure 2.31 [102]. The properties of the deposited coatings depend on the interacting processes in each of these zones. Gas flow over the substrate causes a stagnant boundary layer (zone 1) in the vapor just above the substrate surface. Reactions in this zone may result in a poor quality nonadherent coating. Heterogeneous reactions occur in the vapor/coating boundary (zone 2). Film growth takes place in zones 3–5. Solid state reactions such as phase transformations, precipitation, recrystallization, grain growth occur in these zones. Zone 4 is particularly important for film adhesion.

The most basic type of deposition chamber is the horizontal tube reactor, shown in Figure 2.32, also known as a hot wall reactor or cold wall reactor. This is an efficient geometry for stacking round wafers, as shown in the figure, and is commonly employed for "front-end" IC fabrication steps: polysilicon deposition, silicon nitride deposition, and high-temperature silicon dioxide deposition, as well as deposition of doped oxides such as BPSG or PSG. This type of reactor is not normally employed for deposition of metals such as W, Cu, or TiN due to problems in cleaning (see next paragraph).

This basic design for the hot wall reactor is commonly used as a furnace for oxidation and annealing. The tube is typically made of quartz so it can be heated to temperatures near 1000 °C, with a diameter large enough to accommodate substrates or wafers. The tube is placed in a resistively heated box. Many of these heaters have two or three temperature zones. In this configuration, reactions take place in the gas and as a result, substrates and reactor walls are at the same temperature so the material is deposited on the walls as well as the substrate. One risk here is that material can flake off chamber walls and fall down onto the growing film, causing defects.

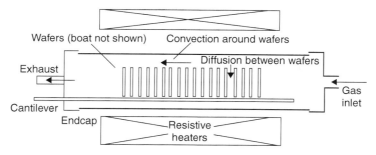

Figure 2.32 Diagram of basic horizontal tube (hot wall) reactor [104].

In the cold wall reactor, only the substrate is heated and no deposition occurs on reactor walls. Homogeneous reactions in the gas are suppressed as a result of the low temperature, and surface reaction become important. There are now steep temperature gradients near the substrate which can cause nonuniform film thicknesses and microstructures. This type of reactor is more popular than the hot wall type. Substrates can be heated by resistive elements placed in the substrate holder (similar to PVD), or by heat lamps, lasers, RF induction, and susceptors [104]. Because this system has a large thermal mass, transient temperature control is challenging. Tens of minutes are usually required to reach stable temperatures, which negates rapid temperature changes. Many systems are now computer controlled.

The reactor can also have vertical orientation, shown in Figure 2.33, which facilitates easier loading and unloading. Substrates are typically stacked vertically in a slotted boat. The gas inlet is at one end of the tube and exhausted at the other end. End caps are used to seal the tube. Operation at low pressures (LPCVD) requires that the ends have o-ring seals that can withstand high temperatures.

Hot wall reactors have a "flat zone" in which the temperature along the axis is nearly constant, and as a result, the coating should be uniform in composition in this zone.

The performance of a tube furnace can be summarized as follows [104]:

- Excellent radial thickness uniformity if pressure is low and the surface reaction is slow
- Axial thickness uniformity depends on substrate placement, operating temperature, gas flow and gas injection geometry
- Large batches are possible

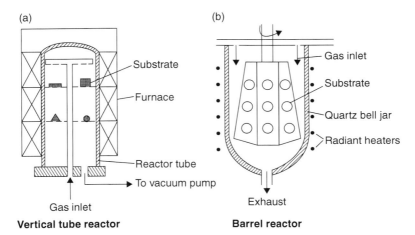

Figure 2.33 Vertical hot wall reactor and barrel reactor [102].

- Process times can be long, but large batch sizes help to compensate for this
- Films can be very pure with minimal contaminants
- High temperature operation possible
- Difficult to implement other processes such as plasma generation

The second type of tube reactor is the showerhead reactor [104], shown in Figure 2.34. Showerhead reactors employ a perforated or porous planar surface to dispense the reactant gases over a planar surface located below. This configuration is a cold wall type since substrates can be heated independently of the walls. The showerhead can also be cooled to prevent deposition on it and clogging of pores. This configuration is amenable to use of plasmas since parallel plates needed to generate the plasma can be readily installed. Also note that the substrate placement is quite different from the tube reactor.

The size of the substrate or batch of substrates essentially determines the diameter of the showerhead. The ceiling height (H_c) is a critical design parameter in that chamber volume, surface to volume ratio, residence time, gas consumption time, and gas radial velocity are all dependent on this dimension. Residence time significantly affects gas phase reactions, and gases must be introduced high enough above the substrate so that a uniform even flow impinges the substrate. H_c, in addition to the showerhead hole configuration,

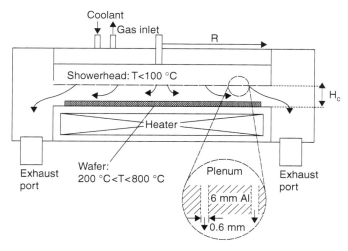

Figure 2.34 Diagram of showerhead CVD reactor [104].

is also important in determining the film's thickness uniformity. It is also possible to use multiple showerheads to avoid mixing gases in the dispensing manifold.

Salient features of this CVD reactor can be summarized as

- Good heat transfer to the substrate
- Zone heating required for good thickness uniformity
- Thickness uniformity depends on temperature and showerhead design
- Amenable to plasma processes
- Easy cleaning
- Amenable to batch processes
- Rignoreapid changes in gases possible (e.g., for multi-layer films)
- If used, plasmas must be carefully contained
- Showerhead pores can easily clog

While a wide range of compositions and microstructures can be deposited by CVD, coatings are generally character-ized by being fine grained, dense, very pure and hard. Virtually any material can be synthesized, including metals, oxides, nitrides, carbides, borides, intermetallic compounds, and some elements.

SiO_2 and Si_3N_4 films are used extensively in semiconductor, optical, and solar applications. SiO_2 can be deposited by several processes:

- Using silane (SiH_4) and oxygen (O_2) precursors with the reaction
  - $SiH_4 + O_2 \rightarrow SiO_2 + 2H_2$
- Or with dichlorosilane and nitrous oxide with the reaction
  - $SiCl_2H_2 + 2N_2O \rightarrow SiO_2 + 2N_2 + 2HCl$
  - or with tetraethylorthosilicate
  - $Si(OC_2H_5)_4 \rightarrow SiO_2 + by\ products$

The choice of source gas depends on the thermal stability of the substrate; Al, for example, melts at 600 °C. SiH_4 deposits between 300 and 500 °C, $SiCl_2H_2 \sim 900$ °C, and $Si(OC_2H_5)_4$ between 650 and 750 °C. SiH_4, however, produces a lower quality oxide than the other methods, and thickness uniformity if poor. Any of these reactions may also be used in LPCVD, but the SiH_4 reaction is also done in APCVD.

The following two reactions deposit nitride from the gas phase:

$$3SiH_4 + 4NH_3 \rightarrow Si_3N_4 + 12H_2$$

$$3SiCl_2H_2 + 4NH_3 \rightarrow Si_3N_4 + 6HCl + 6H_2$$

Si_3N_4 deposited by LPCVD contains ~ 8% hydrogen. Films also have high tensile stress which may cause cracking and delamination in films thicker than 200 nm. An advantage is that the resistivity and dielectric strength are higher than virtually other composition.

Two other reactions may be used in plasma to deposit SiN:H:

$$2SiH_4 + N_2 \rightarrow 2SiN : H + 3H_2$$

$$SiH_4 + NH_3 \rightarrow SiN : H + 3H_2$$

2.2.2   Plasma Enhanced Chemical Vapor Deposition

In the CVD process family plasma enhanced chemical vapor deposition (PECVD) is now used most extensively to deposit virtually any type of thin film material with improved control of properties and improved properties in many cases. It is a valuable method used to deposit optical, tribological, semiconductor thin films, and even composite thin films. Standard CVD processes rely on reaction kinetics, gas transport, and temperature to synthesize various thin film compositions. Similar to PVD processes, introducing plasma into the gas and plasma bombardment of the substrate add another degree of energetics for the formation of thin films, thus the term "plasma enhanced". For example, deposition of Si_3N_4 requires a reactor temperature of 900 °C, which is a major disadvantage and not suitable for many substrate materials. The situation can be mitigated sometimes by using different reactant gases, but the most effective method is to introduce plasma activation. With plasma activation, temperature can be reduced to 350 °C [103]. Reaction gases can be ionized by a number of methods, including DC, RF, microwave and electron cyclotron resonance (ECR) discharges. Another advantage of PECVD is that it is used in hybrid processes, discussed later.

Reactors, typical of the ones diagrammed in Figures 35a and 35b, now have electrodes to supply the plasma activation and chamber pressures need to be lower for the plasma to form [104, 105]. The most often used approaches for generating a plasma are

- A pair of electrodes in a low pressure gas
- A coil wound around a nonconducting tube/reactor that contains the gases
- A pair of electrodes placed on each side of a nonconducting tube containing the reactant gases

As with the conventional CVD reactor, PECVD employs cold and hot wall reactors, but now parallel plates are installed to generate the discharge, as shown in Figure 2.36 for the cold wall reactor. The hot wall reactor is shown in Figure 2.36 [103]. This configuration employs long parallel electrodes to increase batch size. While the DC discharge is the simplest configuration with the plasma created between two conductive electrodes at pressures ~ few Torr, it cannot be used to deposit insulating films. The discharge will quickly extinguish if they are deposited in a DC system. Nonconductive materials require RF activated discharges. A capacitive discharge is excited applying an alternating current (AC) or RF signal between an electrode and the conductive walls

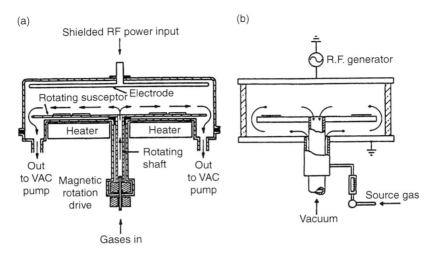

Figure 2.35a and 2.35b Two versions of radial flow PECVD reactor [104, 105].

of a reactor chamber, or between two cylindrical conductive electrodes facing one another. The latter configuration is known as a parallel plate reactor. Frequencies range from of tens to a few thousand hertz, and will produce time-varying plasmas that are repeatedly initiated and extinguished; frequencies of tens of kilohertz to tens of megahertz result in approximately time-independent discharges. The electron cyclotron resonance reactor uses a frequency of 2.45 GHz in a cold wall configuration [108]. A solenoidal magnetic field is used to generate the plasma resonance, hence the cyclotron configuration. This system operates at fairly low pressures ~1 mTorr (similar to PVD) and create a very high degree of ionization.

Frequencies used in PECVD reactors range from DC (0 Hz) to several GHz. Excitation frequencies in the low-frequency (LF) range ~100 kHz, require several hundred volts to sustain the discharge. These large voltages lead to high-energy ion bombardment of surfaces. High-frequency plasmas are usually excited at the standard 13.56 MHz frequency, the same as RF sputtering. At high frequencies, the displacement current from sheath movement and scattering from the sheath assist in ionization, and thus lower voltages are sufficient to achieve higher plasma densities. Thus the chemistry and ion bombardment during deposition can be adjusted by changing the excitation frequency, or by using a combination of low and high frequency signals in a dual-frequency reactor. Excitation

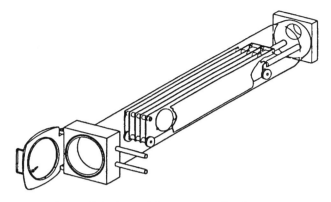

Figure 2.36 Hot wall parallel plate PECVD reactor [103].

power of 10s to 100s W is typical for an electrode with a diameter of 20 cm to 30 cm.

Capacitive (parallel plate) plasmas are typically very lightly ionized, usually resulting in incomplete dissociation of precursors and low deposition rates. Significantly denser plasmas can be created using inductive discharges, generated by an inductive coil excited with a high-frequency signal that induces an electric field within the discharge. Here the electrons are accelerated in the plasma itself rather than just at the sheath edge. ECR reactors and helicon wave antennas have also been used to create high-density discharges. Excitation powers of 10 kW or more are often used in modern reactors.

As stated earlier, frequency and pressure can significantly affect the characteristics of the plasma and properties of the deposited thin film. For example, it has been shown that lower frequency discharges produce films with higher compressive stress because ion bombardment is more intense (note the similarity to ion assist in PVD processes), as shown in Figure 2.37 [106]. More intense ion bombardment has been attributed to a higher sheath potential at lower frequencies. Lighter electrons diffuse out of the plasma first and create a negative bias. There is less time for electrons to diffuse from the plasma between cycles at higher frequencies, which results in less need for the formation of a strong negative bias. Lower energy ions are the result of a weaker negative bias. On the average, the ion experiences only ~1/3 the maximum sheath potential at high frequencies [103].

Based on the above discussion, we see that control of the discharge, electrode geometry, electrode bias, and frequency are the critical first steps in deposition of thin films by PECVD. Process

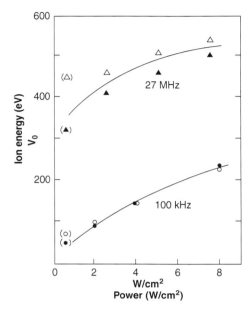

Figure 2.37 Dependence of ion energy with power level and frequency [106].

variables for Si_3N_4, will be used as an example. The process gases for PECVD Si_3N_4 usually are SiH_4, and combinations of NH_3, N_2, H_2, and inert gases. The important run parameters are

- Operating pressure
- Operating temperature
- Discharge frequency
- Reactant gas mixture

The results are for a RF capacitively coupled reactor, such as those shown in Figures 2.35a and 2.35b. The hot wall reactor is typically run at a frequency of 400 kHz and the cold wall at 50 kHz. Dual frequency cold wall reactors are also used. Table 2.12 compares salient properties of thermal CVD and PECVD Si_3N_4 films. The operating temperatures are the first significant difference, 900 °C for CVD and 300 °C for PECVD. Process conditions were as follows:

- Pressure: T = 300 °C, f = 310 kHz, $SiH_4/N_2/NH_3$ = 100/300/1100 sccm

- Frequency: T = 300 °C, P = 13- Pa, $SiH_4/N_2/NH_3$ = 100, 700/700 sccm
- Temperature P = 130 Pa, f = 310 kHz, $SiH_4/N_2/NH_3$ = 100/200/1200sccm

It is desirable for films to have low compressive stress, which improves adhesion and durability. Stress in the PECVD films depends on frequency, pressure, and temperature. For frequencies <4 MHz a temperature of 600 °C will provide films with compressive stress. Although stress can be reduced by decreasing temperature, one must be aware of the temperature sensitivity of the substrate. As a result of the reactive gases used, large amounts of hydrogen can be incorporated into PECVD films. Excess hydrogen can significantly affect many properties of the films. It can also influence stress in the films. Hydrogen content decreases with increased temperature and increased N_2 in the mix [109]. While some of the properties of PECVD films shown in Table 2.12 are not quite as good as those of thermal CVD films, the major advantage is the lower substrate temperature.

Adding another frequency (high + low) to the process provides another dimension of control [110]. The high frequency is used to stabilize the discharge while the lower frequency is used to control ion bombardment. Mechanical stress can be controlled by varying low/high frequency power. Stress can be varied from tensile (T) to compressive (C) by increasing the power to the low frequency supply. Changing the power ratio also changes the way H is bonded into the films, while not significantly affecting the total H content. Si-H bonds can be reduced and N-H bond increased by increasing low frequency power. Again, this can affect a number of important properties, such as stress, refractive index, breakdown voltage, electromigration to name a few.

2.2.3 Atomic Layer Deposition (ALD)

For a more detailed description of ALD please refer to the Third Edition of the *Handbook of Deposition Technologies for Films and Coatings* (Elsevier, 2009) [1]. ALD has particular applications to integrated circuits and semiconductors where film thicknesses are now less than 10 nm and features have high aspect ratios, often greater than 60:1. As IC and interconnect structures decrease in size and thickness, thin films grown by conventional PVD, CVD and PECVD processes generally cannot meet the tight thickness

Table 2.12 Comparison of Si_3N_4 deposited by thermal CVD and cold wall PECVD [107].

Property	Thermal CVD	PECVD
Composition	Si_3N_4:S/N = 0.75	SiN_x:S/N = 0.8–1.0
Density	2.8–3.1 g/cm³	2.5–2.8 g/cm³
Refractive index	2.0–2.1	2.5–2.8
Dielectric constant	6–7	6–9
Dielectric strength	10^7 V/cm	6 X 10^6 V/cm
Bulk resistivity	10^{15}–10^{17} Ω.cm	10^{15} Ω.cm
Surface resistivity	>10^{13} Ω.cm	10^{13} Ω.cm
Intrinsic stress	1.2–1.8 GPa (Tensile)	0.1–0.9 GPa (compressive)
TCE	4 X 10-6/°C	–
Transmitted color	None	Yellow
Water permeability	0	0 – low

requirements. Film thickness often must be less than 10 nm. ALD can meet these challenges, as described below.

ALD is a self-limiting, sequential CVD surface chemistry that deposits conformal thin-films of materials onto a wide variety of substrates. Unlike conventional CVD or PECVD, the ALD reaction breaks the CVD reaction into two half-reactions, keeping the precursor materials separate during the reaction. ALD film growth is self-limited and based on *surface reactions*, which facilitates precise control of atomic scale deposition. By keeping the precursors separate throughout the coating process, atomic layer control of film grown can be obtained as fine as ~0.1 Å/monolayer [111].

ALD has unique advantages over other thin film deposition techniques:

- Films are conformal
- Pin-hole free
- Ultra-thin films possible
- Dense and homogeneous

- Stoichiometric
- Low thermal budget
- Precise and repeatable
- Chemically bonded to the substrate
- Scale up possible

With ALD it is possible to deposit coatings with excellent thickness uniformity, even inside deep trenches, porous media, and around particles. Film thicknesses typically range from 1–500 nm.

ALD can be used to deposit virtually any type of thin film, including metals, ceramics, oxides, nitrides, carbides, insulators, semiconductors, and transparent conductive oxides (TCO). It is particularly important in semiconductor manufacturing processes. The process consists of pulsing gases to produce one atomic layer at a time. For example, metal precursors have the formula ML_x, where M = Al, W, Ta, Si, etc., and ligands L = CH_3, Cl, F, C_4H_{11}, etc. The film is built up by a series of pulses. Consider the pulsed reactions A and B. The four basic steps of the ALD process are [111, 112]:

- Expose the surface of the substrate to the first reactant (A) that reacts with surface sites. This reaction stops when the reaction has occurred at all surface sites.
- Products and the remaining initial reactants are purged from the system, usually with an inert gas.
- The surface is then exposed to the second reactant (B) which reacts with surface sites resulting from the first reaction (A).
- The system is purged again after reaction B is complete.

If surface reaction B returns the surface back to the initial state, then atomic layer controlled growth can be achieved using an alternating pulsed ABAB... reaction sequence. Since gas phase reactants are utilized, ALD does not require line-of-sight for deposition. As a result, very high aspect ratio geometries or porous structures can be easily coated with extremely conformal films. It is critical that the surface of the substrate be prepared to react directly with the molecular precursor. Table 2.13 lists widely used ALD precursors. It is evident that many of the precursors are common to all CVD or PECVD processes.

Table 2.13 ALD precursors [113–123].

Material	Precursors	Substrate Temperature (°C)
Al_2O_3	$Al(CH_3)_3$, H_2O or O_3	20–177
Cu	CuCl, $Cu(thd)_2$ or $Cu(acac)_2$ with H_2, $Cu(hfac)_2xH_2O$ with CH_3OH	200–410
Mo	MoF_6, $MoCl_5$ or $Mo(CO)_6$ with H_2	200–600
Ni	$Ni(acacO_2$, 2 step process NiO by O_3 by H_2	
SiO_2	$SiCl_4$, H_2O	
Ta	$TaCl_5$	
TaN	TBTDET, NH_3	260
Ti	$TiCl_4$, H_2	
TiN	$TiCl_4$ or TiI_4, NH_3	350–400
W	WF_6, B_2H_6 or Si_2H_6	300–350
WNxCy	WF_6, NH_3, TEB (triethylboron)	300–350
ZrO_2	$ZrCl_4$, H_2O	
HfO_2	$HfCl_4$ or TEMAH, H_2O	
ZnS	$ZnCl_2$/purge/H_2S	200–600

For the following reaction to occur, the surface must include hydrogen-containing ligands (-AH) [124]:

Substrate-AH + ML_x → AML_y + HL (note that the − before or after the element implies a ligand)

All initial surface ligands AH are consumed and, because the surface is saturated with AML_y metal precursor, the surface is now covered with L ligands that have not found a metal precursor. All initial AH surface ligands are replaced by ML_y ligands. The next step is to purge the first reaction products from the chamber and start the second reaction by introducing the second precursor. In this step, surface reactivity for another metal precursor is restored by eliminating L ligands and replacing them with AH ligands so the reaction can continue:

$$-ML + AH_z \rightarrow -M - AH + HL.$$

Now the surface is covered again by AH ligands and the element A is deposited. HL ligands are volatile and are removed from the surface. It is critical to restore the original surface for the reaction to be sustained.

Let's examine the ALD of Al_2O_3, shown schematically in Figure 2.38 [111, 124]. The reactants in this case are

- Trimethlyaluminum (TMA, $Al(CH_3)_3$ and water (H_2O)

A thickness of 0.25 molecular layer is deposited in each cycle. The ALD half reactions are [111]

(A) $AlOH + Al(CH_3)_3 \rightarrow AlO - Al(CH_3)_2 + CH_4$

(B) $AlCH_3 + H_2O \rightarrow AlOH + CH_4$

The reaction steps shown in the figure are

a. An –OH terminated surface is exposed to $Al(CH_3)_3$ precursor.
b. $Al(CH_3)_3$ molecules react with surface –OH sites and attach to these sites. CH_4 is produced, is volatile and is purged. The reaction continues until all –OH sites are reacted.
c. The surface is exposed to H_2O.
d. H_2O molecules react with –CH_3 sites, producing volatile CH_4 and –OH ligands attach to each initial –CH_3 site.
e. Running concurrently with the above reaction, adjacent –$Al(OH)_2$ molecules cross link by reacting with an –OH ligand, producing volatile H_2O molecules and a linked -Al-O-Al- network. Thus the Al_2O_3 film is deposited and the initial –OH sites are also restored.

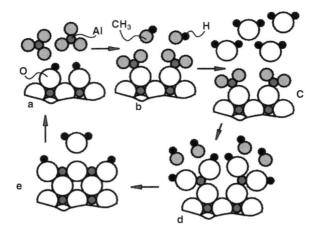

Figure 2.38 ALD of Al_2O_3 (courtesy Sundew Technologies).

In this case the deposition rate is 1.2 Å/pulse at 177 °C. The films are amorphous and insulating.

The main types of ALD reactors are [124]

- Closed system chambers
- Open system chambers
- Semi-closed system chambers
- Semi-open system chambers

Figure 2.39 shows the cross section of a typical ALD reactor [124], and Figure 2.40 shows a plasma enhanced ALD reactor (PEALD) [124]. The ALD system has two temperature controlled baths to provide precursors into the reaction chamber. Reaction rates during gas pulses in ALD systems must be rapid, and purge rates must be equally fast to achieve high deposition rates. This requires the use of high speed valves to pulse the precursors and a vacuum system to effectively purge the chamber. The reaction chamber must be heated in order to initiate surface reactions. Plasma enhancement (PEALD) accelerates reactions and improves overall productivity of the process.

The following materials have been deposited by ALD [126]:

- Oxides: Al_2O_3, TiO_2, Ta_2O_5, Nb_2O_5, ZrO_2, HfO_2, SnO_2, ZnO, La_2O_3, Y_2O_3, CeO_2, Sc_2O_3, Er_2O_3, V_2O_5, SiO_2, In_2O_3
- Nitrides: AlN, TaN_x, NbN, TiN, MoN, ZrN, HfN, GaN, WN

- Fluorides: CaF_2, SrF_2, ZnF_2, MgF2, LaF_3, GdF_2,
- Sulfides: ZnS, SrS, CaS, PbS
- Metals: Pt, Ru, Ir, Pd ,Cu, Fe, Co, Ni
- Carbides: TiC, NbC, TaC
- Compounds: $AlTiN_x$, $AlTiO_x$, $AlHfO_x$, $AlSiO_x$, $HfSiO_x$, TiN_xO_y
- Nanolaminates: $HfO_2/Ta2O_5$, TiO_2/Ta_2O_5, TiO_2/Al_2O_3, ZnS/Al_2O_3, ATO, (AlTiO)
- Doped materials: ZnO:Al, ZnS:Mn, SrS:Ce, Al_2O_3:Er, ZrO_2:Y, ... rare earth metals (Ce_3+ Tb_3+ etc.) also co-doping

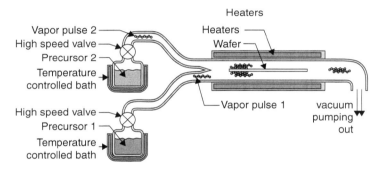

Figure 2.39 Typical ALD reactor chamber [114].

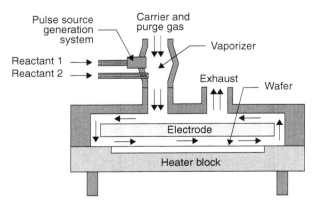

Figure 2.40 PEALD reactor [114].

2.3   Pulsed Laser Deposition

Pulsed laser deposition (PLD) is, in its simplest form, ablation of a target surface in vacuum by high power laser radiation and subsequent deposition onto a substrate. The target material absorbs the energy from the laser beam, heats up, is vaporized and is deposited as a thin film onto a suitable substrate. The process can take place in high or ultra high vacuum or in a reactive gas environment. PLD is used to deposit high quality oxides and nitrides, high temperature superconductors (such as $YBa_2Cu_3O_7$), diamond-like carbon, metallic nano-laminants, and superlattices. The development of new high repetition rate and short pulse length lasers has made PLD a useful process for growth of thin, well defined films with complex stoichiometry [125].

PLD has significant benefits over other film deposition methods, including:

- The capability for stoichiometric transfer of material from target to substrate, i.e., the exact chemical composition of a complex material such as YBCO, can be reproduced in the deposited film.
- Relatively high deposition rates, typically ~100s Å/min, can be achieved at moderate laser fluences, with film thickness controlled in real time by simply turning the laser on and off.
- The fact that a laser is used as an external energy source results in an extremely clean process without filaments. Thus deposition can occur in both inert and reactive background gases.
- The use of a carousel, housing a number of target materials, enables multilayer films to be deposited without the need to break vacuum when changing between materials.

A basic PLD chamber set up is shown in Figure 2.41 [127]. The chamber has the following components:

- Vacuum chamber
- Vacuum pumping system
- Gas supply system
- Pulsed laser source
- Beam guiding optics

- Target manipulation
- Substrate heating and rotation
- Ion source (optional)

While one would think that heating a material with a laser and ablating atoms and molecules would be straightforward, the physical mechanisms of PLD, in contrast to the simplicity of the system shown in Figure 2.41, are quite complex. It involves the physical process of the laser-material interaction on a solid target and the formation of a plasma plume with high energetic species and the transfer of the ablated material through the plasma plume onto a heated substrate surface. The four stages of thin film formation are

1. Laser radiation interaction with a solid target
2. Ablation processes
3. Deposition of ablated material
4. Nucleation and growth of the thin film on the substrate

The resultant film is epitaxially grown, stoichiometric uniform and smooth.

In stage 1, a high power pulsed laser beam is focused onto the surface of the target. At sufficiently high flux densities and short pulse duration, all elements in the target are rapidly heated up to their evaporation temperature. Materials are ablated out with the same stoichiometry as the target. The instantaneous ablation rate

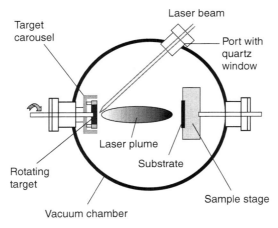

Figure 2.41 Basic PLD chamber configuration [127].

depends on laser fluence. Ablation mechanisms are complex and include collisions, thermal and electronic excitation, exfoliation, and hydrodynamics.

During the second stage, emitted materials are directed towards the substrate in a plume, as shown in Figure 2.42. Much like the distribution of sputtered atoms, the angular distribution of ablated material can be expressed as [128]

$$f(\theta) = ra\cos^m\theta + r(1-a)\cos^n\theta \qquad (2.3)$$

where r is a normalizing parameter and a is the ratio of $\cos^m\theta$ to the entire ablated distribution.

Figure 2.43 shows a plume distribution for $SrZrO_3$ [128]. This plume is highly forward directed, and as a result, the thickness distribution of deposited material is very non-uniform The area of deposited material is also quite small, typically ~1 cm^2, in comparison to that required for many industrial applications. Composition can also vary across the plume (as shown in Figure 2.43) and film. The ablated material contains macroscopic globules of molten material, up to ~10 µm diameter. Much like unfiltered cathodic arc films, these particulates can be detrimental to the properties of the film.

The spot size of the laser and the plasma temperature also have significant effects on the deposited film uniformity. Target-substrate

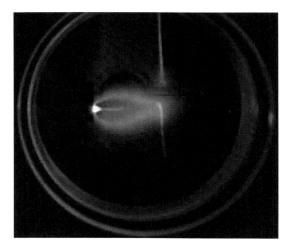

Figure 2.42 Laser plume in pulsed laser deposition.

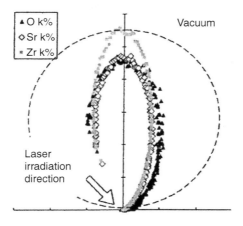

Figure 2.43 Plume distribution for SrZrO$_3$ [128].

distance is another parameter that governs the angular spread of the ablated materials. Masking has been found to be helpful in reducing beam divergence. We will show how thickness uniformity is improved later in the column.

The third stage is important with respect to the quality of the thin film. Again, similar to sputtering processes, ejected high-energy species bombard the substrate surface and may induce various type of damage to the substrate [129]. These energetic species sputter away some of the surface atoms and an interaction region is formed between the incident flow and the sputtered atoms. The film starts to grow after a thermalized region is formed. When the condensation rate of particles is higher than the rate of sputtered particles, thermal equilibrium can be reached quickly and the film grows on the substrate surface.

Nucleation and epitaxial growth of crystalline films depends on factors such as density, energy, degree of ionization, and type of condensing material, as well as temperature and substrate composition and crystallinity. Substrate temperature and supersaturation D_m of the depositing species are important and are related by the following relation

$$D_m = kT \ln\left(\frac{R}{R_e}\right) \qquad (2.4)$$

where k is the Boltzmann constant, R is the actual deposition rate, and R_e is the equilibrium value at temperature T [125, 130] .

Epitaxial growth using PLD is straightforward. Nucleation of the film is not straightforward, and depends on interfacial energies between the substrate, the condensing material and the vapor. Of course, deposition rate and substrate temperature essentially drive the nucleation process. As with all PVD and thin film deposition processes, crystalline film growth depends on surface mobility of the adatoms. Defect free crystalline growth occurs at higher substrate temperature (addressed later) while disordered and amorphous structures tend to grow at low substrate temperatures.

Deposition rate also depends on duration of the laser pulse. Due to the short laser pulsed duration (~10 ns) and the resulting small temporal spread (≤10 ms) of the ablated materials, deposition rate can be very large (~10 mm/s). In addition, rapid deposition of energetic ablation species helps to raise the substrate surface temperature. In this respect PLD tends to require a lower substrate temperature for crystalline film growth.

The nonuniform thickness distribution of the deposited films is definitely a problem. Figure 2.44 shows the thickness distribution of a single plume film [129]. Note that the distribution can be different for each element in the target/film. In order to achieve more uniform thickness distributions necessary for scale up, multi-spot PLD and large area PLD have been developed and demonstrate films with significantly better thickness distributions [128, 131]. Multi-spot PLD is based on the optimized superposition of plumes, which are simultaneously ablated from several evaporation spots. The uniformity of the thickness achievable by superposition of plumes is sufficient for many applications [132].

Another large area technique involves rastering the laser beam across the target while rotating the substrate, much like the method discussed for cathodic arc films [133, 134].

PLD is used to deposit a wide variety of thin film materials, including nonvolatile memories [134], dielectric and ferroelectric films [131], low-k and optical coatings [135], photocatalytic films [136], triboligical coatings [137], low temperature superconductors [138, 139], diamond like carbon [140], and semiconductors [141]. Historically, PLD's first significant success was deposition of $YBa_2Cu_3O_7$ high temperature superconductors.

An advantage of PLD is that stoichiometry of the target material is preserved on the substrate, which is important for ferroelectric

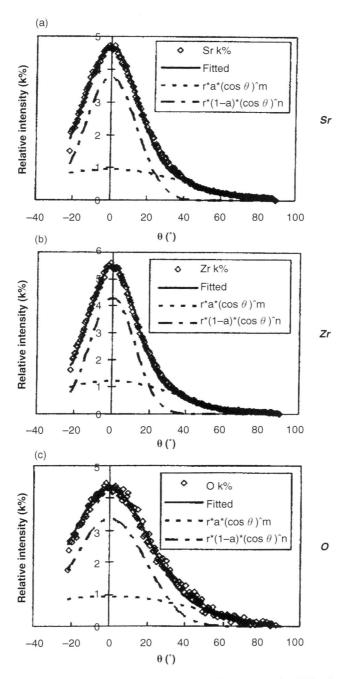

Figure 2.44 Thickness distribution for multiple elements in the PLD plume [128].

and low temperature superconductors that have complex compositions. Materials such as $SrBi_2Ta_{0.8}Nb_{1.2}O_9$ (SBTN) have applications in nonvolatile memories [134, 142, 143]. A SBTN target was irradiated at a 45° angle by 193 nm wavelength beam with a repetition rate of 5 Hz. Pulse duration was 17 ns and pulse energy was 160 mJ. A DC oxygen plasma was also excited by a ring electrode in the region of the plume. All films were crystalline, even without the O_2 plasma. The plasma did, however, affect the dominant crystalline phases present. Growth of c-axis planes, and perovskite phase, was enhanced at low plasma voltages. While films grown with and without plasma bombardment were very dense, films with plasma were significantly smoother than those deposited without.

A wide variety of tribological coatings have been deposited by PLD and by hybrid PLD/magnetron sputtering [127, 137]. ZnO:Al, TiN,TiCN, TiO_2, CrN, CrCN, TiAlN, and DLC all have been demonstrated. PLD coatings generally show high critical loads (>40 N) in scratch tests and tunable tribological properties (friction coefficients between 0.05 and 1.0 and high wear resistance) that depend on coating type.

Thus we see that PLD has many aspects common to all PVD processes, including a source material, high vacuum, reactive deposition, and the capability for ion assist. One of the main advantages, however, is the epitaxial nature of this process and deposition of highly crystalline films. Another is the deposition of complex compositions. PLD does not fit every application. Scale up of PLD is not as straightforward as for magnetron sputtering or e-beam evaporation, and large area coverage (architectural structures, window glazings, vacuum web coating, wear resistant coatings) is much more challenging. High deposition rates also make PLD very attractive. Thus, as with all deposition processes, one must weigh the advantages and disadvantages for each potential application.

2.4 Hybrid Deposition Processes

Multilayer (ML) structures are used extensively in virtually every thin film application, because they provide performance not achievable by single layer systems. ML structures combine two or more materials designed and engineered with a required performance. Many times thin film materials deposited by one process have better or preferred properties compared to the same material deposited by other processes. Hybrid deposition processes combine two or more different deposition processes to take advantage of the optimum performance of each constituent thin film layer. Improvements can be in:

- Deposition rate
- Economics
- Film physical, optical, and chemical properties
- Film elastic and tribological properties
- Permeation/diffusion
- Materials usage
- Safety

The following hybrid systems will be addressed:

- Polymer flash evaporation/PVD
- PECVD/PVD
- CVD/PVD
- PVD/PVD
- Cathodic arc/PVD/CVD

2.4.1   Vacuum Polymer Deposition

Vacuum polymer deposition (VPD), also known as polymer multilayer deposition (PML), combines flash evaporation of an organic monomer and PVD processes to deposit multilayer organic/inorganic thin film structures, often with thousands of layers. This technology was originally developed in the late 1970s by General Electric's capacitor division. One of the first products, which is still being produced by other facilities, was a multilayer aluminum/acrylate capacitor [144]. This capacitor has the advantage of very low dentritic growth which significantly improves performance and reliability. This process produces ultrasmooth, nonconformal, and pinhole-free films at very high deposition rates over large areas. Recent applications include ultra-low permeability gas and water barrier coatings for organic electronics, thin film solar cells, and thin film batteries, multilayer optical coatings, thin film capacitors, nanocomposites and nanolaminates on plastic.

Because of their unique physical, optical, and mechanical properties, thin film polymers have numerous technological and consumer product applications. They offer several advantages over inorganic thin film materials, including low cost, high deposition rates, low optical absorption, high smoothness, wide variety of compositions, formation of composites, high breakdown voltage, and high ductility. Disadvantages are low mechanical strength and wear resistance, low chemical resistance in some cases, high gas permeability, low melting temperature, small range of refractive indices, numerous optical absorption bands between near infra-red and infrared wavelengths. Thus, applications are restricted by many of the above properties, but are also expanded in many areas.

Unlike any other vacuum deposition process, VPD films actually smooth the surface of the substrate [144]. All other vacuum deposition techniques essentially bombard the substrate with species from a source that can bond immediately to the surface (low energy), bond after moving some distance on the surface, or reflect off the surface. These traditional processes "grow" the coating atom-by-atom or molecule-by-molecule from the substrate surface outward, and produce films that tend to reproduce the substrate surface on the length scale of the bombarding species. As a result, substrate surface roughness, from atomic scale upward is replicated by the growing film. Furthermore, this adatom growth can increase surface roughness through mechanisms such as shadowing, dislocations, and grain boundary growth. These growth mechanisms usually increase the roughness as the film grows thicker. A VPD layer, in contrast, does not grow atom-by-atom upward from the substrate; a gas of monomer vapor condenses on the substrate as a full- thickness liquid film that covers the entire substrate surface and its features. The liquid film is then cross-linked into a solid layer by ultraviolet (UV) or electron beam (EB) radiation. The resulting surface is glassy with virtually no defects or pin holes. The VPD layer can be combined with conventional PVD, CVD, PECVD, etc. layers to form low defect, ultra-smooth thin film structures.

The heart of this process is the monomer flash evaporator. Liquid monomer is fed through a capillary tube into an atomizer that distributes a mist of micro-droplets into a heated enclosure. Once they hit the walls of the enclosure, the monomer droplets instantly evaporate and are transported through a series of baffles to the exit slit. The monomer gas exits the slit uniformly and condenses onto a rotating drum, which is held at a temperature lower than the monomer gas. The drum is often cooled. The monomer is now a liquid that spreads uniformly over the substrate due to surface tension

[145, 146] and covers the substrate much like water in a pond covers rocks and roughness in the bottom. Next, the monomer is cured, and crosslinked, either by high intensity ultraviolet radiation or by an electron beam. A polymer is thus deposited onto a web or suitable substrate. The surface of the polymer is specular with virtually no defects or pinholes. A wide variety of monomers can be used, but this process works best for relatively low molecular weight monomers. Acrylates are often used. With this technology, polymer films can be deposited on moving substrates at speeds up to 1000 feet per minute with excellent adhesion to substrates and thickness uniformity of ±2%. The film thickness range known to be obtainable with the VPD process extends from a few tens of angstroms up to tens of microns, with the upper thickness limit largely imposed by the penetration depth of the cross-linking radiation employed.

Virtually every PVD or CVD-related process replicates the surface morphology of the substrate. Polymer films deposited by the VPD process actually smooth the substrate, as shown in Figure 2.45 [146]. This is demonstrated by the SEM photographs that show the surface of virgin polyester, the same surface with a 250 Å-thick sputtered Ta coating and a 1 μm thick VPD layer. Coatings routinely exhibit RMS surface roughness < 0.1 nm.

The web substrate can be subjected to plasma pretreatment before polymer deposition to activate the surface and improve adhesion, or to improve adhesion of subsequent layers [147]. Surface treatments of the linear-chain polymer PET have much higher effects on its surface topography and composition than treatment of the highly crosslinked PML surface. Reactive gases like O_2 and N_2 were found to have a substantially higher impact on the surface topography and roughness of both PET and PML when compared to treatments using inert gases like Ar. O_2, however, also noticeably changes the elemental composition of the VPD surface, and is preferred for adhesion improvement. It was also found that high intensity electron beam corona treatments could interfere with adhesion of metals, metal-oxides, and polar polymers.

Virtually any PVD or vacuum deposition process can be used in conjunction with VPD, including magnetron sputtering, and electron and thermal evaporation. Figure 2.46 shows placement of the polymer evaporator, plasma treatment, sputtering cathodes, and electron beam evaporator around the central drum of a web coater. Since the VPD process requires a higher chamber pressure, differential vacuum enclosures are required for PVD processes that

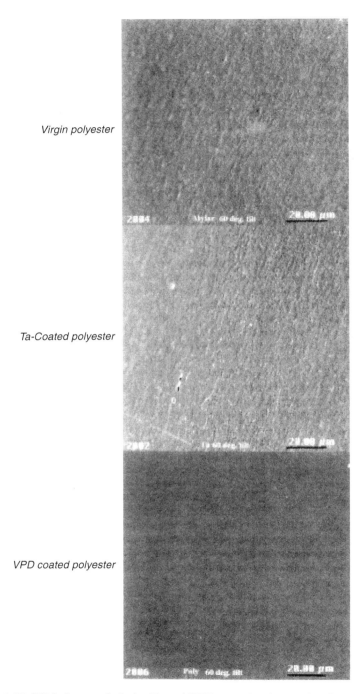

Figure 2.45 SEM picture of virgin, Ta and VPD coated polyester [146].

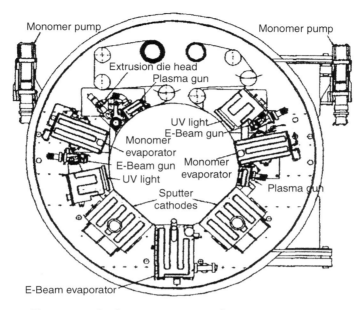

Figure 2.46 Placement of polymer evaporator, plasma treatment, sputtering cathodes and electron beam evaporator around the central drum of a web coater.

operate at lower pressures. The enclosures operate at pressures required for magnetron sputtering (2–5 mTorr) or evaporation (electron beam and thermal). Reactive gases can be introduced into the enclosure for reactive sputtering. This process can be used in roll-to-roll, in-line or planetary substrate configurations.

The VPD process has several distinct advantages compared to conventional organic/inorganic deposition processes [145]:

- Polymer thin films are more cost-effective since polymer layers can be deposited 10 to 100 times faster.
- VPD technology permits polymer, polymer electrolyte, metal, and even oxide and nitride films to be deposited in a single pass through a vacuum coater, as a monolithic structure, in an integrated manufacturing process.
- VPD process eliminates all of the unnecessary handling and laminations involved in conventional processes, and it also results in better adhesion between layers and defect-free interfaces.

- It is amenable to virtually all types of deposition processes, including roll-to-roll, in-line, and batch planetary substrate rotation
- Capital equipment and raw material costs are comparable to or less than those of conventional processes.

This process is not just limited to conventional single or multilayer structures. The polymer can be molecularly doped to increase
its functionality [148]. Insoluble solids can be suspended in acrylate
polymers. Active, luminescent QuinAcridone (QA) MDP can be
produced which allows fabrication of light emitting electrochemical cells, light emitting polymers, and light emitting organic diodes.
Applications for the VPD process include:

- Ultra-barrier materials [144, 149]
- Thin film Li-polymer batteries [150]
- Thin film photovoltaics [150]
- Light emitting polymers [148]
- Multilayer optical coatings on flexible substrates [151]
- Wear and abrasion resistant coatings
- Nanolaminate structures
- Thin film capacitors [144]
- Electroluminescent devices [148]
- Radiation detectors

One of the most successful applications for this hybrid process is
development and marketing of transparent ultra-barrier coatings
for protection of atmospheric sensitive devices and materials. In
addition, this work has spawned an entire new field of thin film
materials [144, 149, 152, 153]. Transparent ultra-barrier coatings
are needed to protect molecular electronics (OLEDs for example),
thin film solar cells, and thin film batteries from degradation due
to water vapor and oxygen permeation. The basic structure of these
coatings consists of alternating layers of polymer/oxide, polymer/
metal, polymer/nitride, shown in Figure 2.47. The coatings are
deposited onto PET, other flexible polymers and plastics to reduce
water vapor transmission rate (WVTR) and oxygen transmission
rate (OTR) by as much as six orders of magnitude. Polymer layer
creates an ultra-smooth, defect free surface for deposition of the
low-permeation oxide layer. A UV cure is generally preferred for
smoothing since an EB cure can increase surface roughness [145].

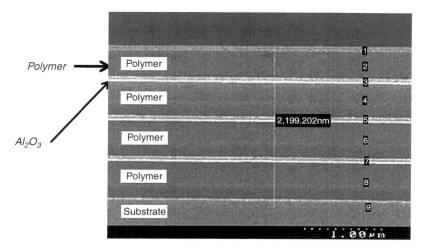

Polymer ———→

Al_2O_3

Figure 2.47 Layer structure of ultra-barrier coating.

This hybrid deposition process is also used to deposit polymer/ oxide and polymer/metal nanolaminate coatings, which are free standing structures consisting of hundreds to thousands of alternating polymer/inorganic layers [153]. The nanolaminate can be transparent, used for light weight windows, or opaque, used for capacitors. In addition to increased functionality, a major advantage of these coatings is their very low surface roughness and low optical scattering. The multilayer plastic substrate can include additional layers, including scratch resistant layers, antireflective coatings, antifingerprint coatings, antistatic coatings, conductive coatings, transparent conductive coatings, and barrier coatings, to provide functionality to the substrate.

Advanced applications for multilayer VPD coatings include contaminant-resistant coatings for flat panel displays, fabrics, fibers and yarns; corrosion-resistant coatings to replace the toxic heavy metals presently used (e.g., chromium); reduction of VOC release; and antibacterial coatings [154]. Contaminant resistant coatings are based on acrylate formulations containing between 49%–65% fluorine, which significantly decreases surface energies and increases hydrophobicity. These coatings are being developed for anti-smudge flat panel displays and windows, increasing chemical resistance for yarns and fabrics. A picture of a coated fiber is shown in Figure 2.48 [154].

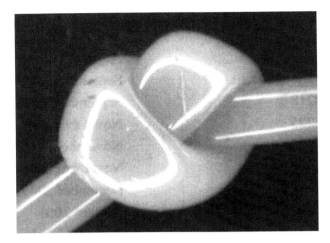

Figure 2.48 Microscope picture of fiber with corrosion resistant coating (Courtesy Sigma Laboratories) [154].

To demonstrate the versatility of this process, PECVD can also be added to the VPD process with little or no modification in vacuum levels [155]. This first example of a hybrid deposition process should give an idea of just how versatile and productive these processes can be and that new types of thin film coatings with increased functionality can be realized.

2.4.2 Magnetron-Based Hybrid Deposition Processes

A number of deposition processes can be combined with the magnetron sputtering process to form hybrid processes, including electron beam and thermal evaporation, cathodic arc deposition, plasma enhanced chemical vapor deposition (PECVD), chemical vapor deposition (CVD), and vacuum polymer deposition (VPD). These hybrid processes can be configured in a simple bell jar, batch coater, vacuum roll coater, or inline coater. To briefly review, the advantage of hybrid deposition processes is that they combine two or more different deposition processes to take advantage of the optimum performance of each constituent thin film layer. Improvements can be in:

- Deposition rate
- Economics
- Film physical, optical, and chemical properties

- Film elastic and tribological properties
- Permeation/diffusion
- Materials usage
- Safety

In fact, it may not be possible to deposit one thin film material by a particular process or the resulting films are of poor quality with poor performance. This is one rationale for combining magnetron sputtering with electron beam (e-beam) processes to deposit multilayer infrared optical coatings. Multilayer IR optical coatings require at least one high refractive index layer material (n_H) and a low refractive index material (n_L) with low absorption often out to wavelengths up to 25 μm. This rules out low index oxides such as SiO_2 and Al_2O_3. Fluorides of magnesium (MgF_2), calcium (CaF_2), hafnium (HfF_4), and thorium (ThF_4) work well at these wavelengths (although ThF_4 has radioactivity issues). However, due to resputtering phenomena, it is difficult to deposit a high quality optical fluoride coating by magnetron sputtering [156]. Films are generally substoichiometric. Optical transmittance of these films depends, for example, on the position of the substrate over the sputtering target, as shown in Figure 2.49. As shown in the figure, transmittance of HfF_4 films located over the target is poor (absorption is high) and

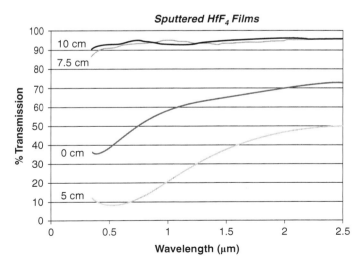

Figure 2.49 Transmission of HfF_4 films placed at various distances from the target axis [156].

improves with increased distance from the center. Additionally, optical properties depend on deposition rate and chamber pressure.

To this end, fluoride layers in multilayer coatings cannot be deposited by magnetron sputtering over an area of reasonable size. Fluorides deposited by the e-beam process, however, display low optical absorption and excellent optical properties, often without ion assist. They also can have good environmental stability. High quality, high index semiconductors, such as silicon (Si) and germanium (Ge), cannot be easily deposited by e-beam evaporation. The hybrid process combining magnetron sputtering for the high index semiconductor and e-beam evaporation for the low index fluoride produces high quality multilayer IR coatings. A schematic of the vacuum chamber configuration is shown in Figure 2.50. The substrate is rotated above the Si target at a suitable chamber pressure for deposition of the Si layers and similarly rotated over the fluoride (CaF_2 or HfF_4) e-beam source at a pressure suitable for evaporation. This sequence is repeated until the required number of layers is deposited. Deposition rate for the fluoride is significantly higher than that of the sputtered Si. All films are deposited in Ar or vacuum. This hybrid process is also well suited for a wide variety of multilayer coatings, and one of the advantages is the ease of scale up.

Magnetrons also work well with filtered cathodic arc (FCAD) processes [157–160]. Figure 2.51 shows a diagram of a magnetron/ FCA process chamber [158]. The FCAD process is described in Section 2.1.10. The magnetron-FCAD process is extremely versatile and combines the best of each constituent system. A diagram of the

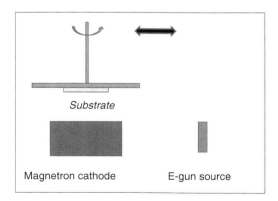

Figure 2.50 Schematic of hybrid magnetron sputtering-e-beam evaporation chamber.

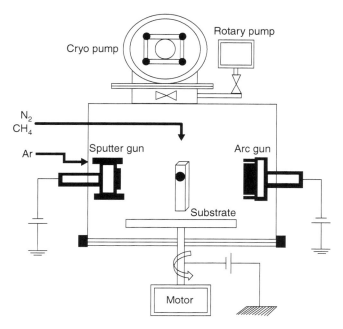

Figure 2.51 Diagram of hybrid FCA/magnetron chamber [157].

chamber and source configuration used to deposit Ti-Mo-Si-N and $CrMoC_xN_{1-x}$ coatings is shown in Figure 2.52 [157, 158]. $CrMoC_xN_{1-x}$ films have improved tribological properties compared to CrN films [157]. To deposit $CrMoC_xN_{1-x}$ coatings Cr is deposited by FCA and Mo is deposited by DC magnetron sputtering. Both metals are reacted with combinations of CH_4 and N_2 to form low friction/wear resistant coatings. Coatings are deposited onto steel and Si substrates. Figure 2.52 shows how the composition of these films depends on the gas volume ratio $CH_4/(CH_4 + N_2)$. Carbon increases and nitrogen content decreases linearly with increased ratio. The lowest coefficient of friction ~0.30 was measured for CrMoC films with the COF of $CrMoC_{0.63}N_{0.37}$ slightly higher.

Ti-Mo-Si-N films were deposited much the same way as $CrMoC_xN_{1-x}$ films, except that Ti is used in the FCA source and Mo and Si were sputtered [158]. Films with thickness ~2 µm were deposited in Ar + N_2 mixtures. Hardness was measured for ternary Ti-Mo-N and quaternary Ti-Mo-Si-N compositions, as depicted in Figure 2.53. Hardness peaked at ~30 GPa for the ternary coatings with 10 at.% Mo. Improved hardness values were obtained

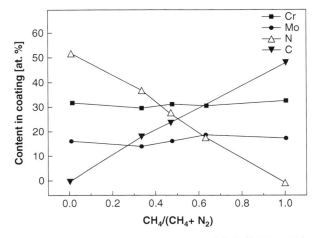

Figure 2.52 Relationship between composition and $CH_4/(CH_4 + N_2)$ ratio for $CrMoC_xN_{1-x}$ films [157].

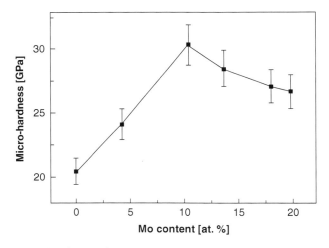

Figure 2.53 Microhardness of ternary Ti-Mo-N and quaternary Ti-Mo-Si-N coatings [158].

for quaternary coatings. Microhardness peaked at ~48 GPa for a Si content of 9 at.% and 10.4 at.% Mo. Microhardness was also found to depend on substrate bias voltage; peaking at 56 GPa at a bias voltage of 100 V. Residual stress was at a minimum ~ −3 GPa near this voltage, so the best of both worlds was achieved. The average coefficient of friction (COF) decreased from ~0.6 to ~0.30 with addition of Si. It is obvious that the addition of Si to the Ti-Mo-N coating significantly increased hardness. This increased hardness was attributed in part to (Ti-Mo)N microcrystallites embedded in an amorphous Si_3N_4 matrix.

Unbalanced magnetron sputtering (UBM) has also been combined with FCAD to deposit nanolaminants and nanostructured TiCrCN/TiCr + TiBC composite coatings [161]. These materials were developed to provide wear resistance, and reduce friction and corrosion in aircraft propulsion elements (gears, bearings, etc.). The system has two dual large area FCAD sources, two UMB cathodes, resistive and e-beam sources, and LPCVD source. Cr and Ti were deposited by LAFCA and B-C was deposited by UBM, all in mixtures of N_2 + 40% CH_4. Figure 2.54 shows the complex layer structure of the nanolaminant coating. Architecture of the coating consists of two segments separated in an intermediate gradient zone. The bottom segment is for corrosion resistance and the top segment is for hardness and low COF. The intermediate gradient zone graded composition from nitride based materials to carbide based materials. Typical thickness of the TiCrN sublayers is 2.5 nm while the total thickness is ~100 nm. Interlayer thickness is ~10 nm. Hardness of the nanostructures is ~28 GPa and elastic modulus) values are near 350 GPa.

FCA has also been combined with CVD for coating building hardware [160]. One of the goals for this work was to replace Cr alloys with more environmentally friendly materials. CVD is used to

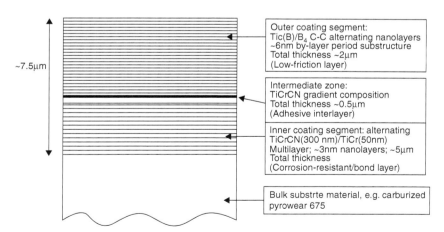

Figure 2.54 Layer structure of TiCrCN/TiCr + TiBC hybrid composite coatings [161].

deposit the organic primer (organosilicone: $Si_iO_xC_yH_z$) for the metal coating, and replaces a Ni/Cr adhesion layer. ZrN is used as the protective coating and an organosilicone top coat may also be applied. This system has the advantages of high hardness, good corrosion, and wear resistance, and just the right color for hardware. Materials and manufacturing costs are low and no toxic waste is generated.

Magnetron sputtering has also been combined with PECVD to deposit corrosion resistant tribological coatings [162]. Batch deposition systems have been designed for high productivity and low operating cost. One of the major goals of this technology is to replace electroplating as a surface treatment. DLC is one of the materials being developed.

While not strictly sputtering, another PVD process, thermal evaporation, has been combined with CVD to produce films with improved properties [163]. This hybrid process, known as hybrid physical chemical vapor deposition (HPCVD) employs a system that consists of a water-cooled reactor chamber, gas inlet and flow control system, pressure maintenance system, temperature control system and gas exhaust and cleaning system. The main difference between HPCVD and other CVD systems is the heating unit also heats a solid metal source. Thus, a conventional HPCVD system usually needs only one heater. The substrate and solid metal source sit on the same susceptor and are heated up inductively or resistively at the same time. Above certain temperatures, the bulk metal source melts and generates a high vapor pressure in the vicinity of the substrate. The precursor gas is then introduced into the chamber and decomposes around the substrate at high temperature. The atoms from the decomposed precursor gas react with the metal vapor, forming thin films on the substrate. Deposition proceeds until the precursor is switched off. The main drawback of single heater setup is the metal source temperature and the substrate temperature cannot be controlled independently. Whenever the substrate temperature is changed, the metal vapor pressure changes as well, limiting the ranges of the growth parameters. The two heater configuration consists of a metal source and substrate that are heated by two separate heaters, thus providing more flexible control of deposition conditions.

HPCVD is very successful in deposition of high quality superconducting MgB_2 films [164, 165]. Films deposited by other processes either have a reduced superconducting transition temperature (T_C) and poor crystallinity, or require ex situ annealing in Mg vapor. HPCVD system can produce high-quality in situ pure MgB_2 films with smooth surfaces, which are required to make reproducible uniform superconducting devices. The HPCVD technique is also an efficient method to deposit carbon doped or carbon alloyed MgB_2 thin films. These films are deposited in the same manner as MgB_2 films, but a metalorganic magnesium precursor (bis(methylcyclopentadienyl) is added to the precursor. The carbon-alloyed MgB_2 thin films by HPCVD exhibit extraordinarily high upper critical field.

It seems obvious that a number of deposition processes can be combined with magnetron sputtering. We are just seeing the beginning of this trend to produce thin films with improved properties. Technologies such as thin film solar cells, electrochromic coatings, organic electronic devices (OLE's, etc.), optical coatings, tribological coatings, and energy efficiency coatings can all benefit from the marriage of deposition processes.

It is not possible to describe all deposition processes used in surface engineering and functional materials engineering application. The ones most often used have been described in this chapter. Other processes not mentioned here will be briefly described where applicable and necessary to fully address the materials and technologies presented in the ensuing chapters.

References

1. *Handbook of Deposition Technologies for Films and Coatings*, 3rd Ed, Peter M Martin, Ed., Elsevier (2009).
2. Donald M. Mattox, *Handbook of Physical Vapor Deposition (PVD) Processing*, William Andrew (1998).
3. Rointan F Bunshaw, *Handbook of Deposition Technologies for Films and Coatings*, Noyes (1994).
4. David Glocker, Ismat Shaw and Cynthia Morgan, *Handbook of Thin Film Process Technology*, Taylor and Francis (2001).
5. M. J. Brett *et al.*, *Proceedings of the 45th Annual Technical Conference of the Society of Vacuum Coaters* (2002) 238.
6. B. A. Movchan and A. V. Demchishin, *Phys Met Metalorg.*, 28:83 (1969).
7. P. Benjamin and C. Weaver, *Proceedings of the Royal Society of London. Series A, Mathematical and Physical Sciences*, Volume 261, Issue 1307, pp. 516–531.
8. T. R. Jensen *et al.*, *Proceedings of the 43rd Annual Technical Conference of the Society of Vacuum Coaters* (2000) 239–243.
9. H. A. Macleod, *Thin Film Optical Filters*, 3rd Ed., Institute of Physics Publishing (2001).

10. H. Zorc *et al.*, *Proceedings of the 41ˢᵗ Annual Technical Conference of the Society of Vacuum Coaters* (1998) 243.
11. A. Carletti and F. Rimediotti, *Proceedings of the 41ˢᵗ Annual Technical Conference of the Society of Vacuum Coaters* (1998) 401.
12. E. Reinhold, M. List, H. Neumann, and C. Steuer, *Proceedings of the 47ᵗʰ Annual Technical Conference of the Society of Vacuum Coaters* (2004) 625.
13. A.L. Brody, *Proceedings of the 35th Annual Technical Conference of the Society of Vacuum Coaters* (1992) 42.
14. G.J. Walters, Proceedings of the 42ⁿᵈ Annual Technical Conference of the Society of Vacuum Coaters (1999) 453.
15. R. Gonzalez-Elipe, F. Yubero, and J. M. Sanz, *Low Energy Ion Assisted Film Growth*, Imperial College Press, London (2003).
16. H. R. Kaufman and J. M. E. Harper, *J. Vac. Sci. Techol.* A 22(1) (2004) 221–224.
17. Dale E. Morton and Vitaly Fridman, Technical Digest Series - Optical Society of America (1998), (9, Optical Interference Coatings), 15–17.
18. D. M. Mattox, *J Electrochem. Soc.*, 115:1255 (1968).
19. D. M .Mattox, R Bunshaw Editor, *Handbook of Deposition Technologies for Films and Coatings*, Second Edition, William Andrew (1994).
20. D. M. Mattox, P. M. Martin Editor, *Handbook of Deposition Technologies for Films and* Coatings, Second Edition, William Andrew, to be published.
21. Donald Mattox, *Handbook of Physical Vapor Deposition (PVD) Processing*, William Andrew (1998).
22. A. Anders, *Proceedings of the 45ᵗʰ Annual SVC Technical Conference* (2002) 360.
23. M. Chhowalla, *Proceedings of the 43ʳᵈ Annual SVC Technical Conference* (2000) 55.
24. *Handbook of Deposition Technologies for Films and Coatings*, R. Bunshaw Ed., Noyes (1994).
25. A. R. Gonzalez-Elipe, F. Yubero and J. M. Sanz, *Low Energy Ion Assisted Film Growth*, Imperial College Press (2003).
26. P. M. Martin and W. T. Pawlewicz, *Journal of Noncrystalline Solids* 45(1), (1981) 15–27.
27. P. M. Martin, J. W. Johnston, and W. D. Bennett, *S.P.I.E. Proceedings* 1323; Optical Thin Films III, New Developments (1990) 291–298.
28. R. DeGryse *et al.*, *Proceedings of the 2007 Fall AIMCAL Conference.*
29. W. D. Sproul and B. E. Sylvia, *SVC 45ᵗʰ Annual Technical Conference Proceedings* (2002) 11.
30. Applied Multilayers Technology Brief.
31. S. L. Rhode *et al.*, *Thin Solid Films*, 193/194 (1990) 117.
32. William D Sproul, PVD Processes for Depositing Hard Tribological Coatings, in *50 Years of Vacuum Coating Technology*, Society of Vacuum Coaters (2007) 35.

33. P. M. Martin *et al.*, *Proceedings of the 50th Annual Technical Conference of the Society of Vacuum Coaters* (2007).

34. P. Eh. Hovsepian, D. B. Lewis and W.-D. Munz, *Surface Coatings and Technology*, 133–134 (2000) 166–175.

35. Hoagland, R.G., R.J. Kurtz, and C.H. Henager, Slip resistance of interfaces and the strength of metallic multilayer composites. *Scripta Mater*, 50(6) (2004) 775–779.

36. C. Morant, Prieto, P. Forn, A. Picas, J. A. Elizalde, E. Sanz, J. M., *Surface and Coatings Technology* (2004), 180–181(Complete), 512–518.

37. Hoagland, R.G., R.J. Kurtz, and C.H. Henager, Slip resistance of interfaces and the strength of metallic multilayer composites. *Scripta Mater.*, 2004. 50(6) 775–779.

38. W. D. Munz, *Proceedings of the 35th Annual Technical Conference of the Society of Vacuum Coaters* (1992) 240.

39. Bernd Schultrich, New hard coatings by nanotechnology. *Jahrbuch Oberflaechentechnik* (2003), 59 128–136.

40. P. Eh. Hovsepian *et al.*, *Proceedings of the 44th Annual Technical Conference of the Society of Vacuum Coaters* (2001) 72.

41. D. Zong *et al.*, *Proceedings of the 47th Annual Technical Conference of the Society of Vacuum Coaters* (2004) 493.

42. D. J. Teer *et al.*, *Proceedings of the 47th Annual Technical Conference of the Society of Vacuum Coaters* (1997) 70.

43. C. Strondl *et al.*, *Proceedings of the 44th Annual Technical Conference of the Society of Vacuum Coaters* (2001) 67.

44. J. M. Anton and B. Mishra, *Proceedings of the 48th Annual Technical Conference of the Society of Vacuum Coaters* (2005) 599.

45. P. Eh. Hovsepian *et al*, *Proceedings of the 45th Annual Technical Conference of the Society of Vacuum Coaters* (2002) 49.

46. P. Eh. Hovsepian *et al.*, *Proceedings of the 47th Annual Technical Conference of the Society of Vacuum Coaters* (2004) 493.

47. T. H. Liu and Y. L. Su, *Proceedings of the 49th Annual Technical Conference of the Society of Vacuum Coaters* (2006) 589.

48. Raymond L. Boxman *et al.*, SVC Summer 2007 Bulletin, 24.

49. A. Madan, *Proceedings of the 42nd Annual Technical Conference of the Society of Vacuum Coaters* (1999) 379.

50. C. Stondl *et al*, *Proceedings of the 44th Annual Technical Conference of the Society of Vacuum Coaters* (2001) 67.

51. M. Wright and T. Beardow, *J. Vac. Technol.* A (1978) 179.

52. D. A. Glocker *et al.*, *Proceedings of the 43rd Technical Conference of the Society of Vacuum Coaters* (2000) 81.

53. D. E. Siegfried, D. Cook and D. A. Glocker, *Proceedings of the 39th Technical Conference of the Society of Vacuum Coaters* (1996) 97.

54. D. A. Glocker *et al.*, *Proceedings of the 48th Technical Conference of the Society of Vacuum Coaters* (2005) 53.

55. A. Blondeel *et al., Proceedings of the 48th Technical Conference of the Society of Vacuum Coaters* (2005) 275.
56. A. Blondeel *et al., Proceedings of the 48th Technical Conference of the Society of Vacuum Coaters* (2005) 187.
57. V. Kouznetsov *et al., Surf. Coat. Technol.,* 122 (1999) 290.
58. Dave Christie, *50 Years of Vacuum Coating Technology and the Growth of the Society of Vacuum Coaters,* Donald M. Mattox and Vivienne Harwood Mattox ed., Society of Vacuum Coaters (2007) 132.
59. G. van Zyle and R. Heckman, *48th Annual Technical Conference of the Society of Vacuum Coaters* (2005) 44.
60. D. J. Christie *et al., 47th Annual Technical Conference of the Society of Vacuum Coaters* (2004) 113.
61. William D. Sproul, *50 Years of Vacuum Coating Technology and the Growth of the Society of Vacuum Coaters,* Donald M. Mattox and Vivienne Harwood Mattox ed., Society of Vacuum Coaters (2007) 35.
62. Karol Macak *et al., J. Vac. Sci. Technol. A,* 18(4) (2000) 1533.
63. J. Almi *et al., J. Vac. Sci. Technol. A,* 23(2) (2005) 278.
64. R Chistyakow and B Abraham, *49th Annual Technical Conference of the Society of Vacuum Coaters* (2005) 88.
65. D. J. Christie, *J. Vac. Sci. Technol. A,* 23(2) (2005).
66. W. D. Sproul, D. J. Christie and D. C. Carter, *47th Annual Technical Conference of the Society of Vacuum Coaters* (2004) 96.
67. K. Sarakinos *et al., Rev. Adv. Mater. Sci,* 15 (2007) 44.
68. S. Konstantinidis *et al., 50th Annual Technical Conference of the Society of Vacuum Coaters* (2007) 92.
69. R. Bandorf *et al., 50th Annual Technical Conference of the Society of Vacuum Coaters* (2007) 160.
70. J. A. Davis *et al., 47th Annual Technical Conference of the Society of Vacuum Coaters* (2004) 215.
71. W.D. Sproul, D.J. Christie, and D.C. Carter, *47th Annual Technical Conference of the Society of Vacuum Coaters* (2004) 96.
72. D. A. Glocker *et al., Proceedings of the 48th Technical Conference of the Society of Vacuum Coaters* (2005) 53.
73. J. Li, S. R. Kirkpatrick and S. L. Rohde, Presentation SE-TuA1, AVS 2007 Fall Technical Conference, October 14–19, 2007, Seattle WA.
74. D. A. Glocker *et al., 47th Annual Technical Conference of the Society of Vacuum Coaters* (2004) 183.
75. A.P. Ehiasarian *et al., Proceedings of the 45th Technical Conference of the Society of Vacuum Coaters* (2002) 328.
76. J. Böhlmark *et al., Proceedings of the 49th Technical Conference of the Society of Vacuum Coaters* (2006) 334.
77. J. Alami *et al., Thin Solid Films 515 (4): 1731.*
78. A. P. Ehiasarian *et al., Surface and Coatings Technology 163–164: 267.*
79. J. Paulitsch *et al.,* Proceedings of the 50th Technical Conference of the Society of *Vacuum Coaters* (2007) 150.

80. U. Heister *et al.*, *41ˢᵗ Annual Technical Conference Proceedings of the Society of Vacuum Coaters* (1998) 187.
81. T. Winkler, *45ᵗʰ Annual Technical Conference Proceedings of the Society of Vacuum Coaters* (2002) 315.
82. D. H. Trinh *et al.*, *J. Vac. Sci. Technol.* A 24(2) (2006) 309.
83. W –M Gnehr *et al.*, *48ᵗʰ Annual Technical Conference Proceedings of the Society of Vacuum Coaters* (2005) 312.
84. D. J. Christie *et al.*, *46ᵗʰ Annual Technical Conference Proceedings of the Society of Vacuum Coaters* (2003) 393.
85. C. Montcalm *et al.*, Proceedings *of the 45ᵗʰ Annual Technical Conference of the Society of Vacuum Coaters*, (2002) 245.
86. John L. Vossen and Werner Kern, *Thin Film Processes*, Academic Press (1978).
87. R. B. Fair, *J Appl. Phys.*, 42 (1971) 3176.
88. R. N. Castellano, *Thin Solid Films*, 46, (1977) 213.
89. C. Weissmantel, *Thin Solid Films*, 32 (1976) 11.
90. S. L. Ream *et al.*, *Proceedings of the 45ᵗʰ Annual Technical Conference of the Society of Vacuum Coaters*, (2002) 412.
91. S. Scaglione and G. Emiliani, *J. Vac. Sci. Technol.* 1 7(3) (1989) 2303.
92. R. Chow *et al.*, *Proceedings of the 49ᵗʰ Annual Technical Conference of* the Society of *Vacuum Coaters*, (2006) 31.
93. See Research Electro Optics: www.reoinc.com
94. E. Quesnel *et al.*, *Proceedings of the 40ᵗʰ Annual Technical Conference of the Society of Vacuum Coaters*, (1997) 293.
95. Makoto Kitabatake and Kiyotaka Wasa, *J Vac. Sci. Technol.* A 26(3) (1988) 1793.
96. T. A. Edison, U S Patent 536, 147.
97. A. Anders, *Proceedings of the 45ᵗʰ Annual Technical Conference of the Society of Vacuum Coaters, (2002) 360.*
98. D. A. Baldwin *et al.*, *Proceedings of the 38ᵗʰ Annual Technical Conference of the Society of Vacuum Coaters*, (1995) 309.
99. John R. Reith and Federick J. Milford, *Foundations of Electromagnetic Theory*, Addison-Wesley (1967).
100. P. J. Martin and A. Bendivid, *Proceedings of the 45ᵗʰ Annual Technical Conference of the Society of Vacuum Coaters*, (2002) 270.
101. X. Yu *et al.*, *49ᵗʰ Annual Technical Conference Proceedings of the Society of Vacuum Coaters* (2006) 536.
102. P.J. Martin, A. Bendavid, and, H. Takikawa, *J. Vac Sci Technol*, A17(4),(1999) 2351.
103. Ben-Shalom, L. Kaplan, R.L. Boxman, S. Goldsmith,and M. Nathan, *Thin Solid Films* 236 (1993) 20.
104. R. Bunshaw, *Handbook of Deposition Technologies for Films and Coatings*, Second Edition, Noyes (1994).
105. A. R. Reinberg, Radial Flow Reactor, U S Patent 3,757,733 (1973).
106. R. S. Rosler *et al.*, *Solid State Technology*, 19(6) (1976) 45.

107. D. L. Smith *et al.*, *J. Electrochem. Soc.*, 129 (1982) 2045.
108. J. R. Hollahan *et al.*, *Proceedings of the 6th International Conf. on Chemical Vapor Deposition*, Electrochem. Soc. (1977) 224.
109. W. A. Claasen *et al.*, *J. Electrochem. Soc.*, 132 (1985) 132.
110. R. Chow *et al.*, *J. Appl. Phys.*, 54 (1983) 7058.
111. E. P. Van de Ven *et al.*, *Proceedings of the 7th Internat. VLSI Multilevel Interconnection Conf.*, IEEE (1990).
112. 111 S. M. George *et al.*, *J. Phys. Chem*, 100 (1996) 13121.
113. 112 M. Ritala and M. Leskela. "Atomic Layer Deposition." *Handbook of Thin Film Materials*, Ed. H.S. Nalwa, Vol. 1, (2002)103–159.
114. See Sundew Technologies.
115. A. E. Braun, *Semicon. Intl.* (2001) 52.
116. Gerald Beyer and Mieke Van Bavel, *Micro.*, (2002) 51.
117. Jerry Gelatos *et al.*, *Sol. State. Technol.* (2003) 44.
118. J. W. Klaus *et al.*, *Thin Solid Films*, 145 (2000) 360.
119. S. M. Rossnagel *et al.*, *J. Vac. Technol.*, B18 (2000) 2016.
120. M. Utrianinen *et al.*, *All. Surf. Sci.*, 157 (2000) 151.
121. P. Martensson, *Acta Univ. Uppsala* (1000) 421.
122. P. Martensson and J. O. J. Carlsson, *Chem. Vap. Deposition*, 3 (1998) 145.
123. R. Solanki and B. Pathongey, *Electrochem. Solid State Lett.*, 3 (2000) 479.
124. P. Martensson and J. O. J. Carlsson, *Chem. Vap. Deposition*, 3 (2000) 45.
125. Nicole Harrison *et al.*, "Atomic Layer Deposition", www.mne.umd/ugrad/course/465.
126. P. M. Martin, *Vacuum Technology Coating*, November 2008, 50.
127. www.planarald.com
128. W. Waldhauser and J. M. Lackner, *Proceedings of the 49th Annual SVC Technical Conference* (2006) 50.
129. I. Konimi *et al.*, *J. Vac. Sci. Technol.* A 26(6) (2008) 1455.
130. K. H. Wong and L. C. Mac, *Applied Physics of the Hong Kong Polytechnic University.*
131. See www.geocities.com/afserghie/LVE.
132. Shigreki Sakai *et al.*, *J. Vac. Sci. Technol.* A 24(4) (2007) 903.
133. J. M. Lackner *et al.*, *BHM*, 149 (2004) 118.
134. A. Anders, *45th Annual Technical Conference Proceedings of the Society of Vacuum Coaters* (2002) 360.
135. Pingxiong Yang *et al.*, *J. Vac. Sci. Technol.* A 25(1) (2007) 148.
136. I. Umezu *et al.*, *J. Vac. Sci. Technol.* A 20(1) (2002) 30.
137. S. J. Wang *et al.*, *J. Vac. Sci. Technol.* A 26(4) (2008) 898.
138. S. V. Prasad *et al.*, *J. Vac. Sci. Technol.* A 20(5) (2002) 1738.
139. *Pulsed Laser Deposition of Thin Films*, Douglas B. Chrisey and Graham K. Hubler ed. John Wiley & Sons (1994).
140. K. H. Wong and L. C. Mac, *J. Phys. D: Appl. Phys.*, 30 (1997) 957.

141. Jaanus Eskusoon *et al.*, *SPIE Proceedings* 6591 (2007).

142. J. Tersoff *et al.*, *Phys. Rev. Lett.*, 78, (1997) 282.

143. C. A. P. de Araujo *et al.*, *Nature* (London), 374 (1995) 627.

144. PLD Brochure, Neocera, www.neocera.com.

145. J. D. Affinito *et al.*, *Society of Vacuum Coaters 39th Annual Technical Conference Proceedings* (1996) 392.

146. M. E. Gross and P. M. Martin, *Vacuum Polymer Deposition*, Third Edition of *Handbook of Deposition Technologies for Thin Films and Coatings*, 3rd Ed, Peter M Martin, Ed, Elsevier, (2009).

147. A. Yializis, G.L. Powers, and D.G. Shaw, in *IEEE Trans.Components Hybrids Manuf. Technol.* 13, (1990) 66.

148. J. D. Affinito *et al.*, *Society of Vacuum Coaters 40th Annual Technical Conference Proceedings* (1997) 210.

149. J. D. Affinito *et al.*, *Society of Vacuum Coaters 41st Annual Technical Conference Proceedings* (1998) 220.

150. M. E. Gross *et al.*, *Society of Vacuum Coaters 41st Annual Technical Conference Proceedings* (2003) 89.

151. M. E. Gross *et al.*, *Society of Vacuum Coaters 49th Annual Technical Conference Proceedings* (2006) 139.

152. P. M. Martin *et al.*, *Society of Vacuum Coaters 38th Annual Technical Conference Proceedings* (1995) 163.

153. G. L. Graff *et al.*, *Society of Vacuum Coaters 43rd Annual Technical Conference Proceedings* (2000) 397.

154. United State Patent 6,962,671; P. M. Martin *et al.*, *Multilayer plastic substrates*.

155. A. Yializis *et al.*, *Society of Vacuum Coaters 41st Annual Technical Conference Proceedings* (1998) 477.

156. J. D. Affinito *et al.*, *Society of Vacuum Coaters 42nd Annual Technical Conference Proceedings* (1999) 102.

157. P. M. Martin *et al.*, *Thin Solid Films* 420–421 (2002) 8–12.

158. Ji Hwan *et al.*, *J. Vac. Sci. Technol.* A26(1) (2008) 146.

159. Jin Woo *et al.*, *J. Vac. Sci. Technol.* A26(1) (2008) 140.

160. S. Anders, A. Anders, M. Rubin, Z. Wang, S. Raoux, F. Kong, I.G. Brown, *Surf Coat Technol* 76 (1995) 167.

161. B Lee and Y Liu, *41st Annual Technical Conference Proceedings of the Society of Vacuum Coaters* (1998) 51.

162. V. Gorokhovsky *et al.*, *Surface & Coatings Technology*, 201 (2006) 3732.

163. H. Delorme *et al.*, *47th Annual Technical Conference Proceedings of the Society of Vacuum Coaters* (2004) 534.

164. X. X. Xi *et al.*, *Physica*, C 456 (2007) 22.

165. X. H. Zeng *et al.*, *Nature Materials*, 1 (2002) 35.

3

Thin Film Structure and Defects

Structure, microstructure, and nanostructure of surface treatments are critical aspects for surface engineering. Structure, including microstructure and nanostructure, influences virtually every aspect of thin film performance, and can be varied over wide ranges by choice of deposition process, deposition conditions, deposition geometry, substrate surface and composition, ion bombardment, and pre and post deposition treatments. Microstructure of the film will also depend on the application. Although the relationship between microstructure and coating properties is often difficult to precisely define due to other factors, there are some general correlations between microstructure and surface properties that can be expressed:

- It is generally desirable for the film to have high density, low porosity, and tightly bound grain structure
- Columnar structure is a strong function of deposition temperature
- A highly columnar or porous film may not be as hard or corrosion resistant as a dense, small grained and defect free film

143

- Stable optical properties are observed for dense, small grained, or amorphous films
- Structural defects tend to increase with increased thickness
- Columnar structure is required for a number of applications, such as piezoelectric and ferroelectric and magnetic thin films
- Compressive stress generally increases microhardness and microfracture toughness
- Interfacial phases and associated microporosity and contaminants all degrade coating adhesion

3.1   Thin Film Nucleation and Growth

To fully understand thin film microstructure it will be instructive to first elucidate film growth processes: how the film evolves during the growth process and the major factors that affect nucleation, growth, and microstructure. Defects are also an integral aspect that must be understood since they can severely degrade film performance.

The microstructure of a thin film evolves as it nucleates and grows on the substrate surface. How a thin film grows is not a trivial process, and depends on a number of factors, including deposition process, substrate surface quality, temperature, energy of incident particles, angle of incidence, etc. We will address a number of structure zone models (SZM) that predict film microstructure shortly These models are important in determining the microstructure and morphology of the thin film, but tell us very little about how growth initiates. Referring to Figure 3.1, the following surface kinetic processes are involved [1]:

- Adsorption
- Desorption
- Surface diffusion
- Cluster formation (nucleation)
- Cluster dissociation
- Step edge adsorption/desorption
- Film/substrate interdiffusion

We will briefly address each process. Figure 3.2 defines the two types of adsorption: physisorption and chemisorption. Surface

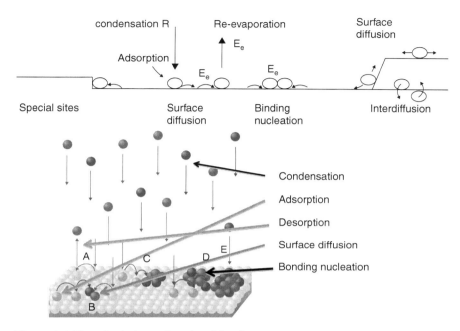

Figure 3.1 Two depictions of surface kinetic processes.

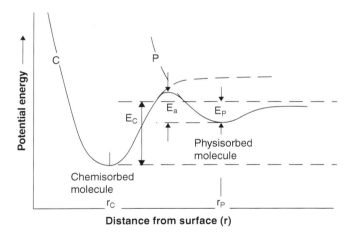

Figure 3.2 Model of physisorption and chemisorption [1].

adsorption occurs when we expose a pristine surface to gas particles. Here impinging atoms and molecules enter and interact within
the transition region between the gas phase and substrate surface
[2]. One of these processes will dominate, depending on the strength
of atomic interactions at the surface. Physisorption will dominate

if the incident particle is stretched or bent but retains its identity. Bonding to the surface will be accomplished by van der Waals forces. Chemisorption will modify the identity (chemical makeup) of the particle through ionic or covalent bonding. These two phenomena can be identified by considering adsorption energies (or heats), E_p and E_C (see Figure 3.2). Typical values are $E_p \sim 0.25$ eV and $E_C \sim 1$–10 eV. Potential energies for these two processes are shown in Figure 3.2 and are plotted as a function of distance from the surface. We see that the chemisorbed particles are bound closer to the surface than physisorbed particles. If both processes occur, energy E_a is the barrier energy that governs the rate of transition from physisorbed to chemisorbed states. Dissociation can occur when the particles dissociation energy E_{dis} is < EC. Energies of desorption and adsorption, E_{des} and E_{ads}, can be related to E_C and E_P.

We see that pressure plays a major role and that uniform coverage is achieved at high pressures (we'll address this below) and/or when more atoms are adsorbed than desorbed. Consider a vapor deposition process at pressure P, the rate of adatom condensation on a substrate is

$$\frac{d\theta}{dt} = k_{ads}P(1-\theta) - k_{des}\theta \tag{3.1}$$

Here $0 < \theta < 1$, with $\theta = 1$ being a monolayer. The first term represents atoms adsorbed at unoccupied surface sites, ~ kads, while the second term represents desorbed atoms. Both rate constants ~ $\exp - (E_{ads\ or\ des}/k_BT)$. Surface coverage as a function of time can then be expressed as

$$\theta = \frac{KP}{(1+KP)}\{1 - \exp[-k_{des}(1+KP)t]\} \tag{3.2}$$

$$K = \frac{k_{ads}}{k_{des}} \tag{3.3}$$

The equilibrium coverage is simply $\theta = KP/(1 + KP)$ and = 1 for KP >>1 [1].

These expressions provide a model for the morphology of vapor deposition on a surface.

Once particles are adsorbed onto the substrate, a number of processes can occur. Particles can diffuse on the surface to initiate cluster formation. This involves lateral motion of adsorbed atoms. During surface diffusion, atoms move between energy minima on the surface of the substrate and diffuse to the lowest intervening potential barrier (E). Thus, surface energies play an important role in how atoms nucleate and the film grows. A vapor-solid interface, with associated interfacial energies, is formed as the next nucleation step.

Nucleation results from a supersaturated vapor. The thermodynamics of nucleation are a major factor in film growth. As with all physical systems, energy minimization is the ultimate goal. Atoms on free surfaces are more energetic than atoms inside the substrate bulk, and are obviously less constrained. The surface energy (γ) is the difference between energy of atoms on free surfaces compared to bound states, and can range from 0.2–3 J/m^2 and is typically ~ 1J/m^2 [1]. There is also a thermodynamic driving force keeping the number of dangling bonds on the surface to a minimum. Metals typically have highest surface energies. The nucleation stage involves condensation from a supersaturated gas (or vapor). Gibbs (G) free energy = U – TS, where U is the internal energy and S is the entropy. In terms of θ, G can be expressed as

$$\frac{G}{n_s} = 4\theta w(1-\theta) - k_B T[\theta \ln \theta + (1-\theta)\ln(1-\theta)] \qquad (3.4)$$

Here n_s is the number of surface sites and w is the bond energy. Thermodynamic equilibrium is achieved when $d(G/n_s)/d\theta = 0$.

The free energies (ΔG) involved are

$$\Delta G = \left(\frac{4\pi}{3}\right)\Delta G_V + 4\pi r^2 \gamma + \Delta G_S \qquad (3.5)$$

$$\Delta G_V = -\left(\frac{kBT}{\Omega}\right)\ln\left(\frac{P_V}{P_S}\right) \qquad (3.6)$$

Here ΔG_V is the volume free energy, ΔG_S (= $1/2Y\epsilon^2$, where Y is the elastic modulus of the growing film and ϵ is the resulting strain) is the strain free energy, r is the radius of the island, P_V is the pressure of the supersaturated vapor, P_S is the equilibrium vapor pressure, and Ω is the atomic volume [4]. ΔG_V decreases upon nucleation due to condensation and γ increases due to creation of new interfaces. Thus we need $\Delta G_V < 0$ for film growth. Figure 3.3 plots these free energies as function of radius [5]. Here ΔG_V decreases, γ increases with and over all formation of a free energy barrier. Ignoring ΔG_S for the moment, the critical free energy ΔG^* and critical radius r^* are defined by $d\Delta G/dr = 0$:

$$r^* = -\frac{2\gamma}{\Delta G_V} \tag{3.7}$$

$$\Delta G^* = \frac{16\pi\gamma^3}{3(\Delta G_V)^2}, \tag{3.8}$$

which is the effective energy barrier for nucleation

From Figure 3.3, note that DG^* is at the radius where we have a maximum or minimum in ΔG.

If we look at a monolayer thick cluster on the surface

$$\Delta G = j\Delta G_V + j(\gamma_{fv} + \gamma_{fs} - \gamma_{sv})\Omega^{2/3} + j^{1/2}\gamma_e \tag{3.9}$$

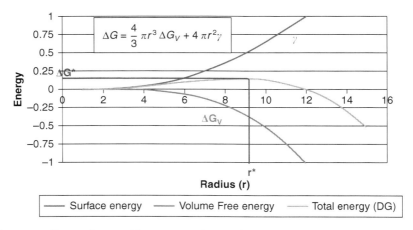

Figure 3.3 Dependence of free energy during film growth on radius of the cluster [5].

j = # atoms/cluster, fs – film substrate, fv – film vapor, sv – surface vapor, and γ_e is the energy/length at the step edge of the cluster.

The mechanism for the effective energy barrier for nucleation can be expressed in terms of these surface energies. The sign of γ determines the mechanism [3]:

- $\gamma_{fv} + \gamma_{fs} - \gamma_{sv} \leq 0 \rightarrow$ a 2-D cluster is stable and layer-by-layer nucleation occurs
- $\gamma_{fv} + \gamma_{fs} - \gamma_{sv} > 0 \rightarrow$ 2-D cluster is not stable and 3-D islands form to minimize substrate contact

The nucleation rate N can then be defined as

$$N = N * A * \omega \text{ with} \tag{3.10}$$

$$N^* = n_s \exp\left(\frac{-\Delta G^*}{k_B T}\right) \tag{3.11}$$

N^* is the number of nuclei at equilibrium, n_s is the total nucleation site density, ω is the rate at which atoms impinge on the critical area A^*.

The three basic models are used to explain thin film growth after nucleation:

- Frank-van der Merwe (FM)
- Three-dimensional island: Volmer-Weber (VW)
- Initially layer-by layer, followed by 3D islands: Stranski-Krastanov (SK)

Recall that the surface energy, γ, is defined as the difference in interatomic energy of atoms at the surface and energy of those in the bulk. When discussing these three growth modes, it will be helpful to refer to Figures 3.4 and 3.5 [6]. Also recall that the surface coverage parameter, θ, is given by

$$\theta = \frac{KP}{(1+KP)}\{1 - \exp[-k_{des}(1+KP)t]\} \; (0 \leq \theta \leq 1 \text{ for growth})$$

$$\tag{3.12}$$

Growth modes

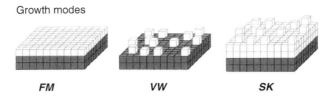

FM   VW   SK

Figure 3.4 Thin film growth modes [1].

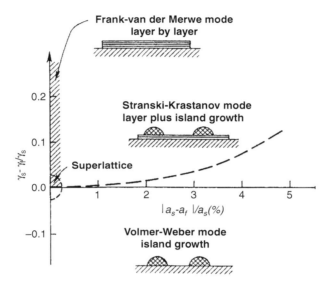

Figure 3.5 Stability regions for three film growth modes in coordinates of surface energy [6].

$$K = \frac{k_{ads}}{k_{des}} \qquad (3.13)$$

Equilibrium coverage is simply $\theta = KP/(1 + KP)$ and $= 1$ for $KP \gg 1$ [1].

In the FM growth mode, the deposited atoms wet the substrate, and interatomic interactions between substrate and deposited materials are stronger and more attractive than those between the different atomic species within the film itself. As we shall see in the following discussion, just the opposite is true for VW growth. SK growth falls somewhere in the middle.

Layer-by-layer FM growth is two-dimensional and results because atoms of the deposited material are more strongly attracted to the substrate than they are to themselves, and can be defined by

$$\Delta\gamma = \gamma_{fv} + \gamma_{fs} - \gamma_{sv} \leq 0 \text{ or} \tag{3.14}$$

$$\gamma_{sv} \geq \gamma_{fv} + \gamma_{fs} \tag{3.15}$$

with fs – film substrate, fv – film vapor, sv – surface vapor. This expression tells us that the sum of the film surface energy and interface energy must be less than the surface energy of the substrate for wetting to occur; it is easier to for layer to layer growth to occur with increased substrate surface energy.

This type of growth is found in the cross hatched region of Figure 3.5. For this type of growth, q ~ 0. A special case of this process is ideal homo- or auto-epitaxy. Superlattices can be formed. In this case $\gamma_{fs} = 0$ since the interface and film is continuous and $\gamma_{sv} \geq \gamma_{fv}$. Strain energy, which is incorporated into γ_{fs}, increases linearly with additional layers. As a result, $\gamma_{fv} + \gamma_{fs} - \gamma_{sv}$ becomes > 0 and FM growth transforms into SK growth.

As mentioned earlier, epitaxy is a direct consequence of FM growth. One of the most interesting examples is the growth of superlattices [8]. Superlattices consist of alternating layers of two different crystalline materials (A/B/A/B/A/B.....), usually semiconductors, with layer thicknesses between 1 and 10 nm [7]. To form laminar and atomically flat the totally free energy must ~ 0. Thus if $\Delta\gamma$ is ≤ 0 for A/B stacking then $\Delta\gamma \geq$ for B/A layers. This condition is usually fulfilled for semiconductor SLs but generally not for metal/semiconductor and metal/metal SLs [1]. In these cases, surfactants must be used to lower surface energy.

A word on the stability of strained layers. Strain energy plays a major role in determining the growth mode and results from the mismatch of film and substrate lattices and layer-layer lattice mismatch. The strain free energy can be defined as

$$G_S = \frac{1}{2} Y \varepsilon^2 \tag{3.16}$$

where Y is the elastic modulus and ε is the strain [6]. Here ε is related to the lattice mismatch strain:

$$f = \frac{[a_0(s) - a_0(s)]}{a_0(f)}, \tag{3.17}$$

where a_0 is the lattice parameter of the film (f) and substrate (s).

There is a critical thickness at which a growing planar film begins to roughen due to island growth. Morphological instability in the form of island growth initiates when the free energy change is [4]

$$\Delta G = \left(\frac{2 \pi r^3}{3} \right) \Delta G_V + \pi r^2 \gamma + \Delta G_S, \qquad (3.18)$$

These parameters are introduced in the above discussion. r is the radius of the island. DGS is essentially the difference in epilayer strain per unit area after the island nucleates relative to conditions prior to nucleation. The critical thickness for island initiation is:

$$h^* = \frac{2\gamma}{Y(\varepsilon^2 - f^2)} \qquad (3.19)$$

for example, $h^* f^2$ for GaAs is 1.8×10^{-10}, the transition from 2D to 3D growth [4].

This leads us to SK growth (layer-plus-island growth), defined by

$$\gamma_{sv} > \gamma_{fs} + \gamma_{fv} \qquad (3.20)$$

Referring to Figure 3.5, here the strain energy per unit area of the film is large compared to γ_{fv}, permitting nuclei to form above initial 2D film layers. The adatom cohesive force is stronger than surface adhesive force. Beyond a critical layer thickness, which depends on strain and the free energy of the deposited film, growth continues through the nucleation and coalescence of adsorbate islands [4]. Figure 3.6 graphically depicts this growth mode.

Coherent island formation during SK growth can be used to fabricate epitaxial nanoscale structures, particularly quantum dots [9, 10, 11, 12, 13]. Techniques such as surface dimpling with a pulsed laser and growth rate control are used to adjust the onset of the SK transition (or even suppress it altogether) [13, 14]. Nanopatterning is used to create templates for SK growth [15]. Geometry and size of the nanostructures can be predictably controlled using the SK transition, which in turn, can alter their electronic or optoelectronic properties (such as band gap). Nanolithographically patterned substrates have been used as nucleation templates for SiGe clusters [12, 16]. Several studies have also shown that island geometries can be altered during SK growth by controlling substrate morphology and growth rate [13, 17]. Pyramidal and dome-shaped Ge islands on Si, shown in Figures 3.7 and 3.8, have been demonstrated on a textured Si wafer. The capability to control the size, location, and shape of these structures could provide invaluable techniques for "bottom up" fabrication schemes of next-generation devices in the microelectronics industry.

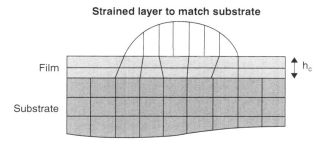

Figure 3.6 SK growth mode.

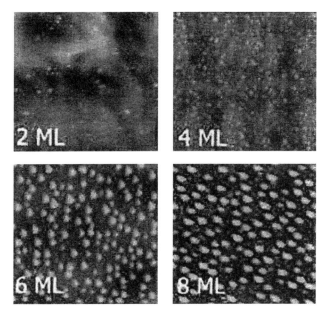

Figure 3.7 Ge nanospheres on Si [15].

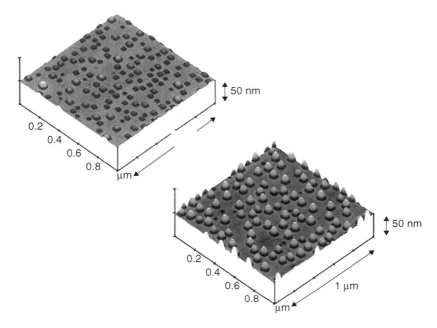

Figure 3.8 Self-assembled quantum dots grown on Si in VW mode [14].

In VW growth, adatom-adatom interactions are stronger than those of the adatom with the surface, leading to the formation of three-dimensional adatom clusters or islands [10]. Growth of these clusters, along with coarsening, described above, will cause rough multi-layer films to grow on the substrate surface. VW growth is defined by

$$\gamma_{sv} < \gamma_{fs} + \gamma_{fv} \tag{3.21}$$

Here the surface tension of the film exceeds that of the substrate. The total surface energy of the film interfaces is larger than that of the substrate-vapor interface. Material "balls up" to minimize interfacial contact with the substrate. Growth is uneven and surface diffusion is retarded. In this growth mode, metals for example, tend to cluster or ball up on ceramic or semiconductor substrates.

Quantum dots can also be grown using the VW mode [18]. Figure 3.8 shows self assembled ZnO quantum dots grown by vapor phase transport on Si. The dots were synthesized by thermally evaporating Zn acetate powders at 500 °C in a 20% O_2/Ar mixture. As shown in the figure, deposition times varied from 2–4 min.

Dot size was controlled primarily by deposition time. The nano-crystals had predominantly the wurtzite phase. Height of the dots increased linearly with diameter, as shown in Figure 3.9 [18]. The optical properties could thus be tuned by controlling dot size.

Thus far we have seen that a wide variety of thin film types and morphologies can be synthesized using the FM, SK and VW growth modes. Growth mode depends primarily on the relative magnitude of surface energies (film-substrate, film-vapor and vapor-substrate) and strain free energy.

3.2   Structure of Thin Films

The structure of thin films can range from amorphous (no long range order) to highly textured with columnar structure. Density and porosity can also vary over wide ranges. Microstructure is also dependent on thickness of film. Figure 3.10 shows the range of thin film microstructures achievable. Figure 3.10a shows an amorphous microstructure typical of many glassy materials, Figure 3.10b shows nanocrystalline solid, Figure 3.10c shows large grains, Figure 3.10d shows columnar structure, and Figure 3.10e shows large columnar structure.

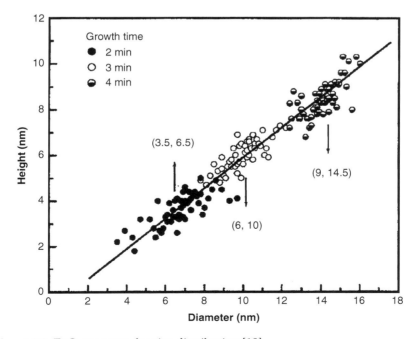

Figure 3.9 ZnO quantum dot size distribution [18].

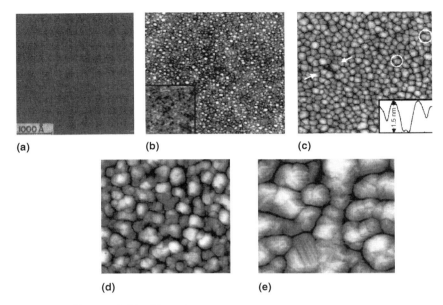

Figure 3.10 Range of thin film microstructures.

3.2.1   Amorphous Thin Films

Amorphous thin films have numerous important applications. Most of the high performance optical coatings (TiO_2, SiO_2, Si_3N_4, Nb_2O_5, Ta_2O_5 for example) used today are amorphous. Amorphous metals show promise of superconductivity. Amorphous thin films are used extensively in microelectronics; in particular, SiO_2 films are used as the insulator above the conducting channel of a metal-oxide semiconductor field-effect transistor (MOSFET). Also, hydrogenated amorphous silicon (a-Si:H) is one of the most important thin film photovoltaic materials.

Virtually every class of thin film material (metal, semiconductor, insulator, semi-metal, chalcogenide) can assume the amorphous state. An amorphous solid is a solid in which there is no long-range order of the positions of the atoms. Amorphous materials have two types of short-range order:

- Topological
- Compositional

Topological short-range order is characterized by an average number of nearest neighbors, coordination number, and mean separation of neighbors. These quantities can be obtained from the radial distribution function (RDF) resulting from x-ray diffraction analysis. Figure 3.11 depicts short-range and intermediate-range order in amorphous carbon (a-C).

Compositional short range order refers to nearest neighbor composition differing from average bulk values. Because they contain a large number of different atomic configurations, amorphous thin films are usually described by a statistical distribution model having two categories [19]:

- Dense random (close) packing (DRP). In this model, atoms tend to fall into regular as well as distorted tetragonal groupings that are further packed into larger units. Atomic clusters have five-fold symmetry and density of ~ 0.64 that of bulk.
- Continuous random network (CRN). In this model, the spacing of atoms is the same as for a crystalline solid, bond length is nearly constant, but there is a significant spread in bond angles about a mean value.

Structure zone models (SZM) were developed to describe the microstructure and morphology of thin films as a function of deposition pressure and temperature and will be described later in this chapter. According to these models, a necessary (but not sufficient) condition for the occurrence of amorphous phases is that $T/T_M < 0.3$. For higher

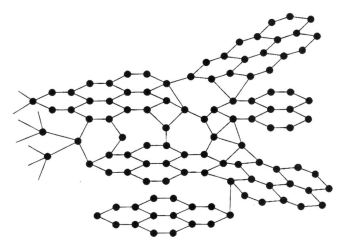

Figure 3.11 Short and intermediate range order in a-C [19].

Table 3.1 Grain sizes and types of structural order in solids [19].

Grain (Crystallite Size)	Type of Structural Order
~ 1–100 mm	Polycrystalline (or single crystal)
~ 1–1000 μm	Microcrystalline
~ 1–1000 nm	Nanocrystalline (nanophase)
~ 0.1–1 nm	Amorphous (noncrystalline)

values, surface diffusion of deposited atomic species allows formation of crystallites with long range atomic order. Table 3.1 shows the classification of structural order with respect to grain (or crystallite) size [19].

Note that disordered films can also be composites and a crystalline alloy can posses long range structural order and still be compositionally disordered. Examples of this type of material include SiGe and CuZn. Composites are discussed in detail in Chapter 4. Figure 3.12 compares atomic arrangements of ordered (top) and disordered alloy (AB) thin films [19]. Figure 3.12a shows A and B atoms arranged periodically in space on a lattice so the alloy is ordered both structurally and compositionally. Figure 3.12b shows that A and B atoms can occupy the same crystalline sites but are spatially disordered.

3.2.2   Grain Growth in Thin Films

The next step in microstructural complexity is polycrystalline solids and thin films. Grain (or crystallite) growth is important since it affects virtually every property of the thin film, including density, mechanical stress, tribological properties, physical and optical properties, and corrosion resistance. Grain and texture evolution depend on a number of factors:

- Deposition process
- Plasma bombardment
- Energy of incident atom flux
- Substrate temperature
- Deposition rate
- Impurities
- Lattice mismatch between film and substrate

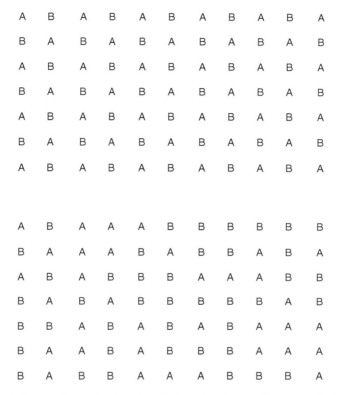

Figure 3.12 Comparison of ordered and disordered crystalline alloy [19].

As we shall see, it's all about *energy*. We are all aware that when a solid is heated, grains or crystallites will grow. When heated, the grain boundary with radius R will expand with a velocity v, which relates expansion to time: $v \sim dR/dt$. It's all about energy:

$$R^2 - R_0^2 \approx kt^{1/2}$$

[21], where R_0 is the initial grain radius and k is a thermal constant. Note that, even though the size of the grain increases, the shape does not change appreciably in bulk solids, as shown in Figure 3.13 [20]. Grain growth models for thin films are complex and involve grain boundary, substrate interface, and upper film surface free energies. Figure 3.14 shows how grain boundaries move and evolve for randomly and highly oriented grains growth [21, 22]. The energy density residing in grain boundaries, substrate-film, and film-air interfaces all drive grain growth in thin films, with surface

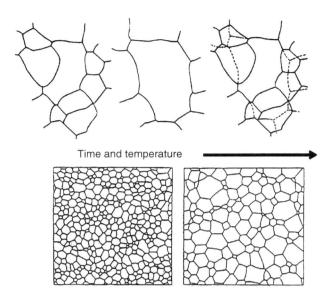

Figure 3.13 Grain Growth by successive heating [20].

free energy variations as the most important factor. The following general observations summarize grain growth patterns in thin films:

- Abnormal grain growth is the "norm" for thin films. Grain size distributions are generally bimodal, and some grains can grow abnormally large.
- As a result of abnormal grain growth, an average crystallographic orientation evolves, which is more pronounced as films get thinner.
- Low strain growth occurs at higher deposition temperatures. In this case, surface and interfacial energy minimization dominates grain growth.
- Grain growth in thin film often stagnates at grain sizes 2–3 times the film thickness. Grooves in the surface form, which essentially absorb or diffuse much of the driving energy.

Figure 3.15 shows the sources of energy available to control microstructure of thin films [23]. *Thermal energy*, expressed as $k_B T$, is probably the most important source for grain growth. At a temperature of 400 °C = 58 meV/atom [7], is the activation source for many grain growth mechanisms. *Energy* released to form the grain is the product of the area of the grain boundaries that are

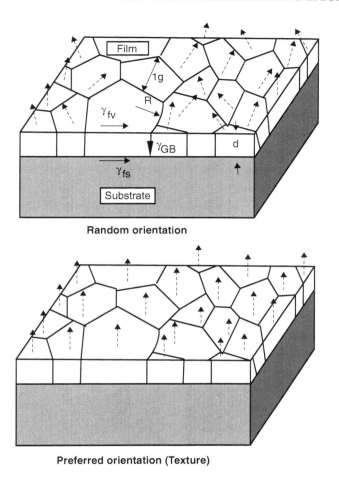

Random orientation

Preferred orientation (Texture)

Figure 3.14 Evolution of grain boundaries for random and highly oriented growth [21, 22].

eliminated and grain boundary free energy (γ_{GB}). If the initial length of a cubic grain is 0.1 μm and $\gamma_{GB} = 0.5 \text{ J/m}^2$, then the energy needed to grow the grain to 1 μm is ~ 15 MJ/m³ or 1.1 meV/atom. *Energy minimization* at the film's air and substrate interfaces promotes grain growth. Interface energy depends on the thickness of the film. For example, the energy density for a 1 μm thick film is ~ 0.15 meV/atom while this value is 10 times higher for a 0.10 μm thick film. Energy is expended when grain growth is associated with solute precipitation. *Discontinuous precipitation* occurs when grain boundaries sweep through grains that are supersatuarated with respect to solute. This process is capable of releasing energies

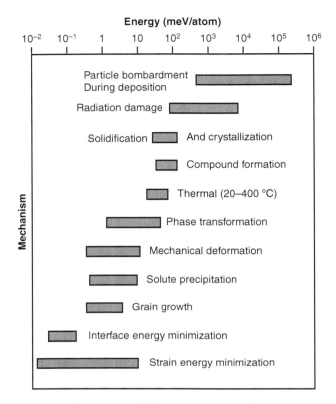

Figure 3.15 Sources of energy available to control thin film microstructure [23].

~ 7 meV/atom. Referring to Figure 3.15, generally, the higher the energy, the higher the capability of driving grain boundary change or grain growth.

One would expect stresses to be generated due to grain growth. Stresses result due to the fact that grain boundaries are less dense than the grain itself, and when grain boundaries are reduced or modified, density modulations occur in the film that create tensile stresses. This is depicted in Figure 3.16 [6]. If we assume spherical grains with initial diameter L_0, grain related stress proceeds as shown in the figure. Length change of the grain can be expressed as

$$\frac{1}{L_0} - \frac{1}{L} = \frac{3(1-\nu)\gamma_{GB}}{2E(\Delta a)^2} \tag{3.22}$$

(a) Vertical boundary

(b) Horizontal boundary

3. Edge dislocation

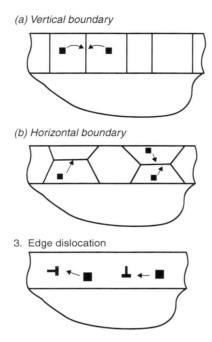

Figure 3.16 Stress build up in grain boundaries [24].

Here v is Poisson's ratio and E is Young's modulus of the film. Stress due to grain growth can be expressed as

$$\sigma = \frac{3\gamma_{GB}}{2\Delta a},$$ (3.23)

where Δa is excess volume per grain boundary [25]. If $\gamma_{GB} \sim 0.3 \, J/m^3$ and $\Delta a \sim 10^{-4} \, \mu m$, then

$$\sigma \sim 5 \, GPa$$

In early stages of film growth, the films can consist of small grains. These grains coalesce as the film grows [25, 26]. As grains, or crystallites, grow the gap, or grain boundary, between them decreases and cohesion initiates between them, as shown in Figure 3.17 [27].

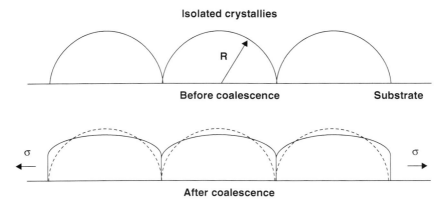

Figure 3.17 Diagram of crystalline coalescence during film growth [27].

Some elastic deformation of grains also begins to occur. The average stress due to this phenomenon is

$$\sigma_{ave} = \frac{2\pi\gamma}{3t_C}, \tag{3.24}$$

where t_C is film thickness and γ is the surface energy (= $\gamma_s - 1/2\gamma_{GB}$) as explained above.

Stresses also form when vacancies are annihilated and diffuse into the grain boundaries. Contributions to total stress, however, are expected to be minimal for this case.

In order to be stable, all forces acting on grain boundaries must be in equilibrium [21]. Forces acting on the grain boundary include:

- Grain boundary energy (= $2\,\gamma_{GB}/L$)
- Surface and interface energy variations
- Grain grooving
- Interactions with second phase particles

After grain growth has stagnated, equilibrium in driving forces for grain growth can be expressed as [5]

$$\frac{2\gamma_{GB}}{l_g} + \frac{\Delta\gamma_{fv}}{d} + \frac{\Delta\gamma_{fs}}{d} = \frac{\gamma^2_{GB}}{d\gamma_{fv}} + \frac{3\gamma_{GB}f}{2r} + \frac{3\gamma_{GB}C_s}{a} \tag{3.25}$$

or
GB energy + variation in surface energy + variation in interface energy = groove force + precipitate force + solute force

Generally, the limiting grain size is ~ 6 times the film thickness [7]. For comparison, for annealed Au and Al films, the largest grains are ~ 10 and 3 times film thickness.

Thus we see that grain growth in thin films is no simple matter and depends on a number of driving forces and free energies. Grains can grow as a result of deposition conditions and post deposition heat treatment, as shown again in Figure 3.13. Grain size of Ni-W films increases from ~ 3 mm to ~ 20 mm with heat treatment at 700 °C. Post deposition heat treatment results can be summarized as

1. A threshold temperature is usually involved with increased grain growth
2. Chemical ordering transition is observed after grain growth at sufficiently high temperatures
3. Grain boundary segregation remains almost constant after heat treatment
4. Relaxation of grain boundaries is manifested through a large heat release in the absence of grain growth
5. Alloying can improve the thermal stability of grain size

No one factor precisely determines how grains will grow in a thin film, but temperature appears to be a dominant player. A number of energies and forces work with and against each other as the grain grows and eventually reaches equilibrium. Because of this fact, the energetics and mechanisms in each material must be considered separately, particularly if alloying or impurities are present.

3.2.3   Columnar Structures

Columnar microstructures comprise the other end of the microstructure spectrum. Columnar structure, shown in Figure 3.18, is by far the most common type of thin film microstructure, and consists of a network of low density material that surrounds an array of higher density parallel rod-shaped columns [7]. The microstructure of a thick sputtered AlN film shown in Figure 3.18 is well behaved with very little deviation from straight rod-like columns. Figure 3.19 shows a highly and densely packed RF sputtered AlON film [20]. Columnar structures are formed when the mobility of the adatoms on the surface of the substrate is low and clustering occurs. Low mobility is caused by low energy resulting from high pressure or low substrate temperature. Columnar structure, due to low adatom

AIN piezoelectric film

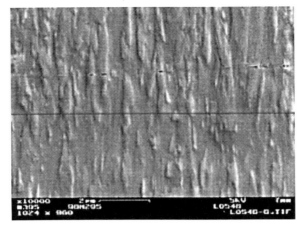

Figure 3.18 SEM picture of well-behaved columnar thin film microstructure.

Figure 3.19 Columnar structure in an RF sputtered ALON film [28].

mobility and low deposition temperature, is often found in amorphous films. Columnar films often have enhanced or anisotropic magnetic, electric, piezoelectric, ferroelectric, and optical properties.

The tangent rule was developed to explain the geometry of columnar grains deposited at an angle α to the substrate [29]. Figure 3.20 summarizes the geometry and angles, and this relation can be expressed as

$$\tan \alpha = 2 \tan \beta \qquad (3.26)$$

Here β is the angle between the column axis and the normal to the substrate. The main structural feature can generally be related

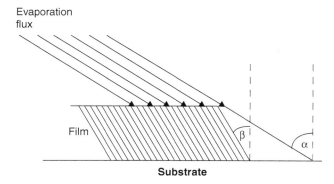

Figure 3.20 Geometry used in the tangent rule [29].

to geometric shadowing. An extreme case of this is glancing angle deposition (GLAD), discussed later. The tilt of the columns is attributed to surface diffusion in that adatoms have a momentum component parallel to the substrate surface and thus tend to migrate along the surface [30]. Oblique incidence even has the potential to shift structure zones. Ion bombardment introduces additional random motion on the surface, and columnar tilting is reduced or even eliminated in this case. Figure 3.21 confirms that ion bombardment inhibits columnar growth [31]. SiC film (a) was deposited with no ion bombardment while SiC film (b) was exposed to ion bombardment during deposition.

While the columnar structure shown in Figure 3.18 is fairly typical of that found in thin films, there are a number of variant, even pathological structures. Under certain conditions (low atom mobility), columnar structure can be further divided into nano-, micro- and macrocolumns [31]. Figure 3.22 shows an example of all this combined structure in a-Ge films [30]. Characteristic sizes of the columns ranged from 1–3 nm up to 300 nm. Five levels of structure have been identified for a-Si:H films [31] with characteristic sizes

- 1–3 nm
- 5–20 nm
- 20–40 nm
- 50–200 nm
- 200–400 nm

We see from Figure 3.22 that the nanocolumns are bundled inside the microcolumns, which are in turn bundled inside macrocolumns.

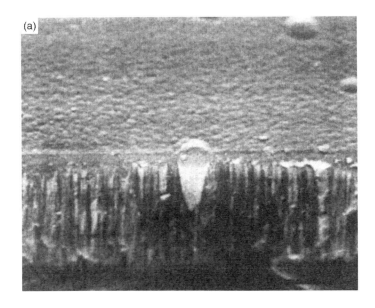

Figure 3.21 SEM micrograph of two SiC films; film (a) had no ion bombardment and film (b) was exposed to ion bombardment [31].

It should be emphasized that low mobility conditions are a requirement for this type of structure.

Pathological structures, some of which are very useful and some of which are detrimental, include nodules, inclusions, GLAD

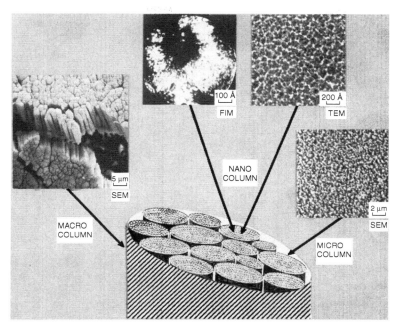

Figure 3.22 Physical structure of nano-, micro and macro columns in a-Ge films [31].

structures, and exaggerated columns. These structures tend to be more frequent in thicker films where they have a chance to nucleate and grow. A nodule, such as that shown in Figure 3.23, is a ubiquitous and annoying manifestation of columnar structure. Nodules are particularly troublesome when they distort surface features but become a disaster when they pop out of the film and form a pit. They need a seed such as substrate pits, scratches, contamination, dust and particulates from the deposition source, to nucleate. The boundary around the nodule is less dense than the nodule itself, which degrades the adhesion of the nodule to the substrate and the rest of the film. It should be noted that the material inside the nodule is as dense as the rest of the film. The best way to avoid nodules is to clean the substrate as thoroughly as possible, make the surface as smooth and defect-free as possible, and keep your deposition source "quiet".

Once distortions start growing in columnar structures (or any type of structure for that matter), they usually cannot be stopped or mitigated. Shadowing enhances all defect formation. Cone formation is a related growth morphology that results from

Figure 3.23 Nodule in and AlON rugate filter [28].

self-shadowing and random competition for cluster growth [31]. Cone seeds nucleate with apexes at the substrate surface and grow geometrically, known as fractal growth. As the cones widen, they die off and others grow at their expense. Obviously surface roughness also increases with cone growth.

Shadowing is used as a tool to control columnar growth in GLAD coatings [32]. Figure 3.24 shows a variety of columnar structures fabricated using the GLAD process. Tilted columns, zig-zag columns, helical (chiral) columns and vertical columns all can be deposited. Columnar growth is best described by the Volmer-Weber mode [33, 34]. Because films are deposited at glancing angles, the tangent rule becomes important. However, it must be modified for increasingly oblique deposition [35]:

$$\beta = \alpha - \arcsin\left\{\frac{(1-\cos\alpha)}{2}\right\} \tag{3.27}$$

or [12] by

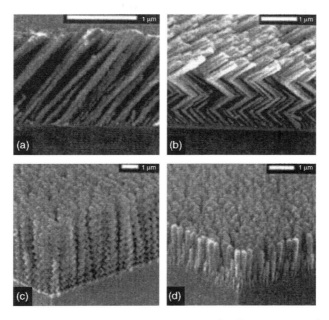

Figure 3.24 Columnar structures fabricated using the GLAD process [32].

$$\tan\beta = \left(\frac{2}{3}\right)\frac{\tan\alpha}{(1 + \Phi\tan\alpha\sin\alpha)} \qquad (3.28)$$

where Φ depends on the diffusivity and deposition rate.

The dependence of β on α in fact has been described by four different models, shown in Figure 3.25 [32]. A word of caution: because columnar structure is sensitive to deposition conditions and material being deposited, each material will behave differently and must be developed independently.

Column evolution starts with growth about a symmetric axis. As columns evolve, they lose their symmetry and tend to fan out in a direction perpendicular to that of the incident vapor. Eventually, the columns will chain together and at glancing incidence, structural anisotropy can be very pronounced, thus forming the basis for the range of column geometries.

In this and the last two sections we see that a wide range of columnar structures are possible in thin films. Structure zone models have been expanded to included non-normal incidence of

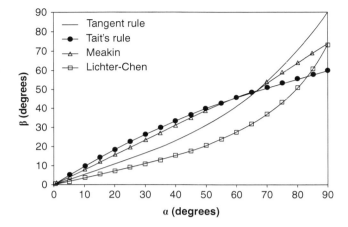

Figure 3.25 Four different curves relating β to α [32].

the incident flux. Structure depends on a number of factors, but most important is angle of incidence of the incident flux to the substrate and subsequent clustering and shadowing. Columnar structure also depends on deposition process, substrate surface morphology, material, and adatom mobility.

3.3 Thin Film Structure Zone Models

Physical vapor deposition (PVD) processes, including evaporation (thermal and electron beam), diode sputtering, magnetron sputtering, and ion plating were first used to modify the surface of metals. Ion bombardment during film growth significantly affects microstructure and increases the density of the film. Second generation processes stress enhanced ionization and ion bombardment of the growing film, and included ion assisted deposition, dual magnetron sputtering, unbalanced magnetron sputtering, cylindrical magnetron sputtering, ion beam sputtering, pulsed laser deposition, and plasma surface modification.

Over the past three decades, a number of models have been formed to predict and explain the structural morphology of vapor deposited thin films. Each successive model further refines and expands previous models based on current deposition technology. The first structure-zone models (SZM) were developed for evaporation and sputtering processes to related deposition parameters to resulting film microstructure [36, 37, 38]. To first order these SZMs are generally valid for most thin film materials. As just described, condensation of a film from the vapor involves bonding of incident atoms into adatoms which then diffuse over the substrate or film surface until they either desorb or are trapped at low energy lattice sites. The process of adatom nucleation and surface diffusion involves a number of processes [7]:

- Shadowing
- Surface diffusion
- Bulk diffusion
- Desorption

All these are dependent on chamber pressure, substrate temperature, deposition rate and substrate placement, and are the basis for SZMs used to predict film microstructure.

The following SZMs and structure zone diagrams (SZD) will be described:

- Evaporated films
- Sputtered films
- Revised SZDfor evaporated metal films
- Revised SZMD for sputtered films
- Ion bombardment

All models are based on the nucleation process described above. One factor that separates them is the energy of incident atoms, which is much lower for evaporated films. Earliest models proposed by Movchan and Demchishin [39] were based on very thick *evaporated* metal films and oxides. Deposition rates ranged from 12,000–18,000 Å/min for Ti, Ni, W, Fe, ZrO_2 and Al_2O_3 films as thick as 2 mm. The model consists of three zones (1, 2, 3), shown in Figure 3.26, determined by the ratio of substrate temperature to melting temperature (T_S/T_M). The model has the following salient features (keep these in mind when reviewing later models):

Zone 1: $T_S < 0.3\ T_M$ (metals), $T_S < 0.26\ T_M$ (oxides) – columns and voids.

Zone 2: $0.3T_M < T_S < 0.45T_M$ (metals), $0.26\ T_M < T_S < 0.45\ T_S$ (oxides) – dense columnar structure.

Zone 3: $T_S > 0.45\ T_M$ (metals and oxides) – polycrystalline structure.

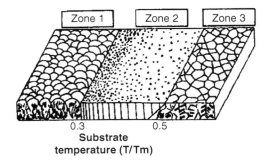

Figure 3.26 Structure zone model of Movechan and Demchishin [39].

The model is based on surface and grain boundary diffusion with columnar grain size increasing with T_S/T_M. Note that no other deposition parameters are included. These will become more important for sputtering.

A five zone model has also been suggested for evaporated films [40]. Three zones similar to those above apply for $T_S < T_M$. Major differences are:

- Zone 1 ($T_S/T_M < 0.1$), atomic shadowing dominates film growth, resulting in an isolated columnar structure
- Zone 2 ($0.1 < T_S/T_M < 0.3$), columnar grains
- Zone 3 ($0.3 < T_S/T_M < 1$), accelerated recrystallization, surface faceting, grain growth, and twinning occur.

Similar SZMs have been developed for sputtered thin films, but now chamber pressure is an important parameter [37]. Figure 3.27 shows the 4-zone SZD for this model, and can be summarized as

- Zone 1 ($T_S/T_M < 0.1$ @ 0.15 Pa to < 0.5 @ 4Pa), voided boundaries, fibrous grains, promoted by substrate roughness and oblique deposition
- Zone T: ($0.1 < T_S/T_M < 0.4$ @ 0.15 Pa, T_S/T_M between 0.4–0.5 @ 4 Pa) – fibrous grains, dense grain boundary arrays.
- Zone 2: (T_S/T_M between 0.4–0.7) – columnar grains, dense grain boundaries.
- Zone 3: (T_S/T_M between 0.6–1.0) – large equiaxed grains, bright surface.

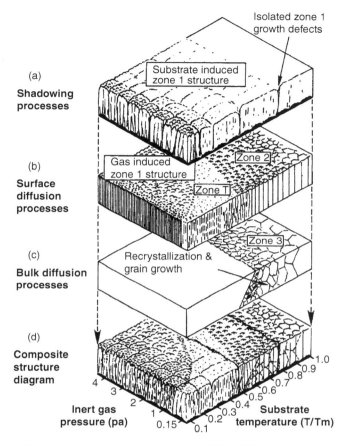

Figure 3.27 Structure zone model for sputtered films [37].

We see that similar structures evolve at slightly lower temperatures for evaporated films compared to sputtered films. Amorphous and crystalline films can form in Zone 1 as a result of shadowing. Surface diffusion is the dominant factor controlling Zone 2 growth. Lattice and grain boundary diffusion dominate in the high temperature Zone 3. Zone T is essentially just a transition region between Zones 1 and 2.

We see for sputtering that inert gas pressure influences film structure through indirect mechanisms such as thermalization of deposited atoms with increased pressure and increased component of atom flux due to gas scattering [41]. Reduction of gas pressure results in more energetic particle bombardment which densifies the film.

These models have been revised and refined over the last two decades [38, 39, 42, 43, 44, 45, 46]. Revised SZMs incorporate ion bombardment, grain structure, film thickness effects, substrate roughness, angle of incidence, and external substrate bias. Additionally, computer simulations and Monte Carlo simulations have enhanced understanding of how film structure evolves. We will address a few of the revised models, which build on SZMs discussed above. Ion bombardment effects are important both in sputtering and evaporation. However, ion bombardment is an integral part of sputtering (DC and RF) processes but must be introduced externally into evaporation processes. Figure 3.28 shows the revised SZD for RF sputtering, which now includes ion bombardment resulting from induced substrate bias [38]. Here the floating potential replaces chamber pressure. This makes sense since particle energy can be related to both pressure and substrate bias (expect more energetic bombardment at low pressures and higher bias: $V_S \sim 1/P$). Zone T is widened compared to zone 1, which can be interpreted as increased ion bombardment enhancing adatom mobility similar to higher substrate temperatures.

Other models address reactive pulsed magnetron sputtering and closed field unbalanced magnetron sputtering, which has become of interest this last decade [45, 46]. Because we are dealing with reactive sputtering, one major variable, particularly for Al_2O_3, is O_2

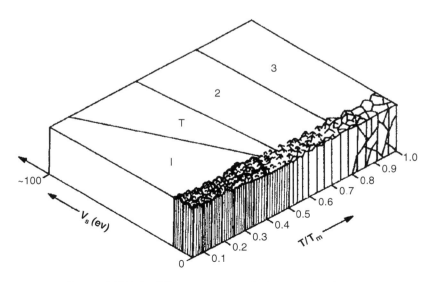

Figure 3.28 Revised SZD for RF sputtering, including ion bombardment.

content in the discharge. The model finds that only Zones 2 and 3 are applicable to this process: Zone 2 applies to substroichiometric oxides with columnar grains and Zone 3 includes stoichiometric films with ceramic properties. The temperature range for Zone 2 microstructures for UBM sputtered films is shifted from T_S/T_M = 0.15–0.45 to T_S/T_M = 0.5–0.77 and Zone 3 is now T_S/T_M between 0.45–0.68. Enhanced ion bombardment in these processes is clearly responsible for the shift to lower temperatures. An SZM has also been developed for ion plating, as shown in Figure 3.29 [47]. Again ion bombardment energy is a major variable.

Figure 3.30 shows the extreme case of employing deposition conditions that would generally be in Zone T for sputtering but still gives a highly columnar structure. The difference here is that the substrate was tubular and rotating above the sputtering target. We see that porosity extends all the way to the substrate. These models must also be extended for glancing angle deposition (GLAD) films.

These models are very useful but do not cover all possible deposition scenarios, such as geometry and shape of the substrate and surface roughness. Among the numerous parameters not addressed by these SZMs are geometry (surface topography, rms roughness, defect density, etc.), physical (residual or background gas pressure, deposition rate), physio-chemical (free energy and enthalpy

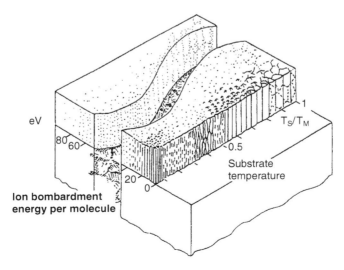

Figure 3.29 SZD for ion plating [47].

Figure 3.30 Enhanced columnar structure resulting from substrate shape and rotation.

of formation, diffusion), and chemical (concentration gradient) in nature [42]. Monte Carlo simulations, such as that shown in Figure 3.31, show step-by step growth of a thin film under various deposition conditions [48]. For an in depth discussion of film nucleation, please refer to Chapter 12 (J Greene) of the Third Edition of the *Handbook of Deposition Technologies for Films and Coatings* (P M Martin, Ed.), Elsevier, 2009. From Figure 3.31 we see that nucleation proceeds with island growth and eventual coalescence.

A word about how residual gas pressure and the resulting co-deposition affect microstructure. Residual gas pressure results from gas molecules adhering to chamber walls, fixturing, and substrates. These gas molecules then desorb and co-deposit on the substrate as the film grows. Resulting effects of co-deposited gas molecules are: [42]

- They occupy surface sites and limit surface diffusion of the deposited coating material

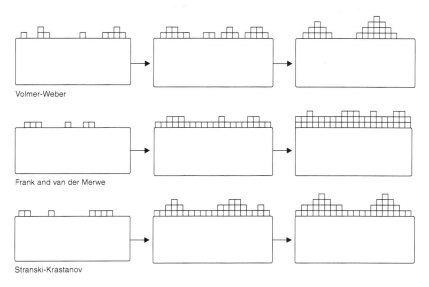

Figure 3.31 Monte Carlo simulation of film nucleation and growth.

- They can be trapped in the growing film
- They can react with the growing film on the substrate surface

The resulting microstructure, particularly for evaporated films, can be nodular, hillocks, or flat topped grains [49].

For all SZMs, the normalized substrate temperature (T_S/T_M) appears to be the most important variable when describing film growth. We can also relate this parameter to encompass total energy (thermal, electronic, kinetic, chemical). Guenther proposed the SZD shown in Figure 3.32 [42]. The model consists of four structure zones, including a vitreous phase, as a function of total particle energy in relation to activation energy for surface diffusion and free energy of solidification (T_S/T_M). The applicable coating processes are shown at the bottom of the figure. We see that the earliest SZMs are reflected in this model, with the exception of the amorphous phase. Guenther describes the next level of detail as a three-dimensional scheme incorporating temperature, particle energy, and second order deposition parameters (residual gas pressure, surface roughness, angle of incidence, defects, etc.).

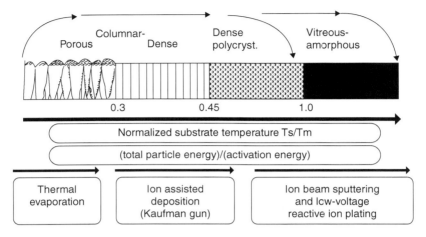

Figure 3.32 Extended and generalized SDM of thin film growth [42].

3.3.1   Zone Structure Model Updates

SZMs presented in the last section address mainly sputtering and evaporation PVD processes. While these models include ion bombardment and substrate bias, an updated model addresses energetic deposition characterized by a large flux of ions for deposition by filtered cathodic arc deposition (FCAD) and high power pulsed magnetron sputtering (HPPMS) [50]. First, let's briefly review these two processes to understand the sources of ion bombardment. FACD and HPPMS are reviewed in Sections 2.1.9 and 2.1.10. Both processes are characterized by very large ion fluxes. FACD uses an electric arc to vaporize material from a cathode target. As a result, a plasma is formed around the target region. The vaporized material condenses on a substrate, forming a thin film.

The arc evaporation process begins by initiating a high current, low voltage arc on the surface of a cathode (target) that creates a small (usually a few μm wide), highly energetic emitting area known as a cathode spot. The temperature at the cathode spot is extremely high (~ 15000 °C), which results in a high velocity jet of vaporized cathodic material, leaving a crater behind on the cathode surface. The cathode spot self-extinguishes after a short time period and re-ignites in a new area close to the previous crater. The arc is rastered over the entire target. For nonconducting and semiconducting targets it is usually necessary to preheat the target material to increase electrical conductivity. A plasma is also created by this process. The arc has an extremely high power density resulting in a high level of ionization (30–100%), consisting of multiply charged ions, neutral particles, clusters and macroparticles (droplets). If a reactive gas is introduced during the evaporation process, dissociation, ionization, and excitation can occur during interaction with the ions and a compound film will be deposited. Thus, the first significant difference between FCAD and PVD processes is the generation of energetic clusters and macroparticles. Particle energies range from 5–25 eV. Among the many designs that filter the ejected particles is one that consists of a quarter-torus duct bent at 90° from the arc source with the plasma guided out of the duct by electromagnetic plasma optics.

HPPMS utilizes extremely high power densities ~ kWcm^{-2} in short pulses (impulses) of tens of microseconds at low duty cycle (on/off time ratio) of <10%. Plasma current density can be as high as 6 Acm^{-2}, while discharge voltage is maintained at several hundred volts [7]. The discharge is homogeneously distributed across the surface of the cathode of the chamber. HPPMS generates a high density plasma of the order of 10^{13} ions cm^{-3} containing high fractions of target metal ions [6], and is typically operated in short pulse (impulse) mode with a low duty cycle in order to avoid overheating of the target and other system components. In every pulse the discharge goes through several stages.

Figure 3.33 shows the updated SZM for FACD and HPPMS [50], which includes contributions from ion bombardment and plasma on film growth. Let's compare this with the most recent SZM for sputtered films shown in Figure 3.28 [37, 38]. This model includes ion bombardment and substrate bias, but no plasma contributions. Recall the 4-zone SZD for this model, which can be summarized as:

- Zone 1 (T_S/T_M < 0.1 @ 0.15 Pa to < 0.5 @ 4Pa), voided boundaries, fibrous grains, promoted by substrate roughness and oblique deposition
- Zone T: (0.1< T_S/T_M < 0.4 @ 0.15 Pa, T_S/T_M between 0.4–0.5 @ 4 Pa)–fibrous grains, dense grain boundary arrays.
- Zone 2: (T_S/T_M between 0.4–0.7) – columnar grains, dense grain boundaries.
- Zone 3: (T_S/T_M between 0.6–1.0) – large equiaxed grains, bright surface.

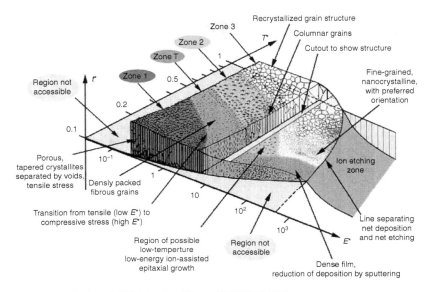

Figure 3.33 Updated SZM for FACD and HPPMS [50].

- Ion bombardment and substrate bias are mainly applicable to Zones 2 and 3.
  ○ The model finds that only Zones 2 and 3 are applicable to this process: Zone 2 applies to sub-stoichiometric oxides with columnar grains and Zone 3 includes stoichiometric films with ceramic properties.
  ○ The temperature range for Zone 2 microstructures for UBM sputtered films is shifted from $T_S/T_M = 0.15$–0.45 to $T_S/T_M = 0.5$–0.77.
  ○ Zone 3 is now T_S/T_M between 0.45–0.68.
  ○ Enhanced ion bombardment in these processes is clearly responsible for the shift to lower temperatures

Figure 3.33 shows that the updated SZM for FACD and HPPMS also maps out stress regions. We see many features described by previous SZMs, but the big difference is that the model's axes are now directly related to factors that directly affect film growth rather than deposition conditions (target current, voltage, pressure, temperature, power pulse length, etc.). The new SZM modifies the axes of the present models to more realistically accommodate actual processes affecting film growth and makes the following changes to previous models [50]:

- Replace the linear T_h axis with a generalized temperature T*, which includes the homologous temperature + the temperature shift caused by the potential energy of particles arriving on the surface (i.e., energetic particle and ion bombardment).
- Replace the linear pressure axis with a logarithmic axis for a normalized energy E*, which encompasses displacement and heating effects caused by the kinetic energy of bombarding particles.
- Label the z-axis with a net film thickness t* which will allow maintenance of film structure while including the effects of thickness reduction by densification and sputtering. Ion etching is included in this quantity.

Comparing Figures 3.33 and 3.28, while the axes are different, the familiar structure zones (Zones 1, T, 2 3) are preserved in this new model, but we see that new sub-zones are added:

- Ion etching zone (in red)
- Region of possible low temperature low energy ion-assisted epitaxial growth
- Transition region from tensile to compressive stress
- Fine-grained nanocrystalline with preferred orientation
- Dense films due to reduction of deposition by sputtering

We see that ion etching can occur at high particle energies and a wide range of T*. Film thickness tapers off when ion etching is involved and several regions are not accessible. Film growth under low particle energies is not allowed in all zones as is growth with high energies in Zone 1.

Again, it all boils down to energetic: the types of energy involved are

- Kinetic energy of bombarding particles (penetrating and non-penetrating)
- Thermal energy that effects adatom mobility

- Substrate bias affects kinetic energy
- Potential energy includes a number of types:
- Heat of sublimation E_C
- Ionization energy E_i

Thus, $E_{pot} = E_C + (E_i - \phi)$. Here ϕ is the work function of an electron needed for neutralization.

The generalized homologous temperatures can be expressed as

$$T^* = T_h + T_{pot} \tag{3.29}$$

with

$$T_h = \frac{T}{T_M} \tag{3.30}$$

$$T_{pot} = \frac{E_{pot}}{kN_{moved}} \tag{3.31}$$

T_{pot} is the characteristic temperature of the heated region where N_{moved} atoms are rearranged.

This model includes atomic displacement due to kinetic energy and associated momentum, which encompassed surface and bulk displacement. Bulk displacement requires energies in the 12–40 eV range. There is also a threshold at which non-penetrating ions can promote surface diffusion, while penetrating particles promote bulk displacement and can have a very short ballistic phase.

While this model takes a significant step in addressing effects of plasma, ion bombardment, and ion etching, i.e., where the ion flux is large, and introduces more realistic factors affecting film growth, there are still several factors to be addressed:

- Growth differences between pure elemental and compound films
- Phase separation
- Nanostructured films
- Amorphous films at low temperatures

While pressure and temperature are still important, factors that more realistically address film growth are generalized temperature, particle energy, and thickness. More has yet to be accomplished but the author admits that deposition processes are so complex that a single model may not be able to entirely predict thin film growth.

It will become evident in the ensuing chapters, particularly in chapter 4, and in virtually every type of surface engineering application, that structure zone models are used to guide deposition of thin films, which directly affects microstructure and properties.

References

1. Milton Ohring, *Materials Science of Thin Films*, Academic Press (2002).
2. M. Konuma, *Film Deposition by Plasma Techniques*, Springer Verlag (1992).
3. tam.northwestern.edu/Lecture 21, Thin Film Nucleation and Structure.
4. B. W. Wessels, *J. Vac. Sci. Technol.*, B(15) (1997) 1057.
5. See Web.utk.edu~prack/thinfilms/thinfilmgrowth.
6. K. N. Tu, J. W. Mayer and L. C. Feldman, *Electronic Thin Film Science for Electrical Engineers and Materials Scientists*, Macmillan (1992).
7. Manijeh Razeghi, Fundamentals of Solid State Engineering, Kluwer Academic Publishers (2002).
8. R. Tsu, *Superlattice to Nanoelectronics*, Elsevier (2005).
9. John Venables, *Introduction to Surface and Thin Film Processes*, Cambridge University Press (2000).
10. K. Oura *et al.*, *Surface Science: An Introduction*, Springer (2003).
11. Y. R. Li *et al.*, *Thin Solid Films*, 489 (1–2) (October 2005) 245–250.
12. C-h Chiu *et al.*, *Physical Review Letters*, 93, 13 (2004) 36105.
13. T. Schwarz-Selinger *et al.*, *Phys Rev B*, 53, 12 (2002) 125317.
14. Fumiya Watanabe *et al.*, *Physical Review Letters*, 94, 6 (2005) 066101.
15. Ted Kamins, Applied Materials Epitaxy Symposium, Santa Clara, CA, September 19, 2002.
16. G. Bauer *et al.*, *Physica Status Solidi*, 203, 14 (2006) 3496.
17. O. E. Shklyaev *et al.*, *Physical Review Letters*, 94, 17 (2005) 176102.
18. J. G. Lu *et al.*, *Appl. Phys. Lett.*, 89 (2006).
19. Joel I. Gertsten and Frederick W. Smith, *The Physics and Chemistry of Materials*, Wiley (2001).
20. H. Hillert, *Mat. Sci. Forum*, 204–206 (1996) 3–18.
21. C. V. Thompson, *Mat. Res. Soc. Sypm. Proc.*, 343 (1994) 3.
22. R. Carel *et al.*, *Mat. Res. Soc. Sypm. Proc.*, 343 (1994) 49.
23. J. M. E. Harper and K. P. Rodbell, *J. Vac. Sci. Technol.*, B15 (1997) 763.
24. Joost Vlassak, Thin Film Mechanics, presented at Harvard University, April 28, 2004.
25. R. W. Hoffman, *Phys. Thin Films*, 3 (1968) 211.

26. R. W. Hoffman, *Phys. Thin Films*, 34 (1976) 185.
27. Freund and Chason, *J. Appl. Phys.*, 89 (2001) 4866.
28. H. A. Macleod, *Thin Film Optical Filters*, 3rd Ed, I.O.P. (2001).
29. J. M. Nieuwenhuizen and H. B. Haanstra, *Philips Tech. Rev.*, 27 (1966) 8.
30. K. Hara *et al.*, *J. Magn. Mater.*, 73 (1988) 161.
31. R. Messier *et al.*, *J. Vac. Sci. Technol.*, A2(2) (1984) 500.
32. Matthew M. Hawkeye and Michael Brett, *J. Vac. Sci. Technol.*, A25(5) (2007) 1317.
33. R. W. Vook, *Int. Met. Rev.*, 27 (1982) 209.
34. K. Reichelt, *Vacuum*, 38 (1988) 1083.
35. R. N. Tait *et al.*, *Thin Solid Films*, 226 (1993) 196.
36. John A. Thornton, *Ann. Rev. Mater. Sci*, 7 (1977) 239.
37. R. Messier *et al.*, *J. Vac. Sci. Technol.*, A2(2) (1984) 500.
38. Russell Messier, *J. Vac. Sci. Technol.*, A4(3) (1986) 490.
39. B. A. Movchan and A. V. Demchishin, *Phys. Met. Metallogr.*, 28 (1969) 83.
40. J. V. Sanders, in *Chemisorption and Reactions on Metal Films*, J. R. Anderson, ed., Academic Press (1971).
41. J. Greene, in *Multicomponent and Multilayered Films for Advanced Microtechnologies: Techniques, Fundamentals and Devices*, O. Auciello and J. Engemann, eds., Kluwer (1993).
42. Karl H. Guenther, *SPIE* Vol. 1324 (1990) 2.
43. H. T. G. Hentzell *et al.*, *J. Vac. Sci. Technol.*, A2 (1984) 218.
44. Q. Tang *et al.*, *J. Vac. Sci. Technol.*, A17 (1999) 3379.
45. P. J. Kelley and R. D. Arnell, *J. Vac. Sci. Technol.*, A17 (1999) 945.
46. P. J. Kelley and R. D. Arnell, *Vacuum*, 56 (2000) 159.
47. E. Moll *et al.*, *Surf. Coating Technol.*, 39/40 (1989) 475.
48. R. W. Vook, *SPIE* Vol. 346 (1982) 2.
49. P. S. McLeod and L. D. Hartsough, *J. Vac. Sci. Technol.*, A14 (1977) 263.
50. Andre Anders, *Thin Solid Films*, TFS-26898, in-press.

4

Thin Film Tribological Materials

The most often used and simplest type of tribological thin film structure is the single layer. More complex multilayer structures and composites (discussed in Section 4.2), however, generally offer improved performance, but at the cost of increased process complexity and cost. There are literally hundreds of thin materials used to improve the tribological properties of solid surfaces. Because it would not be possible to address all these materials, we will describe the following broad categories:

1. Binary thin films, including transition metal based materials, chromium-based materials, boron-based materials, carbon and diamond-based materials, tungsten and molybdenum based materials.
2. Ternary thin film materials
3. Quaternary and higher component thin film materials
4. Nanostructured materials

An attempt has been made to divide these materials into base-metal categories, but the reader should be advised that there is still considerable overlap which could not be avoided.

187

How do thin film coatings improve wear resistance and what are the qualities of a good tribological coating? Can one type of coating reduce all the types of wear discussed above? When are ductile or brittle coatings required? The most important properties of a wear resistant tribological coating are listed in Section 4.1. Wear resistant coatings and thin films are deposited by a number of processes, described in Chapter 2 and can possess a wide range of microstructures, described in Chapter 3. Damage mechanisms have been discussed in Chapter 1.

4.1   Hard and Ultrahard, Wear Resistant and Lubricous Thin Film Materials

Wear resistant thin films are deposited by a wide variety of processes, but most often by PVD and CVD techniques because they can be scaled up to industrial sizes and quantities. Figure 4.1 provides a listing of hard thin film materials and compares the microhardness of PECVD and PVD deposited thin films [1]. Several types of films deposited by PECVD have microhardness in the "super-hard" range while the number is limited for PVD deposited films. Superhardness is defined as microhardness greater than 40 GPa. Hardness is generally improved by increasing energy of deposited atoms using high substrate temperatures, substrate bias, plasma bombardment (ion assist) or a combination of all four. It will be instructive to have this background when reviewing the thin film materials systems described in this chapter.

4.1.1   Titanium Based Thin Films

This family includes titanium diboride (TiB_2), titanium carbide (TiC), titanium boron carbide (TiBC) and nitrides TiCN, TiAlN, AlCrN, TiBCN and many more that it are just not possible to include. All these materials are exceptionally hard and corrosion resistant. Bulk TiB_2 has a gray metallic color and has a number of applications due to its exceptional hardness (25–35 GPa), high melting point (3225 °C), high thermal conductivity (60–120 W/mK) and high electrical conductivity ($\sim 10^5$ S/cm). Thin films typically have hardness $\sim$ 34 GPa, conductivity $\sim$ 2.87 E10^4 S/cm, and reported friction coefficient (COF) between 0.02–0.15 [2, 3]. Because of their high conductivity, films display metal-like reflectivity at NIR and

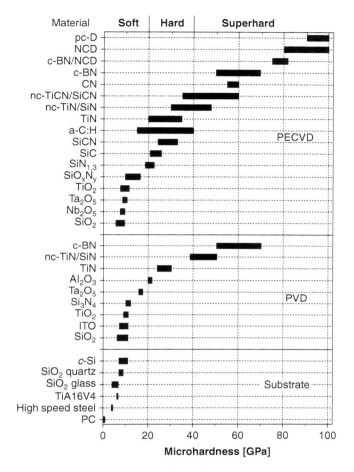

Figure 4.1 Microhardness of PVD and CVD films with commonly used substrates [1].

IR wavelengths and are being considered for solar control coatings on satellite structures [3].

TiB$_2$ is deposited mainly by magnetron sputtering [2, 3]; however, a base coat such as Ti or Cr is used to improve adhesion to a wide variety of materials, including aluminum, stainless steel, magnesium copper alloys, silicon aluminum alloys, titanium alloys, and carbon-carbon alloys [3].

In addition to being extremely hard and wear resistant [4, 5, 6], TiC is being developed for medical implants [7], decorative coatings [8], and photovoltaics [6]. Much of the work on this material took place between 1970 and 2000 with PVD films being preferentially developed in the last decade. Unlike TiB_2, TiC is deposited by a number of techniques, summarized in Table 4.1. Hardness of TiC appears to be typical of most titanium hard coats and peaks in the

Table 4.1 Tribological properties and deposition processes for Ti-alloy coatings.

Material	Deposition Process	Hardness (GPa)	COF	Reference
TiB_2	Magnetron sputtering	34	0.02–0.15	2, 3
TiC	Magnetron sputtering	12, 28	0.15–0.53	13, 14
	Closed field unbalanced magnetron sputtering	23–42	0.20	4
	Reactive pulsed arc evaporation		0.53	6
	Plasma activated electron beam evaporation	35		5
	Ion plating			
	Hollow cathode evaporation	3400 H_V		8
	CVD			9, 10, 11, 118, 19
	PACVD			17
	Pulsed laser ablation	28, 30		7, 12
	Cathodic arc deposition	43	0.15–0.53	17
	Ion assisted ion plating			7

Table 4.1 (cont.) Tribological properties and deposition processes for Ti-alloy coatings.

Material	Deposition Process	Hardness (GPa)	COF	Reference
TiCN	Magnetron sputtering	3648 H_V	0.28–0.35	22, 23, 29, 30
	Filtered cathodic arc deposition	45	0.1–0.25	24, 25, 31, 32
	CVD			28, 31
TiAlN	Magnetron sputtering	26.8 2100–2900 H_V 20–36	0.5–0.65	38, 49, 41, 42, 43
	Filtered cathodic arc	32		44
$Al_{1-x}Cr_xN$	Magnetron sputtering	4000 H_V		53
	Unbalanced magnetron sputtering	31		46
	Closed field UBMS	30.5–40.8		46, 50, 51
	Filtered cathodic arc	28, 31, 38		46, 48, 49, 50
Ti-B-C-N	Magnetron sputtering	5500 H_V, 5100 H_V		54, 55
	Filtered cathodic arc			56

mid 30s GPa. Highest hardness values ~ 42 GPa are reported for closed field magnetron sputtering films [4, 15, 17]. As with other hard and low friction coatings, composition, hardness, COF, and mechanical properties depend to a large degree on energetics of the deposition process. Figure 4.2 shows the dependence of hardness magnetron sputtered films on substrate bias [13]. Highest hardness values are obtained at highest substrate biases, or strongest

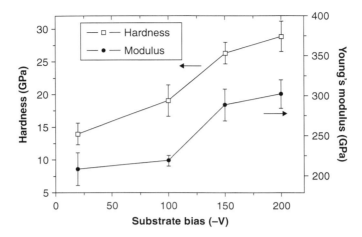

Figure 4.2 Dependence of hardness and Young's modulus of magnetron sputtered TiC films on substrate bias [14].

substrate bombardment. The critical load, however, peaked at a bias of −100 V, indicating a deterioration of adhesion at higher bias (possibly due to increased compressive stress). Films also had a (111) preferred crystal orientation.

As expected, hardness of magnetron sputtered TiC films also depends on deposition rate, substrate target spacing (~ adatom energy). Figures 4.3 and 4.4 show the dependence of hardness on power to the Ti target and target substrate spacing for films cosputtered using Ti and C targets [4]. Lowest COF ~ 0.18 was obtained at highest Ti target powers of 475 W and largest substrate target distance of 180 mm.

Note that *adatom energy* will be a common thread linking hardness and wear resistant properties of triological thin films, regardless of deposition technology.

TiC films deposited by CVD and PACVD use the following gas reactions [18, 19]:

$$TiCl_4 + CH_4 \rightarrow TiC + 4HCl$$

$$TiCl_4 + CCl_4 \rightarrow TiC + 8HCl$$

As a result of using plasma assist, lower substrate temperatures between 350 °C and 600 °C were used with adherent films deposited at temperatures above 400 °C. Without plasma assist or

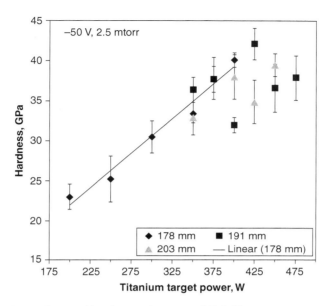

Figure 4.3 Dependence of hardness of sputtered TiC films on power to the Ti target [4].

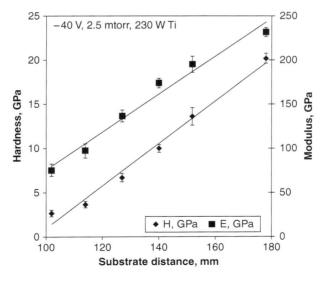

Figure 4.4 Dependence of hardness of sputtered TiC films on substrate target spacing [4].

enhancement, substrate temperatures are in the range 850 °C to 1050 °C [4].

TiC films are also deposited by arc activated deposition [5, 6], ion plating [20], pulsed laser ablation [7, 12], and ion assisted ion plating [7]. Reactive evaporation processes use mixtures of Ar + CH_4 with a Ti source. Wear rates for ion plated TiC films were found to be superior to films deposited by CVD and were lowest for acceleration voltages near 400 V and C_2H_2 pressure of 3 mTorr.

Hard and wear resistant nitride alloys of Ti-B-C include TiCN, TiAlN, and TiBCN. Nitrides are generally less brittle and have a higher fracture toughness than their corresponding carbide [21]. TiCN thus combines the hardness of TiC and the high toughness of TiN. TiCN is harder than and used as a low friction replacement for TiN in moderate temperature applications. It most commonly is copper to bronze colored but, depending on C content, color can also range from pink to blue. TiCN has applications for reusable wear resistant coatings for cutting tools, jewelry, golf clubs, and some scientific instruments. TiCN coatings are deposited by magnetron sputtering [22, 23], cathodic vacuum arc [24, 25], laser ablation [26], plasma immersion, ion implantation and deposition [27], and chemical vapor deposition (CVD) [28]. CVD is used to deposit TiCN coatings in the cutting tool industry due to the high adhesion of deposited coatings. However, as discussed above, the high deposition temperatures of CVD in the neighborhood of 1000 °C limit the selection of substrate materials.

TiCN films are deposited by magnetron sputtering using a high purity Ti target in N_2 + CH_4 mixtures [29, 30]. Films with highest hardness values ~ 3500 kgmm^{-1} are obtained for compositions of 70% CH_4 + 30% N_2. These values were essentially the same as TiC films.

CVD films are deposited using a shower head reactor and decomposing the metalorganic presursor tetrakisdimethylamino titanium (TDMAT) at temperatures between 350 °C–490 °C [31].

Cathodic vacuum arc is a promising physical vapor deposition technique for the deposition of well adhered TiCN coatings at low temperature due to their high ionization rate and high ion energy [27]. Films with hardness of 45 GPa and COF between 0.1 and 0.25 are deposited using a Ti source in mixtures of 95%N_2 and 5%CH_4. The system consisted of one dual filtered arc source, one rectangular plasma-guide chamber, one deposition chamber, auxiliary anodes, heating system, substrate bias system, and vacuum system.

The dual filtered arc source consisted of two primary cathodic arc sources utilizing round Ti targets, which are placed opposite to each other on the sidewalls of the plasma-guide chamber, surrounded by rectangular deflecting coils, and separated by an anodic baffle plate [32]. Substrate bias was –40V with a substrate temperature of 350 °C.

TiAlN and AlTiN are arguably currently the most widely used nitrides in this family and are also used extensively in multilayer and nanolaminate structures. Applications include wear resistant coatings for cutting tools, dies, molds, biomedical implants, decorative coatings components, bearings and machinery parts, as well as nano-scale materials synthesis technologies required for high-precision applications in microelectronics, pharmaceuticals, biomedical, and chemical process industries [33]. TiAlN and AlTiN are especially useful for high temperature cutting operations, and are used to machine titanium, aluminum and nickel alloys, stainless steels, alloy steels, Co-Cr-Mo, and cast irons.

Figure 4.5 shows a picture of TiAlN coated drill bits. Four important TiAlN compositions are deposited in industrial scale by PVD processes:

- $Ti_{50}Al_{50}N$
- $Al_{55}Ti_{45}N$
- $Al_{60}Ti_{40}N$
- $Al_{66}Ti_{34}N$

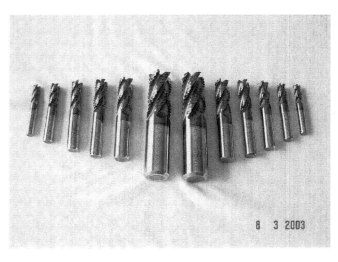

Figure 4.5 Picture of TiAlN coated drill bits.

For example, selected properties of the $Al_{66}Ti_{34}N$ are:

- Hardness = 2600 to 3300 H_V.
- Phase stability to 850 °C, start of decomposition to AlN+TiN.
- Intense oxidation starts at about 800 °C (ca. 300 °C higher than for TiN).
- Lower electrical and thermal conductivity than TiN.
- Typical coating thickness 1 μm to 7 μm.

Coatings are often doped with C, Cr, Si, B, O, or Y in order to improve selected properties for specific applications. The coatings are also used to create multilayer systems (discussed next later in this Chapter), e.g. in combination with TiSiXN, TiN [34, 35]. Nanocomposite TiAlSiN and TiAlN/TiN and TiAlN/WC-C nanol-aminate coatings exhibit superhardness hardness and outstanding high temperature workability [36, 37]. Figure 4.6 shows coating property trends for PVD coatings and Figure 4.7 compares the

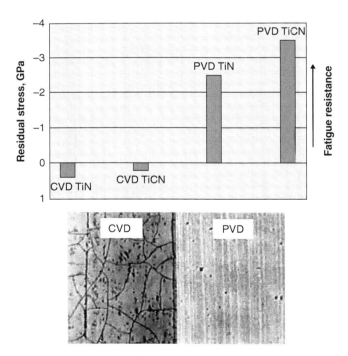

Figure 4.6 Coating property trends PVD coatings [38].

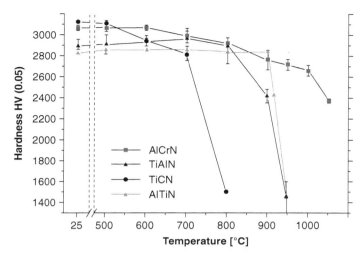

Figure 4.7 Comparison of the hardness of PVD coatings after vacuum anneal [38].

hardness of PVD AlTiN, TiAlN, TiCN, and AlCrN after annealing in vacuum [38]. An excellent review of PVD decorative hard coatings is given by Don Mattox [38]. Nanolaminates and nanocomposites will also be addressed in this Chapter.

Coatings are generally deposited by the magnetron sputtering (including cylindrical magnetron, dual magnetron, unbalanced magnetron [38, 39, 40, 41, 42, 43], and filtered cathodic arc deposition [42]. Magnetron sputtered films are deposited using Ti and Al targets in Ar + N_2 mixtures and DC or RF power sources. Composition depends on relative power to each sputtering target, N_2 partial pressure, and substrate temperature. Adhesion and surface smoothness improves with increased substrate temperatures up to 200 °C [39]. Hardness increased to 3600 H_V with the addition of vanadium (V) and the composition $Ti_6Al_4V{:}N$ [42]. The maximum working temperature for these films was ~ 700 °C .

Films with composition $Ti_{0.23}Al_{0.67}N$ deposited by filtered cathodic arc deposition displayed a hardness of 32 GPa [36, 44, 45]. A phase separation into c-TiN and h-AlN appeared at a temperature of 1100 °C.

Other nitrides that deserve honorable mention are (Ti)AlCrN and TiBCN. AlCrN and (Ti)AlCrN have reported hardness greater than TiAlN or AlTiN films and improved oxidation and temperature resistance [38, 46]. Industrial applications have increased for

AlCrN and include improved wear resistant coatings for cutting tools and operation at temperatures over 800 °C. $Al_{1-x}Cr_xN$ films are deposited by magnetron sputtering (unbalanced, closed field unbalanced), reactive cathodic arc deposition, ion plating, and plasma PVD processes [46]. Hardness ranges from 21.2 GPa to 41 GPa and depends on deposition technique and composition, as shown in Table 4.1. Highest hardness values of 40.8 GPa are achieved for CFUBMS films with composition $Al_{0.71}Cr_{0.29}N_{1.2}$ [47]. Filtered cathodic arc deposition (FCAD), however, appears to be the most widely used deposition technique. Figure 4.8 plots the dependence of hardness on Cr content (x) for FCAD films, showing hardness peaking at a Cr content of ~ 0.40 [46].

Hardness is found to decrease with post deposition annealing temperatures greater than 900 °C [52] as shown in Figure 4.9. The figure shows that films with Al content ~ 0.71 have the highest hardness but are also most affected by high annealing temperatures. Oxidation also increases significantly for temperatures greater than 900 °C, as shown in Figure 4.10 [52]. However, films with higher Al content tend to oxidize at a slower rate.

We will return $Al_{1-x}Cr_xN$ films when discussing multilayer and nanolaminate coatings.

Hardness and oxidation resistance can be enhanced to some extent by alloying elements such as Ti, V, Nb, Mo, and W with

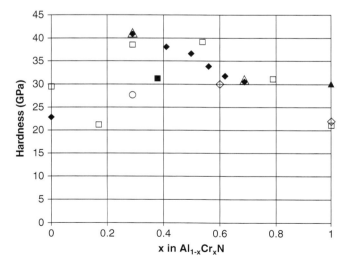

Figure 4.8 Dependence of hardness of FCAD films on Cr content (x) [46].

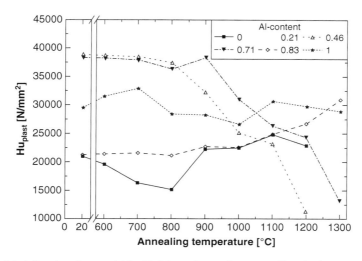

Figure 4.9 Microhardness of $Al_{1-x}Cr_xN$ coatings after annealing in Ar atmospheres [52].

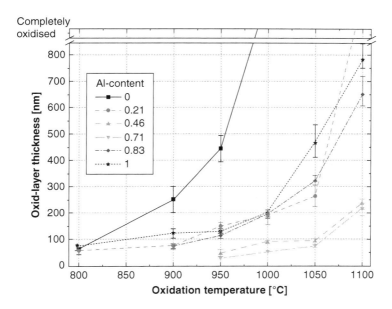

Figure 4.10 Thickness of oxide layer of various coatings after annealing in air [52].

AlCrN. Figure 4.11 plots hardness of AlCrTiN, AlCrVN, AlCtYN, AlCrNbN, AlCrMo, and AlCrWN alloys [48]. The latter two alloys are slightly harder than the first four.

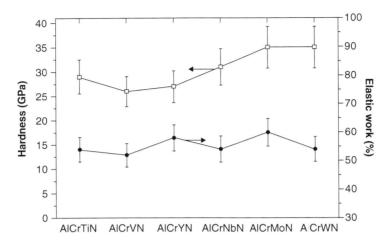

Figure 4.11 Hardness of AlCrTiN, AlCrVN, AlCtYN, AlCrNbN, AlCrMo, and AlCrWN alloys [48].

We end with TiCBN films, although there are still a number of hard coating compositions that will not be addressed (Ti-Hf-C-N, Ti-Al-V, Cr-C-N, MCrAlY, Ti-V-N, and Ti-Zr-N for example). This material appears to have been popular generally before 2000 and is deposited primarily by magnetron sputtering [54, 55] and filtered cathodic arc deposition [56]. Hardness values of 5500 H_V and 5100 H_V have been reported for magnetron sputtered films and depend on composition. Films are deposited using TiB_2 targets in mixtures of Ar + N_2 + CH_4. Measured wear rate is only slightly higher than Ti-B-C films and less than sputtered TiN, TiB_2, Ti-B-N films [54].

4.1.2 Boron Nitride and Related Materials

The boron compound family includes boron nitride (BN), boron carbide (B_xC_y), boron carbon nitride (B_xC_y:N), and boron carbon silicon nitride (BCSiN). Other borides exist but are more appropriately placed in the titanium alloy family (TiB_2, TiBC, TiBN). BN is transparent down to the near UV and cubic BN (c-BN) is extremely hard, second only to diamond, and wear resistant. This makes it an attractive material for protecting optics and, because of its low atomic mass, it is virtually transparent to x-rays. This material is thus an excellent protective optical coating and has high thermal conductivity [11, 57, 58]. Cubic BN is also extremely resistant to oxidation, and, because of its low affinity to steel, is used to protect cast irons and steels [59, 60]. Hexagonal BN (h-BN) is the other common allotrope, but is soft and unstable in oxygen and water environments. We thus focus our attention on c-BN.

Table 4.2 summarizes properties of BN allotropes [11] and Table 4.3 lists deposition processes for these thin films.

Table 4.2 Properties of B-N allotropes [57].

Property	c-BN	h-BN	a-BN
Crystal structure	cubic, zinc blend	hexagonal	amorphous
Lattice constant (Å)	a = 3.61	a = 2.50 c = 6.60	----
Density (g/cm³)	3.48	2.20	1.74
Hardness (GPa)	70		
Thermal conductivity (W/cmK)	2–9	a: 0.36 c: 0.68	

Table 4.3 Deposition processes for B-N, B_xC_y, B_xC_y:N and mechanical properties (when reported).

Material	Deposition Process	Hardness (GPa)	COF (typical)	Reference
BN	CVD			63, 64
	PECVD			65
	ECR pulsed plasma	2.7–10.5		66
	UBM			67
	Ion beam assisted Deposition			68
	Pulsed plasma			69
	RF sputtering			70
	RF magnetron			
	Ion plating			57

Table 4.3 (cont.) Deposition processes for B-N, B_xC_y, B_xC_y:N and mechanical properties (when reported).

Material	Deposition Process	Hardness (GPa)	COF (Typical)	Reference
	Activated reactive evaporation	6200 H_V	0.35	
	Atomic layer deposition	50		71, 74
	Dual ion beam deposition	37, 48		72, 73
B_4C/B_xC_y	Magnetron sputtering	35	0.1–0.9	81
	CVD	5000 H_V		78, 79, 80
	Pulsed laser deposition			82
$B_xC_yN_z$	Magnetron sputtering	8–16		86
	IBAD	31		88, 89
$(BN/CN)_n$	Magnetron sputtering	65		90
$(CN/BN)_n$	Magnetron sputtering	100		90

Cubic BN films have a low COF between 0.18 and 0.22 and a-BN films have an even lower COF between 0.15 and 0.18 [61]. COF of c-BN is stable while that of other allotropes can change over time [57]. High mechanical stress and poor adhesion to most substrate materials (not uncommon for nitrides) are a major limitation of c-BN, limiting thickness to ~ 1 μm on cutting tools. As mentioned earlier, c-BN is an excellent protective coating for many optical components, including plastics [62].

As we have seen for DLC, ta-C and CN_x, formation of sp³ bonds is critical for formation of hybridized structures and films with optimum hardness. This ultimately requires high substrate temperatures or high ion bombardment during deposition. Cubic BN is deposited using CVD processes [63, 64], PECVD [65], ECR microwave plasmas [66], unbalanced magnetron sputtering [67], ion assisted deposition [68], pulsed plasma deposition [69], activated reactive evaporation (ARE) [57], RF sputtering [70], atomic layer deposition [71], dual ion beam deposition [16, 17], and ion plating [1], but best results generally have been achieved using CVD and PECVD processes. CVD and PECVD processes involve mixtures of $B_2H_6 + H_2 + NH_3$, $B_2H_6 + N_2$, or $B_2H_6 + BF_3 + B_3N_3H_6$ at substrate temperatures up to 1800 °C. ALD processes use mixtures of BCl_3 and NH_3 [71, 74]. Hybridized bonding has also been achieved by post deposition heat treating.

Hardness of c-BN films depends on a number of factors, determined primarily by deposition process. These factors include B/N ratio [67], density (low porosity), degree of sp^3 binding (or mixed sp^2/sp^3 bonding), and lack of growth defects. Every "c-BN" film has some percentage of h-BN, again affected primarily by the energy and mobility of deposited atoms. For example, Figure 4.12 shows the dependence of c-BN fraction with substrate temperature for RF magnetron sputtered films sputtered from a h-BN target [75]. c-BN fraction saturates at ~ 91% at a substrate temperate ~ 300 °C. Figure 4.13 shows the dependence on N_2 partial pressure and substrate temperature. As substrate temperature is increased to 500 °C, less

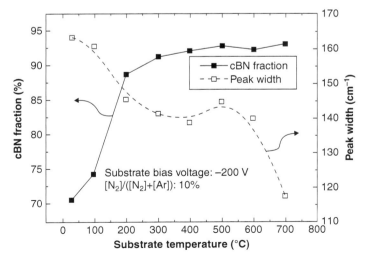

Figure 4.12 Influence of substrate temperature on c-BN fraction for RF magnetron sputtered BN films [75].

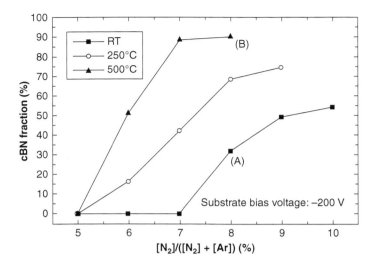

Figure 4.13 Dependence of c-BN fraction on N_2 content in working gas [75].

N_2 is needed to maximize c-BN composition. Significantly lower c=BN fractions are obtained at lower substrate temperatures. Also note that a substrate bias voltage = −200 V was used.

The following conclusions were reported for RF magnetron sputtering [76]:

- Substrate bias plays a decisive role in formation of the cubic phase. There exists a threshold bias above which the c-BN begins to form. Above a bias ~ 150 V resputtering retards film formation.
- Substrate temperature plays an equally important role as substrate bias.
- There exists an optimum working gas composition $(N_2/(Ar + N_2)$ for the formation of the cubic phase (see Figure 4.13).

Similar to DLC and ta-C, the fraction of c-BN in CVD and PECVD films depends primarily on gas mixture, substrate temperature, and substrate bias. The following gas mixtures and substrate temperatures have been reported:

- $B_2H_6 + H_2 + NH_3$ [63, 64]
- $B_2H_6 + N_2$ [63, 64]

- $B_2H_6 + BF_3 + B_3N_3H_6$ [63, 64]
- $B_3N_3H_6$ @ 300–650 °C [77]
- $BCl_3 + NH_3 + H_2 + Ar$ @ [77]
- $BCl_3 + NH_3$ @ 700–1100 °C [77]
- $BF_3 + NH_3$ @ 250–600 °C [57, 71, 74]
- $B_{10}H_{14} + NH_3$ @ 600–900 °C [77]

Because of plasma activation, lower process temperatures are used in PECVD and ECR deposition. A significant substrate bias is needed for predominantly c-BN films.

Boron carbide (B_xC) has stable allotropes ranging from B_4C to $B_{12}C$, however, B_4C (actually $B_{12}C_3$) is the most common and most often used. Figure 4.14 shows the crystal structure of B_4C. In addition to being an extremely hard and wear resistant ceramic, B_4C is used in thermoelectric devices [78, 79]. Table 4.4 lists the various applications for B_4C [78].

B_4C is deposited mainly by CVD [78, 79, 80] and magnetron sputtering techniques [81], although it has also been deposited by pulsed laser deposition (PLD) [25]. In magnetron sputtering films are deposited using a B_4C target and Ar sputtering gas. Films were found to be amorphous with hardness values more dependent

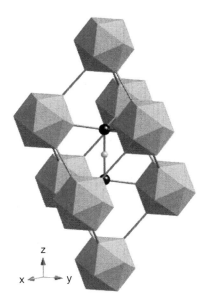

Figure 4.14 Crystal structure of B_4C.

Table 4.4 Wear applications for B$_4$C [78].

Work Material	Tool Types	Life Improvement	Applications
Aluminum alloys (314, 319, 356, 30, 6061, 7075)	Drills, inserts, taps, gun reamers, saws	3–10 X (as high as 20 X)	Automotive, aircraft, saws
Brass–bronze– Copper alloys	Drills, punches, refacing cutters	2–6 X	Electronic parts, spot weld electrode cutters
Cast iron (gray)	End mills, inserts, taps, drills, saws	3–6 X	Engine blocks, piston rings, exhaust manifolds, water pumps
Titanium alloys	Drills, end mills	2–5 X	Aircraft components
Nickel base alloys (inconel, rené, etc.)	Taps, drills, cutters	2–6 X	Aircraft components
Stainless steel	Drills, inserts, reamers	6–20 X	Aircraft components
Composites Graphite–Fiber, Glass–Aluminum Sheet molding compounds Wood products	Router cutters Taps Router bits Cutting routers	2–4 X 5 X 30 X 3–10 X	Aircraft components

upon the physical rather than chemical nature of the film. Substrate bias had a profound effect on the physical microstructure of the films. Those deposited at higher substrate bias were harder than films deposited at lower substrate bias. With no bias applied to the substrate, hardness was ~ 10 GPa. As the bias voltage increased to

80 V, the hardness of the films increased to 18 GPa. At 200 V bias, the hardness of the coating approached 35 GPa. Associated with the change in the hardness, the elastic modulus also increased from 145 GPa at zero bias to 275 GPa at 200 V bias.

Zone T and/or zone 2 morphology (see Chapter 3) with good mechanical properties was achieved at low bias and low substrate temperatures. At low bias, morphology of the coating was columnar. Oxides were found between columns, while the column mainly consisted of boron carbide. At higher bias levels (>200 V), a continuous solid coating was formed. This densification of the coating increases hardness. While the bias affects the hardness, it did not appear to influence the COF of the coating to any great extent.

Figure 4.15 shows a phase diagram plotting carbon concentration of B_xC_y films against CH_4 partial pressure for films deposited by CVD in a low pressure cold wall reactor at ~ 1100 °C using mixtures of $BCl_3 + CH_4 + H_2$ [80]. CVD films are also deposited using mixtures of $BBr_3 + CH_4 + H_2$ at 600 °C [79]. Crystalline phase appears to be sensitive to CH_4 partial pressure with the metastable phases more likely at higher CH_4 partial pressures. As expected, hardness

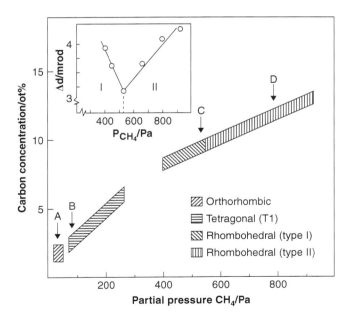

Figure 4.15 Phase diagram for B_xC_y films plotting carbon content against CH_4 partial pressure [24].

depends on C content, as shown in Figure 4.16 [79]. We see that at all substrate temperatures, highest hardness values near 5000 H_V are obtained at ~ 30 at% C. Film morphology also depends on carbon content; For the orthorhombic and tetragonal phases, the crystal facets decrease in size and become less well defined with increasing CH_4 partial pressures, while the opposite behavior is observed for increasing BCl_3. A significant change in grain size is observed when passing a deposition region boundary (i.e., between phases). For example, a fibrous morphology is obtained for the orthorhombic phase at vapor compositions close to the deposition region boundary between the orthorhombic and tetragonal phases.

Boron carbon nitride (B_xC_yN), also known as "heterodiamond" is essentially c-BN that contains between 5 at% and 30 at% C and is generally deposited by PVD and CVD processes [83]. Table 4.5 summarizes hardness values for the B-C-N family. From this table we see that B-C-N has hardness second only to diamond. Thin film deposition processes include magnetron sputtering [86], plasma assisted CVD (PACVD) [87], ion beam assisted deposition (IBAD) [88], and ion assisted evaporation [89]. Magnetron sputtered films were deposited cosputtered using high purity graphite and boron targets in 50/50 Ar + N_2 mixtures with formation of hybridized B-C-N bonds. Elemental composition depended on power to the carbon target with power to the boron target held constant at 150 W.

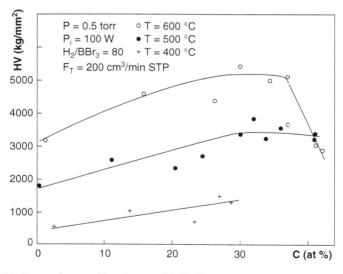

Figure 4.16 Dependence of hardness of B_xC_y films on carbon content [79].

Table 4.5 Summary of hardness values for B-C-N family [84, 85].

Material	Diamond	c-BC$_2$N	c-BC$_5$	c-BN	B$_4$C	ReB$_2$
Vickers hardness (GPa)	115	76	71	62	38	22
Fracture toughness (MPam$^{1/2}$	5.3	4.5	9.5	6.8	3.5	

Composition of the films varied with power to the carbon target as follows:

- 60W: B = 36at.%, C = 32at.%, N = 32at.%
- 180W: B = 23at.%, C = 53at.%, N = 23at.%
- 240W: B = 21at.%, C = 56 at.%, N = 23 at.%

Hardness increased from 8 GPa @ 60 W to 16 GPa @ 240 W target power.

As with all films discussed in this chapter, energy of incident atoms and ion bombardment plays a significant role in bonding and film properties. To this end, ion beam and plasma assisted deposition processes would be expected to produce B-C-N films with hybridized bonding and high hardness values. Films are grown by ion beam assisted processes using B$_4$C in N$_2$ gas using different ion accelerating currents (fluencies) and voltages (energies) [88, 89]. The hardness of films with compositions of 49 at.% B, 42at.% C, and 9at.% N peaked at 31 GPa at an ion current density of 60 mA/cm^2 and a voltage between 1.5 and 2 kW. COF showed a mild dependence on ion bombardment and ranged from ~ 0.60–0.76. Figure 4.17 shows a ternary phase diagram indicating relative compositions of BCN films for different IBAD conditions and a range of N$_2$ + CH$_4$ mixtures [89]. The composition of the samples, for ion assistance with N$_2$ and N$_2$ +CH$_4$, is illustrated in Fig. 4.17. Two clear trends are apparent from the figure: for N$_2$ bombardment N is incorporated in the films to the detriment of the cC content and, therefore, BN compositions are preferred. However, with N$_2$ +CH$_4$ bombardment, films become C rich but the incorporation of nitrogen is less effective, resulting in the production of B$_x$C films. This is not surprising since BN and B$_x$C compounds are stable as bulk compounds, while no CN$_x$ bulk material has yet been synthesized.

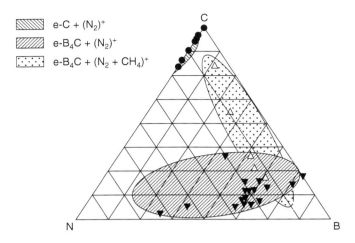

Figure 4.17 Ternary diagram showing relative compositions of B-C-N films for different IBAD conditions [89].

B-C-N films are deposited by PACVD in a horizontal furnace using mixtures of $BCl_3 + CH_4 + N_2$ or tris(dimethylamino) boron (TMAB) and N_2 source gases [87]. Chamber pressure is 130 Pa with RF powers up to 90 W and substrate temperature up 730 °C. High temperature was not sufficient to dissociate TMAB, but this gas was readily dissociated when RF power was introduced. Apparent from the source gas, films contained some H (not quantified). B/N ratio increased with increased power while C decreased at powers up to 50 W and leveled off at higher powers.

An interesting example of obtaining superhard films combining B-C-N is multilayer BC/BN films deposited by [90]. Nano-period $(BN/CN)_n$ and $(CN/BN)_n$ films are deposited by RF magnetron sputtering using graphite and h-BN targets in 5:1 mixtures of $Ar:N_2$. Deposition was typical of multilayer processes: 2-, 4-, 8- and 10-nm-period $(CN/BN)_n$ multilayer films, in which the top layer was CN, were initiated by placing the substrate first over the h-BN target, and then rotated over the opposite the graphite (C) target. The total numbers of layers in the 2-, 4-, 8-, and 10-nm-period multilayer films were 200, 100, 50, and 40, respectively, assuring that the top layer was CN. Nanometer-period $(BN/CN)_n$ multilayer films in which the top layer is BN were deposited in a similar fashion. Thickness of the layers was controlled by calibration of deposition rate. Figure 4.18 shows the layout of the deposition chamber. The objective for using this multilayer, as with other nanolaminates, was to dissipate the energy of deformation layer-by-layer [91, 92, 93]. Figure 4.19 compares hardness of CN, BN, and nano-periodic films. $(CN/BN)_n$ films with a 4 nm period displayed impressive hardness values ~ 100 GPa while $(BN/CN)_n$ films with 4 nm periods displayed equally impressive hardness ~ 65 GPa. It was determined that virtually all energy was dissipated in nano-periodic films.

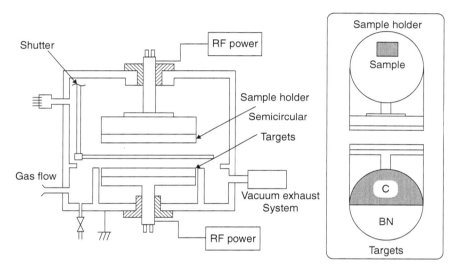

Figure 4.18 Layer out of deposition chamber for nanometer-period $(BN/CN)_n$ and $(CN/BN)_n$ films [90].

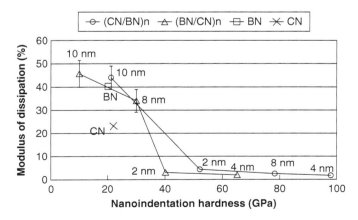

Figure 4.19 Hardness and dissipation modulus of $(BN/CN)_n$ and $(CN/BN)_n$ films [90].

4.1.3 Chromium Based Thin Film Materials: Chromium, Chromium Nitride, Titanium Nitride, and Titanium Carbide

Although Cr and Ni metal electrodoposited coatings were (and still are) used extensively to protect softer and more reactive metals and as decorative coatings, transition metal-based thin films, and diamond-like carbon (DLC) were the first significant breakthroughs in hard tribological coatings [94]. As we shall see, chromium has environmental problems. Table 4.6 presents the wide range of tribological thin film materials available. Nanocomposites and low dimensional materials are now at the forefront of tribological materials development. We begin with simple two-component Ti and Cr based thin films and move to more complex compositions.

It would not be possible to discuss hard coatings without first addressing chromium (Cr). Chromium plating has been used extensively to protect and beautify automobile parts, aerospace components, plumbing fixtures, lighting fixtures, and hardware. Depending on the application, it is deposited in thickness ranging from 0.2–300 µm, predominantly by electroplating processes. Trivalent and hexavalent are the most common forms of plated Cr. Until about a decade ago, hexavalent hard Cr (HCr) was used extensively to provide lubricity, high hardness (900–1100 VH), good corrosion and wear resistance, high heat resistance (stable up to ~ 540 °C), and anti-galling properties. It is usually deposited on various grades of steel, stainless steel, beryllium copper, brass-lead alloys, and copper bronze, and is deposited over a Ni under layer. Decorative chromium is deposited on steel, zinc-based die castings, plastics, stainless steel, aluminum, brass, and bronze for its high reflectivity, excellent tarnish resistance, and good scratch, wear, and corrosion resistance [95]. Corrosion resistance results primarily from the barrier effect of the thick nickel plate under the Cr. It is also used to build up worn or undersized parts. In the thicker applications it may be impervious but is subject to microcracks.

There are, however, a number of serious health concerns associated with hexavalent Cr. Cr is required in small quantities for normal human metabolism of fats, carbohydrates, and protein. High levels of specific forms of Cr can be deadly. Hexavalent Cr is readily absorbed into the lungs, through broken skin, and in the gastrointestinal tract through accidental poisoning. It is rapidly distributed through the body and stored in red blood cells, particularly through industrial exposure. Exposure to hexavalent Cr, particularly in Cr production and pigment industries, is associated with lung cancer and has been designated a confirmed human carcinogen. Hexavalent Cr is also corrosive and causes chronic ulceration and perforation of the nasal membrane and skin. Trivalent and divalent forms of chromium have a fairly low toxicity level. These materials can cause skin problems (such as dermititis), but not fibrosis and scarring.

As a result, a concerted effort has been made to replace HCr with less toxic alternatives. Trivalent Cr (TCr) is one alternative. Plating processes are much more environmentally friendly and physical properties are comparable. In addition to being nontoxic and of comparable cost, TCr offers a number of advantages over HCr [95]:

- Increased plating rates
- Better coverage
- Reduced tendency for burning

TCr also meets requirements for corrosion, adhesion, hardness, wear and appearance. Disadvantages are:

- Slightly different color
- Plating process is difficult to control
- Thick, hard coatings are not possible

Co-based alloys are a viable alternative to HCr, and are found to provide superior corrosion resistance. Ni-W-B is also used to replace HCr, and provides barrier-type protection and excellent corrosion resistance to acids. Untreated hardness is 600 VH while treated alloys have V_H between 950–1050, and 950–1050V_H after heat treatment.

Titanium nitride (TiN) and titanium carbide (TiC) are two of the first PVD-deposited tribological thin film materials developed to replace HCr [95, 96, 97, 98, 99, 100, 101, 102, 103]. TiN is also widely used as a decorative coating, low friction coating, for medical implants, and optical heat mirror applications [104, 105, 106]. These materials were first deposited by ion plating, followed closely by planar magnetron, unbalanced magnetron dual magnetron sputtering. Other deposition techniques, such as cylindrical magnetron sputtering, pulsed magnetron sputtering, ion beam sputtering, hollow cathode, metal-organic plasma spray, filtered cathodic arc, atomic layer deposition (ALD), MOCVD, and PECVD are now used [99, 107–114]. PVD is preferred for steel parts because deposition temperatures do not affect phase composition steel. It should also be noted that TiN can be formed on steel surfaces by the nitriding process.

TiN thin films are applied to cutting tools, milling cutters and other machine tools to significantly improve edge retention and corrosion resistance. Figure 4.20 shows a picture of a TiN coated die set (TiN coated die sets and drill bits are readily found at any industrial or commercial tool supplier). In addition to steel parts, TiN is also deposited onto a variety of higher melting point materials such as stainless steels, titanium, and titanium alloys. The hardness of TiN coatings, as well as electrical properties, optical properties, and corrosion resistance depend on deposition process and resulting microstructure.

As stated above, TiN is also replacing HCr and TCr as a top-layer coating, decorative and protective, usually with Ni or Cr plated substrates, on consumer plumbing fixtures, and door hardware. TiN is also used in medical devices such as scalpel blades and orthopedic bone saw blades where sharpness and edge retention are important and medical_implants, as well as aerospace and military applications [115]. Coatings are used in implanted prostheses, such as hip replacement implants. Useful layer thicknesses range up to ~ 3 μm.

Stoichiometric and highly conductive TiN films have a metallic Au color and are used to coat costume jewelry and automotive trim for decorative purposes. Substoichiometric films have a grey–black appearance. Because of its relatively high conductivity,

Figure 4.20 TiN coated die set.

~ 30–70 µΩ·cm, TiN is used in semiconductor production as a conductive barrier between a silicon device and the metal contacts and barrier coating. Table 4.6 displays highest hardness values, lowest values for the coefficient of friction (when available), and elastic constant (when available) achieved for various deposition techniques. Single layer films typically have a columnar structure and display strong (111) preferred orientation [116, 117]. Hardness values as high as ~ 39 GPa have been reported for single layer films deposited by pulsed magnetron sputtering [116].

TiN is often deposited over Cr and Ni adhesion layers. Figure 4.21 shows a typical layer structure for TiN on an Al part [121]. TiN thin films are deposited by magnetron sputtering in N_2/Ar

Table 4.6 TiN, CrN, TiC hardness and resistivity values achieved for various deposition processes.

Material	Deposition Process	Hardness	Coefficient of Friction	Young's Modulus (GPa)
TiN	Pulsed magnetron [116]	39 GPa	0.03–0.25	
TiN	Dual unbalanced magnetron [117]	14.2–30.3GPa		
TiN	Multiarc ion plating [120]	35–45 GPa		
TiN	Inductively coupled plasma assisted DC magnetron [118]	7000 KH		250–390
TiN	Cylindrical magnetron [107]	2780 VH		
TiN	Filtered cathodic arc [119]	3500 VH		
CrN	Reactive magnetron [122]	24–31 GPa		
CrN	Reactive magnetron [123]	3.4–31.6		94–299

Table 4.6 (cont.) TiN, CrN, TiC hardness and resistivity values achieved for various deposition processes.

Material	Deposition Process	Hardness	Coefficient of Friction	Young's Modulus (GPa)
CrN	Reactive magnetron [126]	18 GPa		
CrN	Reactive ion plating [124]	1300–2200 V_H		
TiC	Magnetron sputtering [103]	2900 VH		
TiC	Cylindrical closed field magnetron [127]	3–42 GPa	0.15–0.53	10–200 GPa
TiC	Reactive magnetron [129]	34.4 GPa	0.15–0.50	
TiC	Reactive magnetron [130]	15–28 GPa		210–300 GPa
TiC	Plasma immersion ion implantation [128]		0.10–0.40	

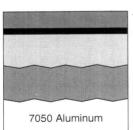

TiN (1.5 μm)
Cr plating (0.25 μm)
Bright Ni (~5 μm)

Electroless Ni
(~12 μm)

7050 Aluminum

Figure 4.21 Layer structure for TiN protective coating [121].

mixtures and a variety of deposition rates, power formats (DC, RF, pulsed, HIPIMS), chamber pressures, N_2/Ar ratios, substrate geometries, and substrate bias.

TiC and CrN thin films have not been explored quite as extensively as TiN films. CrN is most often deposited by reactive magnetron sputtering and ion plating [122–125]. As with TiN and all other hard thin films, hardness and other mechanical properties depend on deposition process and deposition conditions. Cr-N has CrN and β-Cr_2N chemical structures [124]. Deposition of Cr-N films is more difficult than TiN films due to the lower reactivity of Cr with N_2 compared to Ti. As with TiN, Cr-N films are deposited primarily by reactive magnetron sputtering techniques in N_2/Ar mixtures [29, 30, 32, 33, 125]. Figure 4.22 shows the dependence of hardness on N_2 partial pressure of magnetron sputtering CrN films [126]. Preferred crystal orientation can range from (111), similar to TiN, to (200) depending on the energy of ion bombardment [123, 126]. Hardness was also found to depend on substrate bias, Cr/N ratio, and average crystalline size, as shown in Figures 4.23, 4.24, and 4.25 [122, 123]. Maximum hardness = 31 GPa and lowest residual stress = –4.0 GPa (negative = compressive) was found for Cr/N = 1.

TiC is one of the hardest substances known to man, and is even harder than diamond. Similar to the other materials described here, TiC is primarily deposited using magnetron sputtering and CVD techniques [103, 127, 128, 129]. Hardness values and other tribological properties depend on deposition process and process parameters. Magnetron sputtered films are deposited by cosputtering Ti and graphite (C) targets and reactively using a Ti target + Ar/CH_4 mixtures [127, 129]. Hardness of cylindrical magnetron sputtered TiC increases significantly from 21–42 GPa with increased power to the Ti target. Reactively sputtered films have hardness ~ 34.4 GPa.

Sample	N_2 parial pressure	Average hardness
1	0.7 mTorr	18 ± 2 Gpa
2	1.0 mTorr	17 ± 2 Gpa
3	1.5 mTorr	16 ± 2 Gpa
4	2.0 mTorr	16 ± 1 Gpa
5	2.5 mTorr	19 ± 1 Gpa

Figure 4.22 Dependence of hardness of magnetron sputtered CrN films on N_2 partial pressure [126].

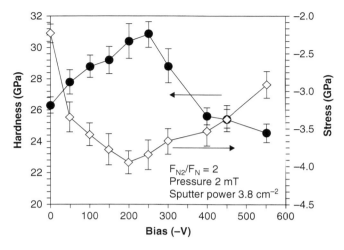

Figure 4.23 Hardness and residual stress on magnetron sputtered CrN films for various substrate bias voltages [122].

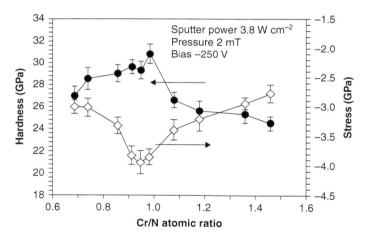

Figure 4.24 Dependence of hardness and residual stress on magnetron sputtered CrN films on Cr/N ratio [122].

Friction coefficients range from 0.0.15–0.50, again depending on deposition conditions.

TiC typically has a NaCl fcc crystal structure, shown in Figure 4.26. Preferred orientation of sputtered films can range from (111) to (200) with columnar microstructure. Microstructure, shown in Figure 4.27, can be columnar porous to highly dense, depending on

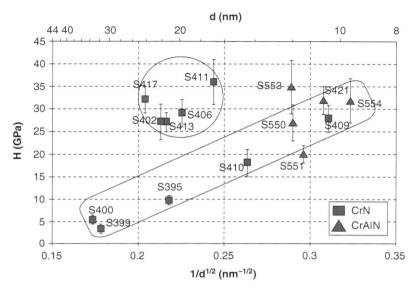

Figure 4.25 Dependence of CrN and CrAlN hardness on average crystallite size [124].

Figure 4.26 FCC crystal structure of TiC.

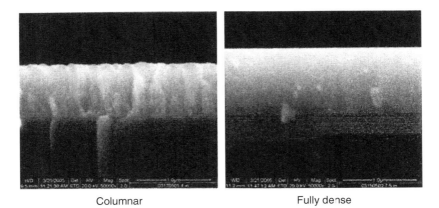

Columnar   Fully dense

Figure 4.27 Comparison of microstructure of columnar and fully dense TiC films [128].

substrate bias and deposition geometry [127]. Films deposited by dual source metal plasma immersion generally have strong (111) and weaker (200) texture [128].

4.1.4 Binary Carbon Based Thin Film Materials: Diamond, Hard Carbon and Amorphous Carbon

The carbon-based family of thin film materials includes diamond-like carbon (DLC), tetrahedral amorphous carbon (ta-C), amorphous carbon (a-C), and hydrogenated amorphous carbon (a-C:H). DLC and ta-C are arguably two of the most widely developed and applied thin film materials used to increase wear resistance and lubricity over the past two decades, and volumes have been published on their properties. DLC is also deposited in microcrystalline, nanocrystalline, and multilayer microstructures. Nanocomposite DLC films are now being developed that have lower wear and friction and significantly higher load carrying capability [36, 131]. High quality DLC films have been demonstrated for a number of applications, including low friction and wear, optical and antireflection coatings, and barrier coatings. DLC is also an excellent abrasion-resistant optical coating, particularly for infrared windows. DLC and ta-C have other extensive applications, such as protecting mechanical components, recording heads, medical implants, race-car engines, optical components, cutting tools, electron, emitters and heat sinks for microelectronics and even razor blades. A number of books have extensively addressed this family [11, 36, 131–135].

Bonding configurations are at the heart of DLC's and ta-C's tribological and electrical properties. Pure diamond has covalent bonding with a cubic crystal structure. Each C atom is covalently attached to four other C atoms in tetrahedral bonds 1.54 Å long [59]. Diamond has the highest atomic density of all semiconductors [133]. DLC, however, exists in seven different forms of amorphous carbon (a-C) materials that display some of the unique properties of diamond. All seven contain combinations of sp^3 and sp^2 bonding. sp^2 bonding is purely graphitic with little wear resistance. Figure 4.28a shows sp^2 bonding and Figure 4.28b shows sp^3 bonding. Diamond and graphite are the major allotropes of carbon; sp^3 bonding is pure diamond while sp^2 bonding is purely graphitic. The reason that there are different types is that even diamond can be found in two crystalline polytypes. The usual one has its carbon atoms arranged in a cubic lattice, while a very rare form (lonsdaleite) has a hexagonal lattice (with c-axis). The various forms of DCL can be combined to form a variety of nanoscale structures that are amorphous, flexible, and yet purely sp^3 bonded "diamond". The hardest, strongest, and most lubricious is such a mixture, known as tetrahedral amorphous carbon, or ta-C. For example, a coating of only 2 μm thickness of ta-C increases the resistance of common (i.e., type 304) stainless steel against abrasive wear, changing its lifetime in such service from one week to 85 years [134]. Such ta-C can be considered to be the "pure" form of DLC, since it consists only of sp^3 bonded carbon atoms.

Figure 4.29 shows a DLC coated dome. Table 4.7 summarizes salient properties of DLC, ta-C, and hydrogenated a-C (a-C:H).

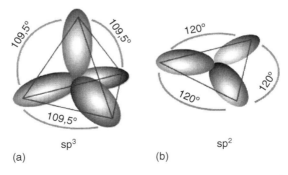

(a)   (b)

Figure 4.28 (a) sp^3 bonding and bond angles. (b) sp^2 bonding and bond angles.

Figure 4.29 Picture of DLC coated dome.

Fluorine has also been added to amorphous carbon films to improve stability, increase dielectric constant, and reduce the COF.

DLC is generally deposited as nanocrystalline diamond. Deposition processes for DLC, ta-C, and a-C:H, described in Chapter 2, are shown in Table 4.8. Formation of high quality DLC requires some bombardment by energetic species, usually ions. As a result, deposition processes such as PECVD, PACVD, MPCVD, ion beam sputtering and filtered cathodic arc deposition have the capability to produce DLC and ta-C. Sputtering, while providing some energetic particle bombardment, typically produces lower quality films. As a result, some form of ion assist is generally required for PVD processes. The Table 4.8, while not entirely complete, gives general ranges for deposition parameters and bonding. The table reflects the fact that DLC, ta-C, and a-C:H have different applications; hard films may not have lowest COF and alternately, films with low COF may not have highest hardness and wear resistance.

Again, the properties of DLC and ta-C films depend on the amount of sp^3 and sp^2 bonding present, which can ultimately be related to the deposition process [59, 135]. Figure 4.30 plots the percentage of sp^3 bonding for the various forms of carbon [134]. The figure shows the two extremes, diamond and graphite, and every combination of bonding in between. Note also that bonding configuration is related to hydrogen content, and eventually back to the deposition process. Microstructures range from purely amorphous to nanocrystalline . Hardness definitely increases with increased

Table 4.7 Properties of diamond and DLC materials.

Property	Thin Film: ta-C	Thin Film: a-C	Thin Film: a-C:H	Bulk Diamond	Graphite
Crystal structure	Cubic (a_o = 3.561 Å)	Amorphous ($sp^3 + sp^2$)	Amorphous ($sp^3 + sp^2$)	Cubic (a_o = 3.567 Å)	Hexagonal (a = 2.47 Å)
Morphology	Faceted crystals	Smooth–rough	Smooth	Faceted crystals	-
Hardness (GPa)	80–88	~3	10–20	90–110	-
Refractive index (@ 550 nm)	2.8–3.5	1.6–2.2	1.2–2.6	2.42	2.26
Electrical resistivity(Ωcm)	$> 10^{12}$	$> 10^{10}$	10^6–10^{14}	$> 10^{16}$	0.4
Thermal conductivity (W/mK)	1100	-	-	2000	3500
Chemical stability	Inert	Inert	Inert	Inert	Inert
H content (H/C)	-	-	0.25–1	-	-

Table 4.8 Summary of DLC, ta-C, and a-C:H deposition processes.

Process	Type(s)	Gases/Precursors	Substrate Bias/Ion Assist	Bond Type	%H	Max Reported Hardness (GPa)	COF
DC magnetron sputtering	a-C, a-C:H	Ar, $Ar + CH_4$		sp^2	0–10	28	
Closed field unbalanced magnetron sputtering	a-C:H	$Ar + C_2H_2$		$sp^3 + sp^2$		14	
Ion plating						30	
Evaporation	a-C			sp^2	0	3	
Atomic layer deposition	ta-C, a-C:H			$sp^3 + sp^2$			
PECVD	ta-C, a-C:H	$Ar + CH_4$, C_2H_4		$sp^3 + sp^2$		40	0.003–0.04
CVD	a-C:H	CH_4, C_2H_4		$sp^3 + sp^2$		19	
PACVD	ta-C, a-C:H	CH_4, C_2H_4	0–-400V	$sp^3 + sp^2$		30	0.10
ECR-CVD	ta-C, a-C:H	CH_4, C_2H_4		$sp^3 + sp^2$			

Method	Material	Precursor	Bias	Bonding			
MPCVD	ta-C, a-C:H	CH_4, $CH_4 + B_2H_6$		$sp^3 + sp^2$			
Plasma decomposition	ta-C, a-C:H	CH_4		$sp^3 + sp^2$			
Ion beam sputtering	ta-C						
Pulsed laser deposition	ta-C, a-C:H			$sp^3 + sp^2$			49
Plasma source ion implantation	a-C:H	C_2H_2, Ar + C_2H_2		$sp^3 + sp^2$			
Filtered cathodic arc deposition	a-C, ta-C		0—2500V	$sp^3 + sp^2$	0	45	
Hollow cathode	a-C:H		-50V	$sp^3 + sp^2$		14	0.08

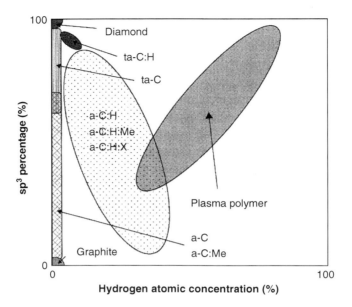

Figure 4.30 Percentage of sp³ bonding vs H content for the various forms of C [134].

sp³/sp² ratio [136]. Table 4.9 shows properties for the various forms of carbon [137].

We will address the following properties:

- Hardness
- Lubricity
- Wear resistance
- Thermal conductivity

Table 4.8 shows that CVD and related processes incorporate some amount of hydrogen in the films. Hydrogen can also be bonded into sputtered films if a hydrocarbon gas is used along with the sputtering gas (usually Ar). Hydrogen replaces some of the carbon atoms in the film and also passivates dangling bonds. Referring to Figure 4.31, C-H σ bonds are thus formed at the expense of π bonds (sp² hybridization) [131]. Hydrogen can also be unbounded and trapped inside the film. This unbounded H can affect the density of states and degrade virtually every property.

As mentioned above, hardness depends primarily on amount of sp³ bonding. In Figure 4.32 we see that hardness increases with

Table 4.9 Mechanical properties of the various forms of C [134, 137].

Type	Hardness (GPa)	Young's Modulus (GPa)
a-C	12–24	160–190
a-C:H	7–30	60–210
ta-C	28–80	210–757
ta-C:H	28–60	175–300
Diamond	90–108	1000
C60	−0.2	
C fibers		200–600
Graphite	0.2–2	10
Carbon nanotube		1000

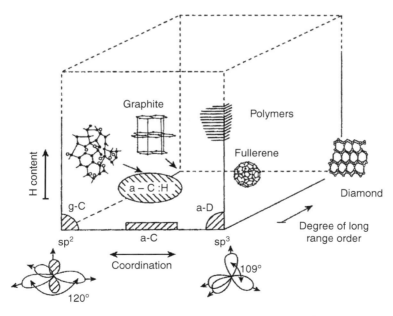

Figure 4.31 Schematic representation of different bonding configurations of C resulting from C coordination, degree of long range order and H content [131].

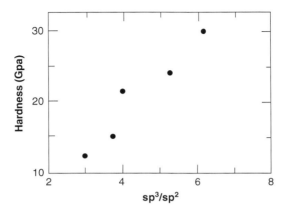

Figure 4.32 Dependence on hardness of a-C films with sp^3/sp^2 ratio [136].

increased sp^3/sp^2 ratio for ion beam assisted magnetron sputtered films [136]. Referring to Table 4.8, we see that sp^3 bonding can occur when H is bonded into the film, and, thus, hydrogen content cannot be ignored when discussing hardness values. As a result of precursors used, CVD and related processes incorporate some amount of H in to the films. Diamond films contain no H, posses 100% sp^3 bonding and have hardness ~ 100 GPa. ta-C films have next highest hardness values ~ 80 GPa with 80–88% sp^3 bonding. PVD processes such as magnetron sputtering, evaporation, and ion beam sputtering do not incorporate H into the films but deposit films with low hardness up to 15 GPa [138]. Amorphous carbon films with bonded H generally have intermediate hardness values with sp^3 bonding ranging from 40–70%. Figure 4.33 shows that hardness of magnetron sputtered a-C:H films decreases with increased H content [138].An extreme example is polyethylene with 100% sp^3 bonding, but also with 67% H and a hardness of only 0.01 GPa.

According to Buhshah [11], diamond films deposited by PECVD exhibit hardness, elastic modulus COF, and wear properties similar to those of natural diamond [135]. Figure 4.34 shows a typical PECVD deposition system used for DLC optical coatings [131]. Thin film coatings were deposited at ambient temperature using mixtures of Ar + CH_4 (methane) and a RF frequency of 13.56 MHz. For comparison, DLC films were also deposited by ion beam CVD. Mixtures of Ar + C_2H_4 were used for IBCVD, and the ion source was located ~ 18 cm below the substrate. In this case, ion bombardment was supplied by the ion source instead of RF excitation of the

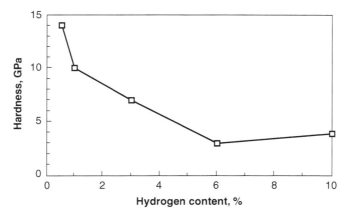

Figure 4.33 Effect of H content on hardness of magnetron sputtered a-C:H [138].

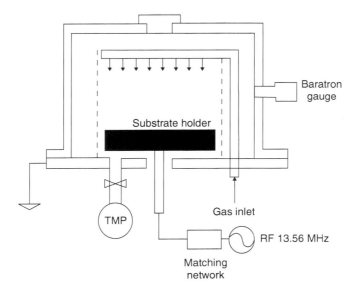

Figure 4.34 PECVD reactor used for DLC films [131].

plasma. Films were characterized for mechanical and optical properties. PECVD films displayed slightly higher hardness values than IBCVD films: ~ 18 GPa average compared to 16.4 GPa and slightly higher elastic modulus; ~ 150 GPa compared to 130 GPa. Typical load-displacement curves are shown in Figure 4.35. The somewhat lower hardness for IBCVD films was attributed to higher hydrogen content.

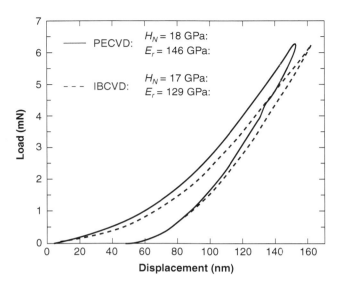

Figure 4.35 Load-displacement curves for PECVD and IBCVD films [134].

Another significant advantage of DLC is its very low coefficient of friction (COF) [139, 140, 141]. Hydrogen bonding (sp^2–sp^3), hydrogen content, and microstructure required for ultra-low COF films can be very different from that found in optical or ultra-hard films. Hydrogen-free films function better in humid environments while hydrogenated films function better in dry and inert gas environments [140]. DLC films with ultra-low COFs between 0.001 and 0.003 and wear rates between 10^{-9}–10^{-10} mm^3/Nm (inert gas) were deposited in pure CH_4 or mixtures of $CH_4 + H_2$ at room temperature [140, 141]. COF depended strongly on H content and H/C ratio of the source gas, as shown in Figure 4.36 [10]. We see that COF is lowest for highest H/C = 10 (25% CH_4 +75% H). COF ranges from 0.7 to 0.003 with increase in C/H ratio from 0 to 10. Note that the lowest values were achieved only in an inert gas or vacuum environments. COF increased to ~ 0.028 in at 50% RH. Wear rates for films with no H were very high (consequence of high COF) at 9×10^{-9} mm^3/Nm while wear rates for C/H = 10 were the lowest at 4.6×10^{-9}.

COF and wear properties are also influenced by surface morphology, grain size, and amount of non-diamond phases present in the film [141–152].

DLC nanocomposite is a new engineered form of this material [132]. The nanocomposite consists of glass-like (a-Si-O)

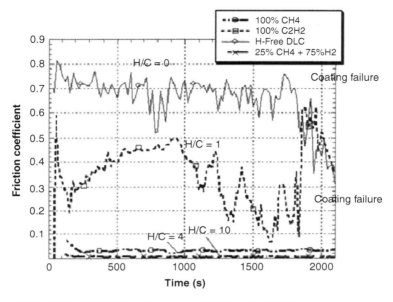

Figure 4.36 COF of DLC films for various H/C ratios [140].

nanoparticles combined with a DLC matrix. There are several Si oxides present in the matrix. One of the advantages of this structure is very low stress, ~ 1 GPa, and hardness in the range 12–17 GPa. DLC films typically have much higher stress. The source gas is polyphenylmethylsiloxane with a combined DC/RF excitation. The COF of the nanocomposite is ~ 0.04 in 50% RH (compare to the above 0.025 for DLC). The wear rate, however, is considerably higher than DLC alone, about 2×10^{-6} mm^3/Nm. Again we see that some properties have to be compromised for the benefit of others.

A wide range of microstructures for DLC, ta-C, a-C, and a-C:H is possible. Microstructure depends primarily on deposition process. Morphology and microstructure depend to a large degree on ion bombardment during deposition [19]. As deposited ta-C and DLC typically have coarse-grained microstructures, but fine grained as shown in Figure 4.37 [135]. Figure 4.38 shows an AFM micrograph of the surface of a hard disc with and without a ta-C film [153]. Figure 4.39 shows the surface morphology of sp^3 bonded ta-C. Films must be polished for several applications, especially optical and biomedical implants [154, 155]. In addition to hydrogen bonding, friction and wear characteristics are directly related to grain size and surface roughness [157]. Friction

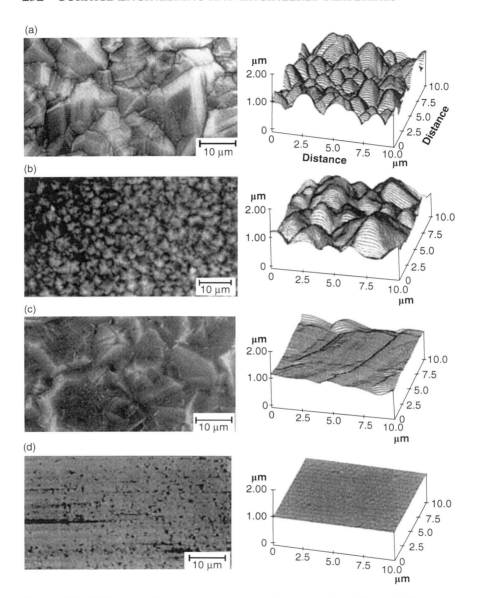

Figure 4.37 CVD-deposited coarse grained to fine grained ta-C films [135].

and wear of polished DLC and ta-C films are comparable to those of natural diamond [147].

Diamond has thermal conductivity greater than that of the most conductive metals, ranging from 900–3500 W/mK [157]. Only

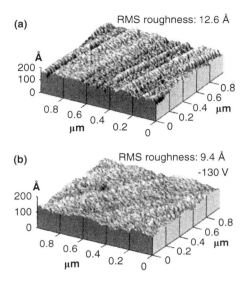

Figure 4.38 AFM micrograph of the surface topography of a hard disk with and without a ta-C film [23].

graphite has higher thermal conductivity [156]. The thermal conductivity of metals is in the range 300–430 W/mK. This high value makes DLC and ta-C excellent heat sinks for microelectronics, while still being good electrical insulators.

4.1.5 Binary Carbon and Silicon Carbide Materials and Multilayers

In this section, we complete our review of carbon and carbon related hard, ultrahard, and wear resistant thin films. Materials addressed are carbon nitride (CN_x), fluorine doped amorphous carbon (a-C:F), silicon carbide (SiC), silicon carbon nitride (SiCN), and titanium carbon nitride (TiCN).

CN_x films are very damage tolerant, and elastic, while retaining the wear resistance and beneficial tribological properties of pure carbon films and have replaced them as overcoats on read and write as well as hard drives in magnetic recording. CN_x films have a history going back to early 1979 [158] and can be synthesized by a variety of methods, including magnetron sputtering, ion beam assisted deposition, PECVD, filtered cathodic arc, and ion plating [159, 160, 161]. Films have high hardness, low coefficient of friction, excellent

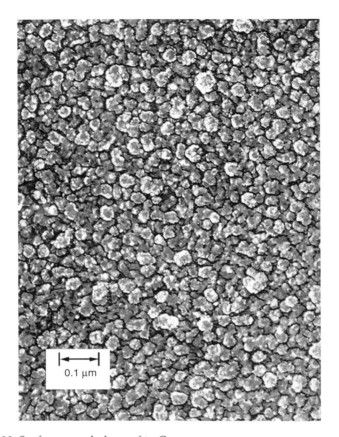

Figure 4.39 Surface morphology of ta-C.

wear resistance, good resistance to chemical attack and are biocompatible. CN_x films are typically amorphous. While it has not yet been successfully synthesized, β-C_3N_4, which has the same structure as β-Si_3N_4, has a theoretical hardness of 427 GPa (diamond = 440 GPa), low intrinsic stress, and an elastic modulus greater than diamond [162]. Nonetheless, CN films display impressive hardness values which depend on deposition process. Hardness and COF results are summarized in Table 4.10.

Hardness, elastic modulus, and COF all depend on nitrogen content and how nitrogen is bonded in the film, all of which depend on deposition process and deposition conditions [161]. However, DLC and ta-C films still have higher hardness and lower COF than any values reported for CN_x. Similar to DLC films, CN_x films contain a mixture of sp^2 and sp^3 bonding, but in this case the bond type is

Table 4.10 Hardness and COF values for CN_x, a-C:F, SiCN and SiC films.

Material	Deposition Process	Hardness (GPa)	COF (typical)	Reference
CN_x	IAD	64		163
	Pulsed bias arc ion plating	32		161
a-CNx	Reactive RF magnetron sputtering	12	0.4	165
a-CN:H	Unbalanced magnetron sputtering		0.03	166
$FLCN_x$		7–43	0.2	164
CN_x	IAD	15–20	0.2	167
CN_x	Ion beam assisted pulsed laser ablation		0.2–0.3	170
a-C:N	Pulsed cathodic arc discharge	75		171
Si-C-N	Magnetron sputtering	4434 -H_V		187
Si-C-N	PECVD	25		186
SiC	RF magnetron sputtering	40		206
SiC	Active Reactive evaporation	3800 H_V		205

C-C(N). For films deposited by the pulsed bias ion plating process, hardness peaks at 32 GPa for nitrogen content ~ 0.081 (Figure 4.40), which also corresponded to maximum C-C(N) sp^3 bonding. Hardness values also peaked for substrate bias voltage near –300 V. Note that these films contained no hydrogen. The elastic modulus showed a similar dependence on composition and deposition conditions, peaking at 450 GPa.

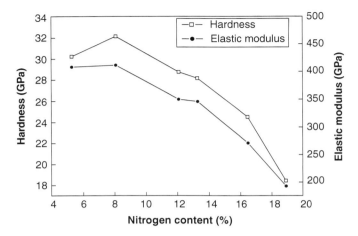

Figure 4.40 Dependence of hardness and elastic constant of CN_x films with N content [161].

CN$_x$ films have also been deposited by electron cyclotron resonance PECVD [168], RF-PECVD [169], N$_2$ beam ion assisted laser ablation [170], and pulsed cathodic arc discharge [171]. Amorphous carbon nitride (a-CN$_x$) films deposited by reactive RF magnetron sputtering show less impressive hardness values [165]. Hardness values for films deposited by pulsed cathodic arc discharge, however, show an impressive 75 GPa hardness [171].

An interesting allotrope of CN$_x$ is fullerene-like carbon nitride [164]. Nitrogen is substituted into predominantly sp^2 hybridized graphite sites, which deforms bond angles in the fullerene and inhibits plastic deformation. Fullerene-like materials are compound materials which have the structural incorporation of odd-membered rings, and are characterized by heavily bent and intersecting hexagonal basal planes. Other fullerene-like materials include C [172], BN:C [173], BN [173], MoS$_2$ [174], TiO [175], NbS$_2$ [176], and WS$_2$ [177]. However, out of these material systems only C, CN$_x$, and BC:N exhibit cross-linking, which enables the material to extend the strength of the covalently bonded two-dimensional hexagonal network into three dimensions.

Figure 4.41 shows that FL-CN$_x$ consists of substitutionally bonded N in bent, intersecting, and cross-linked graphite sheets. Thus the majority of carbon is incorporated in a three-coordinated planar sp^2-hybridized state, which is actually stronger than the hybridized sp^3- bond of diamond [178, 179]. The inset in Figure 4.41 shows that

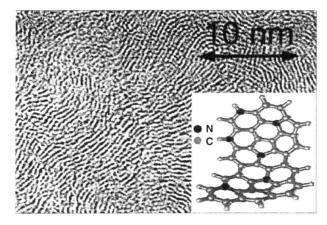

Figure 4.41 TEM picture of curved and intersecting fullerene like sheets in a FL-CNx film. Inset shows distortion and buckling of two-dimensional graphite lattice into three dimensions [164].

the two-dimensional structure of graphite, which deforms easily, is distorted by the bonding of nitrogen [180, 181, 182]. This lattice buckling enables the material to extend the extraordinary strength of a planar sp^2-coordinated carbon network into three dimensions.

Fluorinated amorphous carbon (a-C:F:H) films have been synthesized using plasma decomposition of CF_4-CH_4 mixtures [183]. While these films have lower hardness than other carbon films, they have applications in wear resistant overcoats for magnetic recording media [184], and corrosion and abrasion resistant coatings for infrared optics [185]. In order to incorporate F into the films, they are deposited by CVD processes. No nanoindenter hardness numbers are available, but Vickers hardness ranges from 14–6 with an increased F/C ratio from 0–0.26 [183]. Some H must obviously be incorporated into the films and is generally in the neighborhood of 15 at. %.

While silicon carbide is known as a hard and wear resistant material, Si-C-N films have combined properties of SiC and Si_3N_4 and show promise for wear and corrosion resistance, and excellent chemical stability, wide band gap, and promising mechanical and thermal properties [186, 187]. The Si–C–N system is expected to have different superhard phases namely SiC, β-Si_3N_4 and β-C_3N_4 (see above regarding CN_x) phases, which is a superhard material with theoretically predicted hardness near to diamond (but not yet synthesized) [5]. β-C_3N_4 and β-Si_3N_4 have similar crystal structures and are expected to be miscible in each other and form a Si–C–N phase with excellent hardness. [188–194].

Si-C-N films are deposited by essentially the same methods as CN_x films, which include microwave plasma CVD [31, 32], PECVD [190, 191], ion beam implantation [192], and magnetron sputtering [193, 194]. PECVD films were deposited in mixtures of SiH_4, C_2H_2, and NH_3 [186]. Magnetron sputtered films were deposited using a sintered SiC target in Ar + N_2 mixtures [187]. Targets could also (preferably) consist of pyrolytic SiC. For all deposition processes, hardness generally decreased with increased N content but smoothness increased with increased N content. Not unexpectedly, hardness of PECVD films decreased with increased Si content (or SiH_4/C_2H_2 ratio) as shown in Figure 4.42 [186].

While hardness increased with increased substrate temperature for both magnetron sputtered and PECVD films, hardness peaked at a substrate temperature ~ 500 °C and decreased at 700 °C for magnetron sputtered films. As shown in Figure 4.43 , maximum hardness for PECVD films was obtained at ~ 650 °C.

While sputtered films were predominantly amorphous, some β-Si_3N_4 and β-C_3N_4 phase formation was detected by AFM in films

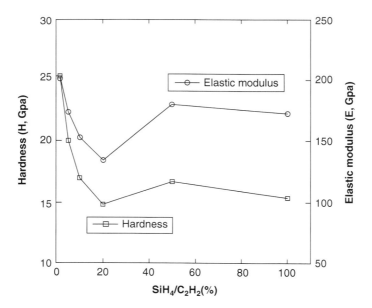

Figure 4.42 Dependence of hardness and elastic modulus with SiH_4/C_2H_2 ratio for PECVD Si-C-N films [186].

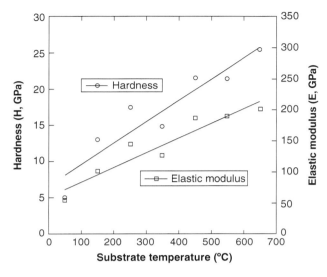

Figure 4.43 Dependence of hardness and elastic modulus with substrate temperature for PECVD Si-C-N films [186].

deposited at 1.0 Pa and 5.0 Pa. The β-Si_3N_4 phase was more prevalent in films deposited at lower pressures (0.4 Pa), most likely due to increased ion bombardment. Films deposited at high pressures ~ 10 Pa exhibited graphitic phases.

Si-C-N films have also been deposited by ion beam sputtering using a Si target doped with $C_5N_5H_5$ (adenine) and bombarding the target with 750V–1250V Ar ions [195]. Amorphous films were obtained using Si rich targets and low accelerating voltages ~ 750 V. Microcrystalline films were deposited when Si content in the sputtering target was decreased.

SiC thin films have a wide variety of useful properties, including good mechanical wear resistance [160], high hardness [196], high thermal conductivity, and very high thermal stability [197]. Applications range from protective coatings against the corrosion of steel [198] to microelectronic devices [199], and from X-ray mask materials [200] to the protection of thermonuclear reactor walls [201], infrared optical coatings, and photovoltaic cells. Films are deposited by a variety of techniques, such as laser-assisted deposition [202], dynamic ion mixing [203], plasma-enhanced chemical vapor deposition [204], magnetron sputtering [205], and reactive evaporation [206].

As with CN_x and Si-C-N films, magnetron sputtering and other PVD processes are most often used [204, 205, 206]. SiC films are deposited by RF magnetron and other sputtering techniques using either a SiC target (pyrolytic SiC preferable but sintered acceptable) or using a Si target in mixtures of Ar + hydrocarbon (CH_4, C_2H_6, etc). While substrate temperature was a factor with hardness of Si-C-N films, there appears to be little dependence for magnetron sputtered SiC films, as shown in Figure 4.44 [206]. There is a significant dependence on RF power (~ deposition rate) and Ar pressure, however [206]. Hardest films were obtained at relative low RF powers ~ 100 W– 150 W and hardness peaked at 40 GPa for an Ar pressure ~ 0.1 Pa. Here we see that this is another example in which adatom mobility is critical and is enhanced by low deposition rate and high ion bombardment of the growing film.

Hardness of SiC films deposited by active reactive evaporation (ARE) showed significant dependence on substrate temperature and bias [205]. Figures 4.45 and 4.46 show hardness dependence on these parameters. Note that films with highest hardness have the β-SiC phase, as discussed above.

While impressive hardness values have been achieved for diamond-related materials, DLC and ta-C thin films still are the

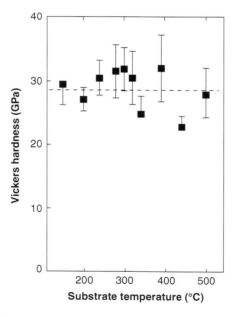

Figure 4.44 Dependence of hardness on substrate temperature for magnetron sputtered SiC films [205].

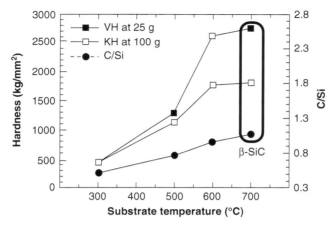

Figure 4.45 Dependence of hardness of ARE SiC films on substrate bias [205].

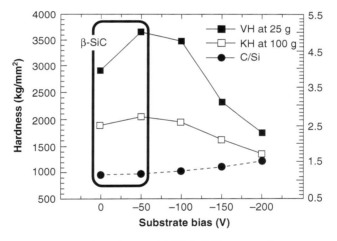

Figure 4.46 Dependence of hardness of ARE SiC films on substrate temperature [205].

"gold standard" for hardness and lubricity. CN_x, Si-C-N and SiC offer advantages of smoother morphologies, large area deposition, and lower synthesis costs which may offset some of the tribological advantages of DLC and ta-C. These diamond cousins also have additional optical, electronic, opto-electronic, and biomedical applications and their properties can be adjusted through modifications in composition.

4.1.6   Tungsten Carbide, Molybdenum Sulfide, Silicon Nitride, and Aluminum Oxide

Several tribological coatings do not fit into one family. In this section, we address the tribological properties of tungsten carbide (WC), molybdenum disulfide (MoS_2), silicon nitride (Si_3N_4 or Si_xN_y), and aluminum oxide (Al_2O_3 or AlO_x). Tungsten carbide melts at 2,870 °C, has a hardness in the range with low electrical resistivity ($\sim 2 \times 10^{-5}$ Ohm·cm), comparable to some metals. As with most of the materials addressed here, WC coatings are used to improve the wear resistance of cutting tools, but are also used in jewelry, particularly wedding rings, and are often used to improve the flow of ink on ball point pen tips. WC coated drill bits and a wedding ring are shown in Figure 4.47. Although not the hardest or most wear resistant of the thin film materials addressed here, it is possible to polish WC and improve the overall surface finish of the coated part. WC oxidizes at temperatures between 500 °C–600 °C.

WC thin films are mainly deposited by magnetron sputtering [207, 208, 209, 210], filtered cathodic arc deposition (FCAD) [211], CVD [212, 213], and plasma spray [214]. Magnetron sputtered coatings are deposited form a high purity W target in mixtures of Ar + CH_4 or Ar + C_2H_2 [207, 208]. Substrate temperatures range from 150–350 °C. Hardness of magnetron sputtered WC films peaks at 40 GPa for C contents ~ 34 at % [207]. Wear rates also appear to be

Figure 4.47 WC coated drill bits and wedding ring.

minimal at this composition ($W_{0.66}C_{0.34}$) [2]. COFs range from ~ 0.10 to 0.20 under dry conditions and are < 0.1 at C content ~ 10 at.% [1]. WC films have considerably lower wear rates than SiC and Si_3N_4 films, as shown in Figure 4.48 [207].

Films are deposited by FCAD using targets with composition $W_{0.30}C_{0.70}$ with substrate bias ranging from 0 to –600V [211]. Surface morphology, crystalline structure, internal stress, hardness, and elastic (Young's) modulus were found to be strongly dependent on substrate bias. Films were smoothest for bias near –300 V. At a substrate bias below –200 V, the increase of substrate bias was found to result in densification of the film with an accompanying change in film structure from single hexagonal a-W_2C phase to a mixture of hexagonal a-W_2C + cubic + WC_{1-x} + cubic W. High density and the strain in films deposited at this bias were attributed to a lattice mismatch between the different phases and resulted in the highest internal stress = 7.2 GPa, hardness = 26 GPa, and elastic modulus = 27 GPa. As shown in Figure 4.49, hardness and elastic modulus decreased with further increase in substrate bias, which was thought to cause relaxation of strain and return film structure to the single a-W_2C phase.

CVD methods that have been investigated include [212, 213]:

- $WCl_6 + H_2 + CH_4 \rightarrow WC + 6\ HCl$ @ 670 °C
- $WF_6 + H_2 + CH_3OH \rightarrow WC + 6\ HF + H_2O$ @ 350 °C

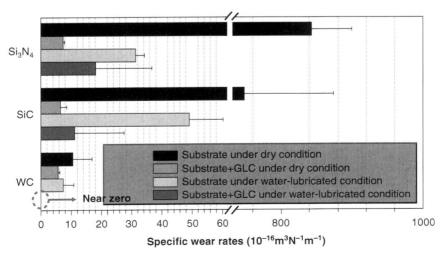

Figure 4.48 Comparison of wear rates for WC, SiC, and Si_3N_4 films [207].

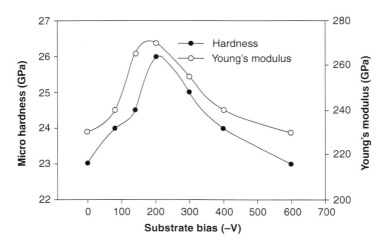

Figure 4.49 Hardness and elastic modulus for FCAD WC films [211].

Mixtures of $WF_6 + H_2 + C_3H_8$ @ 650 °C–1000 °C are also used. Films with a single crystalline phase were deposited at temperatures between 900 °C–1000 °C. A mixture of WC and W_2C was obtained at lower substrate temperatures.

Although not common among other hard wear-resistant coatings, WC and combinations of WC-Co are often plasma sprayed onto cutting tools to achieve high wear resistance and ductility [212, 215, 216, 217]. For films deposited by this process, the W_2C composition is found to be harder than WC films; 3000 H_V compared to 2400 H_V [216], although hardness as low as 1000 H_V have been reported for WC-Co films [215]. Nanostructured WC/Co powders with 50 μm–500 nm size are plasma sprayed. At high temperatures ~ 2200 °C, WC decomposes to tungsten and carbon and this can occur during high-temperature thermal spray, e.g., high velocity oxygen fuel (HVOF) and high energy plasma (HEP) methods [218]. Plasma chemistry and decomposition of the constituent powders is extremely complex and will not be addressed here. Results for plasma spayed nanostructured WC–Co coatings can be summarized as follows [215]:

1. It is possible to produce nanostructured WC–Co powders without non-WC_yCo phases although a high percentage of non-WC_yCo phases is frequently reported in conventional WC-Co powder.

2. Nanostructured WC–Co powder experiences a higher temperature during spraying than that of the corresponding conventional powder. Therefore, if the same spraying parameters are used, WC particles in nanostructured powder will suffer from severe decomposition, which would subsequently degrade the performance of nanostructured WC–Co coatings.

By controlling agglomerate size of feedstock powder, fuel chemistry and fuel–oxygen ratio, near nanostructured WC–Co coatings with a low amount of non WC_yCo phases can be successfully synthesized.

3. Increased hardness, toughness, and wear-resistance can be obtained in near nanostructured WC–Co coatings that are properly synthesized.

Molybdenum disulfide (MoS_2) is used primarily as a solid lubricant but is also combined with wear resistant layers in multilayer thin films to form wear resistant lubricious surfaces [131]. MoS_2 is the most widely used lamellar compound solid lubricant material for space applications, [219] and is used in release mechanisms, precision bearing applications, main weather sensor bearings, and gimbal bearings. MoS_2 coatings by themselves are not suitable for most terrestrial applications, because their tribological properties degrade in humid atmospheres. However, unlike fluid lubricants, MoS_2 is suitable for use in a vacuum environment because of its low outgassing pressure and lack of migration. Before the advent of thin films, MoS_2 powder was applied to the surface and burnished with a soft tissue onto the low friction surface. MoS_2 thin films are now mainly deposited using PVD processes; coatings with best performance (low COF, high adhesion, wear resistance) are deposited by magnetron sputtering [219–226]. Compared to previous methods for the deposition of MoS_2, RF sputtering produces coatings with low COF and enhanced adhesion. Films are deposited using a MoS_2 target or a Mo target in a reactive gas atmosphere [220, 221, 222, 226, 227, 228, 229, 230]. RF sputtered MoS_2 coatings show lower friction coefficients, better adherence, and lower wear rates than coatings produced by other techniques [221]. COF as low as 0.4 are routinely reported. MoS_2 films are also deposited by pulsed laser evaporation [231].

MoS_2 has been combined with Ti to further reduce friction and improve wear resistance [225]. Graphite and Cr were combined in the same study. Films are deposited by closed field magnetron sputtering using MoS_2 and Ti targets, or three graphite targets and a Cr target. MoS_2 in this configuration retains its low COF even in humid environments but wear resistance is significantly increased. Figure 4.50 shows that COFs for MoST (molybdenum-sulfur-titanium) films as low as 0.025 are achieved with wear rates as low as 0.63 mm/h [221].

Recently, nanolaminate and nanocomposite coatings based on MoS_2 have been developed [131, 220, 232, 233, 234, 235]. Nanocomposite, nanoparticle, and nanolaminate coatings involving MoS_2/Au, MoS_2/Ti, MoS_2/(Ni,Pb,Ag,C,TiN, Sb_2O_3,PbO) and MoS_2 fullerines and with very low COF and high hardness have been achieved. MoS_2/Au films are deposited by two different methods: MoS_2 films coated with a thin Au layer (~ 80 nm) are deposited by dc unbalanced magnetron (UBM) sputtering [233] and MoS_2 nanocomposites are deposited by cosputtering MoS_2 and Au targets in an UBM system [234]. While the film initially had a high COF of 0.15, once the Au was integrated into the MoS_2, COF decreased to 0.035. These films also displayed good COF in humid atmospheres. Nanocomposites with an Au content of 16 at.% had the best combination of low electrical resistance and low COF.

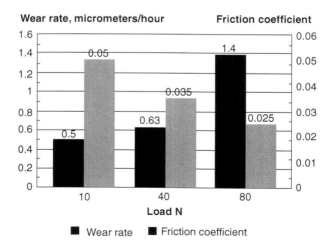

Figure 4.50 Wear rate and COF for MoST films [221].

Si_3N_4 and Al_2O_3 are well known and widely used optical coating materials and are used extensively to provide wear and abrasion resistance to optical materials. Deposition processes and optical properties for these materials are described in Chapter 5.

WC is used in numerous nanolaminate and nanocomposite structures, which will be discussed in Section 4.2.

4.1.7 Transparent Oxides and Nitrides

The tribological properties and wear resistance of transparent oxides and non-transition metal nitrides are generally poorer than corresponding carbides and borides, but the big advantage of these materials is that they are transparent at visible wavelengths. This allows them to be used in protecting optical components, plastics, and metal structures without changing the original color, reflectivity, or transparency. These materials, however, are also used extensively in multilayer optical coatings, thermal barrier coatings, microelectronic devices (e.g., insulators), gas and water permeation barriers, and a variety of other applications. The most commonly used wear resistant oxides are Al_2O_3, Al_xO_y, SiO_2, SiO_x, ZrO_3, Y_2O_3, Ta_2O_5, Nb_2O_5. The most commonly used transparent nitrides are Si_3N_4, Si_xN_y, AlN, and BN. Oxinitrides such as AlO_xN_y and SiO_xN_y are also used extensively because of intermediate optical and dielectric properties.

This group of materials, however, is too extensive to treat each material individually. We will therefore focus on Al_2O_3, Al_xO_y, SiO_2, SiO_x, ZrO_3, and Ta_2O_5, oxides and Si_3N_4, Si_xN_y, and AlN nitrides. A wide variety of deposition processes are used for oxides; the most commonly used processes are magnetron sputtering (all types), high power pulsed plasma magnetron sputtering (HIPPMS), electron (e-beam) evaporation, CVD, PECVD, ALD, and filtered cathodic arc deposition.

Aluminum oxide (alumina, Al_2O_3, Al_xO_y) and silicon oxide (SiO_2 and SiO_x) are arguably two of the most widely used thin film oxide materials. Aluminum oxide films are used extensively in optical applications, permeation barriers for organic devices and solar cells, cutting tools, automotive lighting, biomedical implants, protection of mirror surfaces, and passivation of reactive surfaces. Al_2O_3 thin films are transparent from UV (~ 200 nm) to MWIR (~ 5 μm) wavelengths with a refractive index ~ 1.65 at visible wavelengths. In addition to being hard and wear resistant, this material, although an insulator, has a relatively high thermal conductivity of 30 W/mK. The alpha (α), kappa (κ), and(zetta) (ζ) series of crystalline phases has the hexagonal closed packed (hcp) structure while the gamma (γ), eta (η), and delta (δ) phases have the face centered cubic (fcc) structure. Hardness of alumina films depends on the crystalline phase and ranges from 12.6–22 GPa [11].

The most common crystalline allotrope is the α phase, which is hard and most wear resistant of all phases and also the most thermodynamically stable [236]. This phase is used as an abrasive and to enhance the wear resistance of cutting tools. Amorphous films can also be deposited.

Alumina is deposited by a wide range of deposition conditions. Common to all deposition processes, even with some form of ion and plasma bombardment, is the need for high substrate temperature to achieve optimum performance, whether it is for triboligical, optical, or gas permeation applications. Figure 4.51 shows the general trend in properties with the various crystalline phases with substrate temperature [236]. Films are deposited by magnetron sputtering processes [237, 238, 239], high power pulsed magnetron sputtering [240], PECVD [241, 242], CVD [243, 244], hollow cathode discharge (HAD) [236], plasma activated reactive evaporation [245], atomic layer deposition (ALD) [246], spray pyrolysis [247], and filtered cathodic arc deposition (FCAD) [248].

Oxide films deposited by sputtering techniques generally use

- DC or pulsed power supply with a high purity metal target in mixtures of Ar + O$_2$. RF power can also be used.

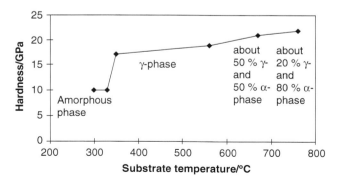

Figure 4.51 Influence of substrate temperature on structure and properties of aluminum oxide [236].

- RF or pulsed power supply with a ceramic target in AR or another inert gas. A small amount of O_2 may have to be added to achieve stoichiometric films.

Therefore, alumina films are deposited by RF or pulsed magnetron sputtering techniques using ceramic Al_2O_3 target, or DC or pulsed reactively sputtered using a metal Al target in mixtures of Ar + O_2. Planar magnetron, unbalanced magnetron, closed field magnetron, cylindrical magnetron, dual magnetron pulsed, all are used to deposit alumina films. Films are deposited in the "metal" mode in the reactive sputtering process to achieve stoichiometric compositions with minimal arcing and high deposition rate. Amorphous films are generally deposited without substrate heating. Crystalline films are deposited at substrate temperatures between 290 °C and 770 °C [249]. The relationship between substrate temperature and crystalline phase is given in Table 4.11 [250]. Figure 4.52 demonstrates that hardness of alumina films deposited by pulsed magnetron sputtering increases with increased substrate

Table 4.11 Summary of crystalline phase of magnetron sputtered alumina films with substrate temperature [249].

Substrate Temperature (°C)	11 kW	13 kW	16 kW	17 kW
290–350	amorphous	amorphous (+50V bias)	amorphous	γ-Al_2O_3
450–480	amorphous (+50V bias)		γ-Al_2O_3	
550–560	γ-Al_2O_3	γ-Al_2O_3	γ-Al_2O_3 (+50V bias)	
670–690	10-% α-Al_2O_3 90% γ-Al_2O_3 (+50V bias)	α-Al_2O_3 90% γ-Al_2O_3	α-Al_2O_3 90% γ-Al_2O_3 (+50V bias)	
750–770	α-Al_2O_3	60% α-Al_2O_3 40% α-Al_2O_3 (+50V bias)	α-Al_2O_3	

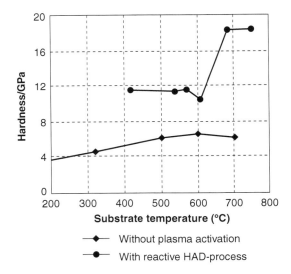

Figure 4.52 Dependence of hardness of magnetron sputtered alumina films on substrate temperature [249].

temperature and amorphous films have lowest hardness values. Substrate bias also has an influence on the fraction of α and γ phases. Again, related to *adatom energetics*. We see that increased plasma density shifts formation of crystalline phases to lower temperatures.

Hardness of magnetron sputtered films also depends on deposition conditions such as O_2 partial pressure and substrate bias [250]. Figures 4.53 and 4.54 show the dependence of hardness of amorphous films with O_2 partial pressure and substrate bias. Similar to films deposited at higher substrate temperatures, hardness peaked at ~ 20 GPa at substrate bias of –50 V. Hardness steadily decreased with increased O_2 partial pressure starting at 0.3 Pa. This was explained by higher packing density at lower O_2 pressures.

CVD and related processes are the other techniques most often used to deposit alumina films. Films are generated by the decomposition of trimethylaluminum ($Al_2(CH_3)_6$) at pressures ~ 0.5 mTorr and temperatures between 300 °C and 600 °C in a shower type reactor [245, 246, 247, 248]. Aluminum acetonate (Al(acac)) is also used in a hot wall, low pressure reactor [251]. Substrate temperatures range up to 600 °C in this process. Dense, pore free crystalline films are obtained at temperatures above 500 °C.

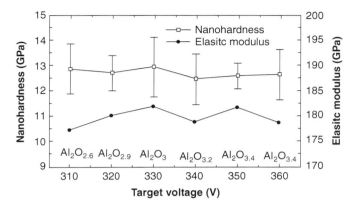

Figure 4.53 Dependence of hardness of unbiased amorphous alumina films on O_2 partial pressure [250].

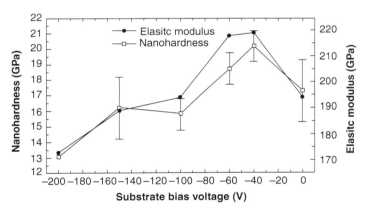

Figure 4.54 Dependence of hardness of unbiased amorphous alumina films on substrate bias [250].

In addition to numerous optical applications as a result of its low refractive index (~ 1.48 @ 550 nm), silicon oxide (silica, SiO_2, SiO_x) films have applications as insulators in microelectronics, transparent barriers, optical coatings, and protecting plastics. SiO_2 films are transparent from the UV (~ 200 nm) to MWIR (~ 2.5 μm) wavelengths. SiO_2 and SiO_x films are primarily deposited by magnetron sputtering techniques [252, 253, 254, 1], CVD and PECVD processes [255, 256, 257, 258]. Atmospheric high frequency plasma [259], atomic layer deposition (ALD) [260], and ion plating [261] are also used to deposit silica thin films. Deposition of high quality films is challenging using evaporation processes without adatom energy enhancement and reacting with O_2, although high quality films have been deposited using ion assist.

Silica films are deposited by DC and RF reactive magnetron sputtering processes (planar, unbalanced, pulsed) using a high purity (often doped to improve conductivity) Si target in mixtures of Ar + O_2 while RF magnetron sputtering techniques are used to deposit from a fused silica target in Ar. Substrates are typically not heated but can reach temperatures near 100 °C due to plasma bombardment. Low temperatures are important for deposition onto temperature sensitive substrates (plastics, etc.). Films are generally amorphous and very smooth.

Deposition of silica thin films was problematic for CVD and related processes several decades ago, but this technology has now become widely used for high quality films. Films are typically deposited from mixtures of SiH_4 + O_2, SiH_4 + N_2O, or from hexamethyldisiloxane (HMDSO) or tetraethoxsilane (TEOS) [1, 257]. Hardness and COF numbers are generally unavailable. Films usually contain 5–15 at.% H. One problem is that particulates can form when SiH_4 is used, which can form nodules and large voids in films. This problem can be partially mitigated by operating at reduced pressures, diluting the SiH_4, heating the electrode, and using a pulsed discharge [1].

In addition to their numerous optical applications, silicon nitride (Si_3N_4 or $SiN_{1.3}$) thin films have applications in passivation and diffusion barriers in microelectronics, surface passivation for microcrystalline Si solar cells, gas and water vapor permeation barriers, high frequency piezoelectric transducers, and biomedical applications. Silicon nitride is also used in nanocomposites (see Section 4.2). Similar to silica, magnetron sputtering [262, 263, 264, 265] and CVD techniques (CVD, PECVD, LPCVD) [1, 59, 266, 267] are most often used to deposit this nitride. Microwave plasma, thermal spray, and ion plating processes are also used [268, 269, 270]. As with SiO_2 films, Si_3N_4 films are generally amorphous but have a columnar microstructure (which makes them useful in piezoelectric applications [262]. Magnetron sputtering involves a high purity Si sputtering target and mixtures of Ar + N_2. Ceramic targets can also be used but, as with sputter deposition of other ceramic thin films, reactive sputtering in Ar + N_2

is still required [264]. Substrates are often unheated but are also heated to temperatures as high as 1300 °C and are often biased to densify the film. Properties depend on N_2 partial pressure, total pressure, substrate temperature, deposition rate, and substrate bias. Density decreases and porosity increases with increased total pressure, which is not surprising when structure zone models are considered. Refractive index decreases to near bulk levels with increased density. Hardness data is hard to find for these films, but Figure 4.55 shows the relationship between hardness and N_2 flow rate [263]. Hardness decreased from 14 GPa to 9.5 GPa with increased N_2 flow, indicating that stoichiometric films are harder than substoichiometric films (nitrogen content also decreased with increased N_2 flow). Hardness values ~ 19 GPa have been reported by Vila *et al.* for unheated substrates and 23.4 GPa for substrates heated to 850 °C [271].

CVD processes normally react SiH_4 with N_2 or NH_3 in an Ar plasma [1, 59]. PECVD processes can also polymerize 1,1,3,3,55-hexamethylcyclotrisilazane [266]. As a result, films can have as much as 30% incorporated H. Films deposited at temperatures near 900 °C are stoichiometric with little incorporated H. Table 4.12 compares properties of CVD silicon nitride.

Aluminum oxynitride (AlO_xN_y) and silicon oxynitride (SiO_xN_y) have intermediate properties between alumina and silica, which

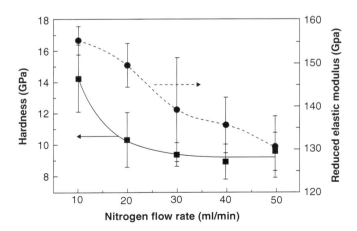

Figure 4.55 Hardness and elastic modulus of Si_3N_4 films deposited with various N_2 flows [263].

Table 4.12 Physical properties of CVD silicon nitride [59].

Property	Si$_4$N$_4$ 1 atm CVD (900 °C)	Si$_4$N$_4$ (H) LPCVD (750 °C)	SiN:H PECVD (300 °C)
Density (g/cm³)	2.8–3.1	2.9–3.1	2.5–2.8
Refractive index (@550 nm)	2.0–2.1	2.01	2.0–2.1
Dielectric constant	6–7	6–7	6–9
Stress @ 23 °C (GPa)	0.7–1.2(T)	0.6(T)	0.3–1.1(C)
S/N	0.75	0.75	0.8–1.0
Thermal stability	Excellent		Variable (>400 °C)

is particularly useful in optical applications. Films are generally deposited by magnetron sputtering and CVD techniques. The refractive index of PECVD SiO$_x$N$_y$ films depends on N$_2$O/(N$_2$O + NH$_3$) flow rate in a mixture with SiH$_4$ [1]. Refractive index decreases from that of Si$_3$N$_4$ (~1.9) to that of SiO$_2$ (~1.46) increased oxygen content. Similar behavior is observed for AlO$_x$N$_y$ films [272].

Although nanolaminates will be addressed in other chapters, because there are very few oxide nanostructures reported in the literature, it is appropriate to discuss AlN/Si$_3$N$_4$ nanolaminates here [254]. Superlattices with 300 alternating layers were deposited by reactive layers deposited by reactive magnetron sputtering. Layer thickness was ~ 3 nm. This coating is deposited by alternately moving the substrate over Al and Si sputtering targets in mixtures of Ar + N$_2$.

Figure 4.56 shows the microhardness of this structure, which is significantly harder than either constituent material. Hardness measured on fused silica substrate was ~ 30 GPa. Hardness of single layer coatings ranges from 10–18 GPa. One of the advantages of this nanolaminate is that it is deposited with no substrate heating, thus permitting deposition on temperature sensitive materials such as plastics and some semiconductors. This structure is also transparent from the UV to NIR wavelengths.

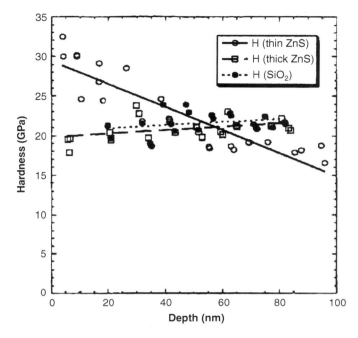

Figure 4.56 Microhardness of nanolaminate on fused silica and zinc sulfide substrates [254].

4.1.8   Zirconium Dioxide and Yttria Stabilized Zirconium Dioxide

Zirconium dioxide (zirconia – ZrO_2) thin films are used extensively as thermal barrier coatings in gas and jet turbine engines. They also have applications as membranes in solid oxide fuel cells (SOFC), optical coatings, security devices, wear resistant coatings for knife blades, high-k dielectrics, and oxygen sensors. Pure ZrO_2 has a monoclinic crystal structure at room temperature and transitions to tetragonal and cubic structures at increasing temperatures. The volume expansion caused by the cubic to tetragonal to monoclinic transformation induces very large stresses, and will cause pure ZrO_2 to crack upon cooling from high temperatures. Several different oxides are thus added to zirconia to mitigate this problem and stabilize the tetragonal and/or cubic phases: yttrium oxide (Y_2O_3) is most often used but MgO, CaO, and Ce_2O_3 are also used [273].

Cubic ZrO_2 also has a very low thermal conductivity, which has led to its use as a thermal barrier coating (TBC) in jet and diesel engines. Ytttria stabilized zirconia (YZO) is used in oxygen sensors and fuel cell membranes because of its high ionic conductivity for oxygen at high temperatures. ZrO_2 and YZO films are deposited primarily by all forms of magnetron sputtering [274], electron beam evaporation [275], CVD [276], PECVD [277], ALD [278], and gas jet processes [279].

4.2 Multifunctional Nanostructured, Nanolaminate, and Nanocomposite Triboligical Materials

During the 1980's materials described in this chapter were used extensively for tribological applications. However, as early as the 1990s and particularly during the last decade, attention has shifted to multifunctional nanostructured, nanolaminate, superlattice and nanocomposite triboligical materials. Applications include optical coatings, nonlinear optics, thermoelectrics, fluorescent sensors, magnetoresistance sensors, C-C structures, transparent conductive coatings, and improving mechanical and structural properties (stiffness, ductility, hardness, wear resistance, strength, stress, weight). These materials can have superior wear resistance, known as "superhard" properties and enhanced lubricity, which has increased and expanded biomedical, transportation, aerospace, and manufacturing applications [36, 280, 281, 282, 283].

Composites can consist of very dissimilar materials, such as polymers and ceramics, polymers and metals and magnetic materials, metals and ceramics, semiconductors and ceramics, polymers and semiconductors, etc. The particles do not have to be compatible with the host, they just have to mix and be distributed uniformly in the host. Composites can consist of metal/metal, metal/ceramic, ceramic/ceramic, polymer/metal, and polymer/ceramic structures. Thin film composite materials can be deposited by physical vapor deposition, MBE, CVD, PECVD, and vacuum polymer evaporation.

Nanocomposites can also have high fracture toughness in addition to high mechanical hardness and wear resistance. All involve heterostructures. It will be instructive to first define these terms:

1. Nanocomposite: A multiphase solid material where one of the phases has one, two, or three dimensions

of less than 100 nm or structures having nano-scale repeat distances between the different phases that make up the material [284]. It is typically the solid combination of a bulk matrix and nano-dimensional phase(s) differing in properties due to dissimilarities in structure and chemistry. The mechanical, electrical, thermal, optical, electrochemical, catalytic properties of the nanocomposite differ markedly from that of the component materials.

2. Superlattice and nanolaminate: A periodic structure of layers of two (or more) materials, typically, thickness of one layer is several nanometers and generally < 10 nm. For mechanical applications, materials are typically carbides, nitrides and oxides.

3. Nanostructure: This category can include nano-scale flakes, fibers, particles, tubes, wires and rods, as well as chiral and sculpted structures [285].

Composites can have unique structural, optical, electrical and magnetic properties not possible with a simple single component material. One of the best known composite materials is fiberglass, which is composed of glass fibers in a polymer matrix. The polymer matrix gives the glass fibers increased flexibility and formability, while preserving their insulating and structural properties. This family of materials (including thin films) is highly disordered and inhomogeneous on a macro- and microstructural scale. Inhomogeneities can be fibers, clusters of atoms or molecules, grains with different crystalline phases (nanocrystalline clusters), inclusions with different electrical and magnetic properties. Note that the particles can have the same composition as the host material, but will have a different structural geometry. Carbon-carbon composites are a good example, where carbon fibers or threads are incorporated into a carbonaceous resin.

Due to the large number of nanostructured coatings, it will not be possible to address all materials systems. Table 4.13 lists the large number of nanocomposites and nanolaminates and provides a extensive that have been studied. This list, although incomplete, describes nanostructured materials and triboligical properties where available. Data for each property, however, was not available for all materials.

Table 4.13 List of nanostructured tribological materials.

Nanostructure Type	Materials	Hardness (GPa)	COF (Dry Air)	Reference
Nanocomposite	Zr-Cu-N	55		36
	Zr-Ni-N	57		36
	CrCu-N	35		36
	Al-Cu-N	47		36
	Zr-Cu-N	23, 54		36, 325
	nc-TiN/ a-Si_3N_4	40–60		36
	nc-TiN/ a-Si_3N_4/a-TiS_2	>80		36
	TiC_xN_y/SiCN	57	0.11	111, 323
	Au/MoS_2		0.035	288
	MoS_2/Ti		0.02	289
	WC/DLC	27, 42	0.05–0.2	290, 291, 318
	WC/TiAlN	38–50		292
	TiC/DLC/WS_2	30	0.01	286
	WC/Co/ Cr/CrN		0.30	293
	TiN/Mo	31		306
	DLC/Cu	38		309
	DLC/Ti	30		309
	TiN/ a-Si_3N_4			311
	W_2N/ a- a-Si_3N_4	51		312, 325
	VN/ a-Si_3N_4			312
	TiN/c-BN	54		312
	TiN/a(TiB_2 + TiB)			313
	TiN/TiB_2			314

Table 4.13 (cont.) List of nanostructured tribological materials.

Nanostructure Type	Materials	Hardness (GPa)	COF (Dry Air)	Reference
	TiC/TiB_2			314
	TiC/DLC			315
	$TiAlN/AlN$	47, 40		313, 325
	$AlCrTaTiZr(N_xC_y)$	32		319
	TiZrAlN	35		320
	TaN/Ag	25.2		321
	$nc-TiC/aC(Al)$	20		322
	$TiAlN/C$	18	0.25	324
	AlCuN	50		326
	TiAlVN	42		327
	$nc-TiN/BN$	69		325
	TiBC	71		325
	$W_{86.7}Ni_{8.3}N_5$	55		325
	$W_{68}Si_{14}N_{18}$	45		325
	$ncMoC/$ $a-(C + Mo_2N)$	49		325
Nanolaminate	$TiAlN/VN$	39		294, 295, 296
	TiAlNYN-VN	78		294
	$TiAlN/CrN$	55		294
	TiCN-TiN	25		297
	$CN/TiCN/TiN$	35		297
	$TiN/TiC_xN_y/$ $Si_3N_4/SiCN$	55		297
	Fe/Ni			297
	CoCu			297
	CoNi			297

Table 4.13 (cont.) List of nanostructured tribological materials.

Nanostructure Type	Materials	Hardness (GPa)	COF (Dry Air)	Reference
	Ni/Ti			297
	Cu/Ti			297
	Ti/TiN			297
	Hf/HfN			297
	AlAlN			297
	Mo/NbN			297
	TiN/CrN	40		298
	TiN/VN	56		299
	TiN/NbN	51		300
	TiAlN/CrN	37.5		301
	$TiN/(V_xNB_{1-x})N$	48		302
	CrN/NbN			303
	CrN/CrAlN			304, 305
	TiAlN/CrAlN			306
	NiAlN/TiBCN	25	0.32	307
	TiC/TiN		0.53	308
	DLC/TiC/TiN/Ti			310
	DLC/TiC/ TiCN/Ti			310
	TiN/Nb	52		325
	TiN/CN_x	45–55		325
	ZrN/CN_x	40–45		325
	$TiN_x/C-N$	20–50		325
	Cu/Al	6.5		328

Figure 4.57 shows the anatomy of a nanocomposite coating [285]. A superhard coating is defined as having a microhardness > 40 GPa, which is based on a combination of nanocrystalline and amorphous phases in composite structures to suppress ductility and increase strength, using grain boundary effects [286, 287]. To qualify as a nanocomposite, nanocrystalline grains must be 3–10 nm in size and separated by 1–3 nm within an amorphous matrix, which consists of other ceramics, metals, carbon, etc. (see Figure 4.57). It has been suggested that the nanocrystals should be oriented in a common direction (have low angle boundaries) in order to provide inter-action across the amorphous matrix and maximize the desirable super-hardness effect [316].

The increased hardness of composites compared to that of single phase coatings is based on the suppression of dislocations by using small 3–5-nm grains and inducing grain incoherence strains when using 1-nm thin matrix for grain separation [282, 283]. The composite strength σ can be defined with respect to crack propagation as [317].

$$\sigma = \left(\frac{4E\gamma_s}{\pi a} \right)^{\frac{1}{2}} \text{ where}$$

(4.1)

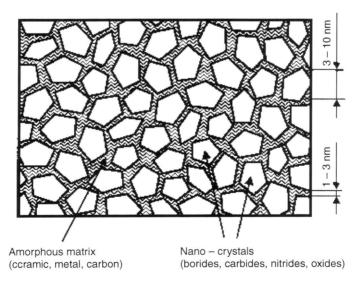

Amorphous matrix
(ceramic, metal, carbon)

Nano – crystals
(borides, carbides, nitrides, oxides)

3 – 10 nm

1 – 3 nm

Figure 4.57 Design schematic of an amorphous/crystalline nanocomposite with high-strength characteristics [290].

E = elastic modulus, γ_s= surface energy of the grain/matrix interface and a = initial crack size. Thus we see that composite strength can be increased by increasing elastic modulus and surface energy of the combined phases, and by decreasing the crystalline grain sizes. For superhard composites, in addition to the selection of an appropriate material system, the elastic modulus artificially increases with decreasing grain sizes due to the lattice incoherence strains and high volume of grain boundaries [286]. Below this limit, the strengthening does not occur because grain boundaries and grains become indistinguishable, and as a result, the stability of the nanocrystalline phase is greatly reduced.

Nanolaminates and superlattices function in much the same way as nanocomposites in that energy of crack propagation is diffused at each layer interface [295, 296]. Figure 4.58 shows how hardness increases with decreasing layer thickness for several

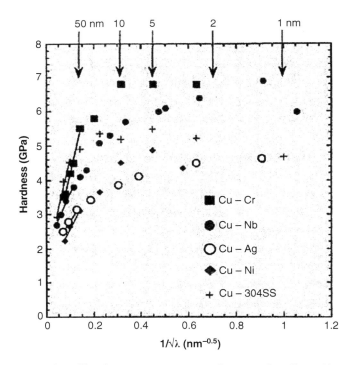

Figure 4.58 Results of hardness measurements showing the effect of layer thickness [295].

bimetal systems [295]. At layer thicknesses on the order of 10 nm, where the hardness approaches maximum, the mechanical properties of nanolaminates and superlattices are strongly influenced by the nature of the mismatch of their crystal lattices at the interfaces as well as the very high ratio of interface volume to total material volume.

Nanostructured coatings are primarily deposited by magnetron sputtering (planar, unbalanced, closed field, HIPPMS, cylindrical, pulsed), CVD (PACVD, PECVD, conventional CVD) and filtered cathodic arc deposition (FCAD). Electron beam evaporation, dual ion beam sputtering, and pulsed laser deposition are also used but are harder to scale up to industrial processes. We will examine four examples involving magnetron sputtering, PECVD, and FCAD. As with all thin film processes, coating properties are primarily determined by deposition parameters. Virtually all magnetron sputtering processes involve cosputtering or alternative sputtering of two different materials, and often reactive sputtering is involved [326, 328]. Common to all deposition processes, size and crystallographic orientation of the grains must be controlled. With respect to sputtering, this involves low energy bombardment and mixing processes and using structure zone models (see Chapter 3) to obtain the correct microstructures [326, 327]. Ion bombardment can restrict grain growth. Mixing is determined by relative sputtering rates of the constituent materials and reactive processes. Additionally, no substrate bias and heating are required to form nanocrystalline structures; however, these parameters are used to fine tune the structure of the film.

In order to demonstrate the effects of deposition conditions, Figure 4.59 plots microhardness of CrNiN nanocomposites against N_2 partial pressure, CrNi target composition, substrate bias (ion energy) and combined substrate temperature, substrate bias, target composition, and N_2 partial pressure [326]. Amorphous films (first box on left) have lowest hardness while nanocrystalline films (third and fourth boxes from left) have highest hardness, peaking at 45 GPa for a N_2 partial pressure of 0.1 Pa, substrate temperature of 200 °C, substrate bias of –200 V, and ion current of 1 mA/cm².

Although structure zone models are sufficient in explaining microstructure of single phase films, they fall short in describing two- phase nanocomposite films. A model based on two phases is shown in Figure 4.60 [329]. The SZM for low Cu (or impurity) content looks much like the single phase SZM. However, as Cu content increases, dense fine grained films without columnar structure form.

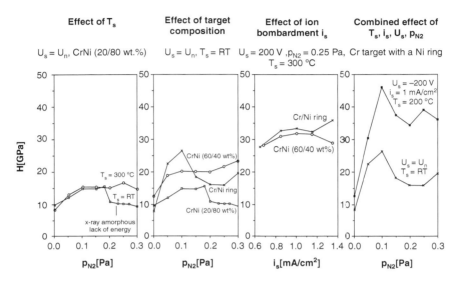

Figure 4.59 Effect of different process parameters on microhardness of Cr-Ni-N nanocomposites [326].

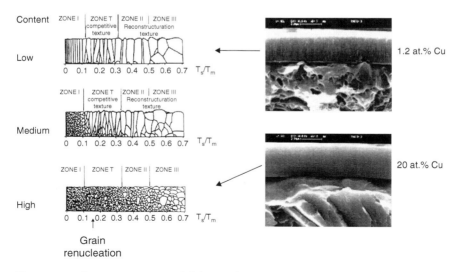

Figure 4.60 Structure zone model for Zr-Cu-N nanocomposite films [329].

The following observations have been made for magnetron sputtered ncMeN/soft metal two- phase films [326]:

1. Mechanical properties of the nanocomposite coatings strongly depend not only on the content of the soft metal but also on the element which forms the hard nc-MeN phase.

2. The superhard (>40 GPa) nanocomposite coatings which exhibit the same hardness can have a very different structure. The same hardness can be achieved not only when the films are composed of larger (> 10 nm) grains (well developed X-ray reflections), but also in the case when films are composed of smaller (< 10 nm) grains (broad, low-intensity X-ray reflections) . In the first case, the maximum hardness of the film can be achieved only when all grains of the hard phase (nc-MeN) are oriented in the same direction and the size of these grains has an optimum value ranging from approximately 10 to 30 nm, see for instance the system Zr-Cu-N or Zr-Y-N.

3. The superhard (> 40 GPa) nanocomposite coatings with large grains (d > 10 nm) exhibit a large macrostress σ (see above) of several GPa. On the contrary, superhard nanocomposite coatings with small grains ($d < 10$ nm) exhibit a low macrostress $\sigma < 1$ GPa .

There is a direct relationship between hardness and reduced elastic modulus ($E^* = E/(1-v^2)$ where v = Poisson's ratio), as shown in Figure 4.61. Hardness generally increases linearly with E. This plot also demonstrates that hardness is a function of the hard phase material (AlN, ZrN, TiN). The parameter H^3/E^2 is a measure of resistance to plastic deformation when plotted against hardness H [111, 325, 326]. Higher values of H^3/E^2 indicate a higher resistance to plastic deformation (lower E). Thus we would expect ceramic/metal composites to have higher values. Figure 4.62 shows that indeed, ceramic/metal nanocomposites have an increased resistance to plastic deformation.

Observations based on this figure are:

1. Films with a given hardness can have different values of the reduced Young's modulus $E^* = E/(1-v^2)$.

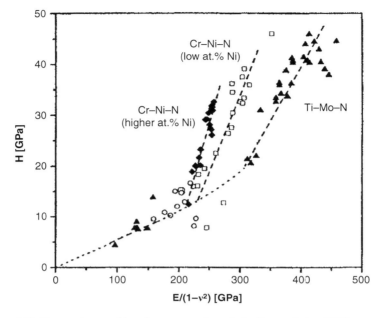

Figure 4.61 Dependence of hardness on reduced elastic modulus [326].

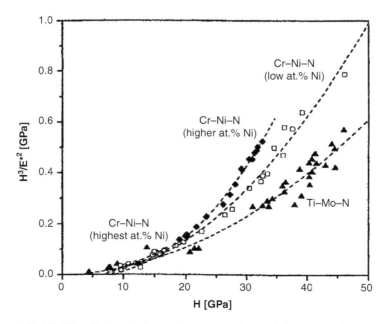

Figure 4.62 H^3/E^2 vs H, indicating resistance to plastic deformation [326, 327].

2. The value of E* can be controlled by the chemical composition of the nanocomposite coating, i.e., by the choice of the elements forming the hard ncMeN phase and soft metal phase.
3. Nanocomposite films composed of two hard phases (Ti-Al-N, Ti-Mo-N) and the nanostructured hard nitrides of transition metals (TiN, ZrN) exhibit a higher E* compared with nanocomposites composed of one hard and one soft (metal) phase. In the last case, the value of E* strongly depends also on the element which forms the hard nc-MeN phase.
4. The resistance to plastic deformation increases with decreasing value of the reduced Young's modulus E*.

This study also concluded that three phase films are needed to achieve hardness > 70 GPa.

Nanocomposite TiC_xN_y/SiCN superhard coatings are deposited by PECVD [111, 323]. Figure 4.63 shows the structural evolution of

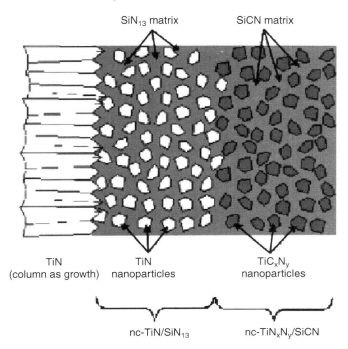

Figure 4.63 Schematic evolution of TiC_xN_y/SiCN superhard nanocomposite [111, 323].

the nanocomposite. nc-TiC$_x$N$_y$/SiCN films with average grain size of ~ 10 nm were deposited in a radio frequency (RF, 13.56 MHz) PECVD system in which the substrates were placed on the RF powered substrate holder, which developed a self-induced DC bias voltage. Substrate temperature was 500 ^{0}C. The working gas mixture consisted of TiCl$_4$, SiH$_4$, CH$_4$, N$_2$, H$_2$ and Ar. The fabrication process consisted of a short, 15-min surface pretreatment in Ar at a DC bias V$_b$ = – 600 V and p = 50 mTorr (6.66 Pa), and was followed by a 3h deposition process performed at V$_b$= – 600 V and p = 200 mTorr (26.66 Pa), using a gas mixture of TiCl$_4$/N$_2$/SiH$_4$.

Figure 4.64 shows that hardness depended on C content and peaked at 55 GPa at C = 10 at.%. COF increased from 0.11 to ~ 0.17 with an increase of C from 0 to 25 at%. Above a C content of 15 At.% phase separation was suppressed and individual phases were found to intermix, thus degrading the nanocomposite. The films also displayed a high resistance to plastic deformation, as shown in Figure 4.65. Impressive values ~ 1.8 are achieved for the PECVD nanocomposites (compare these to Figure 4.61).

Table 4.14 shows industrial applications for transition metal nitride based superhard nanocomposite coatings [295].

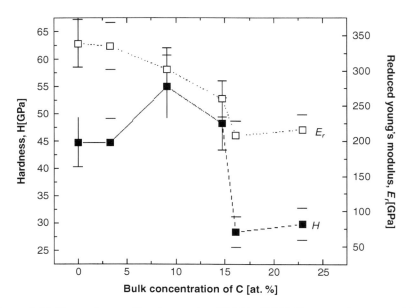

Figure 4.64 Dependence of hardness of nc-TiC$_x$N$_y$/SiCN films as a function of C content [111, 323].

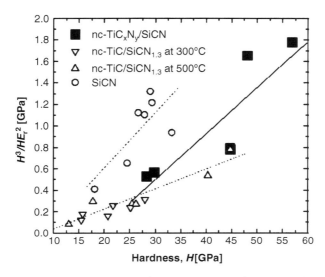

Figure 4.65 H^3/E^2 vs H for nc-TiC$_x$N$_y$/SiCN, nc-TiC$_x$N$_y$/SiN$_{1.3}$ and SiCN films [323].

The most technologically important multilayer and nanolaminate ceramic/ceramic coatings are TiN/VN, TiN/NbN, TiN/CrN, TiN/(VxNb1–x)N, TiN/AlN, TiAlN/TiN, TiAlN/CrN, CrN/NbN, CrN/CrAlN and TiAlN/CrAlN [297]. Figure 4.66 shows the structure of a generic nanolaminate coating. Table 4.15 lists coating systems and applications of transition metal nitrides [297]. Note the importance of TiN in these coatings.

Magnetron sputtering is most often used to deposit multilayer and nanolaminate structures. TiAlN/CrN nanolaminate coatings are deposited by unbalanced magnetron sputtering [330, 331]. As with virtually all multilayer coatings, the substrate was rotated from sputtering target to target to deposit individual layers. Substrates were rotated between Ti$_{50}$Al$_{50}$ and Cr targets in Ar + N$_2$ gas mixtures for a total of 800 layers. Hardness peaked at 4000 kg/mm^2 for layer thickness of 6.5 nm. Nanoindentation measurements carried out after heat treatment of the films showed that the multilayer films retained hardness as high as 2600 kg/mm^2 even at 800 °C, while hardness of TiAlN films decreased sharply to 2200 kg/mm^2 at 700 °C. Hardness of TiN and CrN film decreased to 1000 kg/mm^2 and 700 kg/mm^{2,} respectively.

Figure 4.67 shows a TEM micrograph of a magnetron sputtered Cu/Au nanolaminate film deposited on PET. With addition of high

Table 4.14 Industrial applications for transition metal nitride based superhard nanocomposite coatings [295].

Coating System	Applications
TiSiN	Diffusion barrier liners for gigascale Cu interconnect applications High speed cutting tool machining
TiAlSiN	High speed machining of various steel alloys
AlSiTiN	High speed machining of various steel alloys
TiAlBN	Machining of steel
TiBN	Machining of Ti based alloys
$(Ti_{1-x}Al_x)N/a\text{-}Si_3N_4 + TiN$	Machining of hardened steel
CrSiN	Green machining of brass High speed micro drilling of circuit boards
CrAlSiN	High speed micro drilling of circuit boards Dry turning of case iron
AlCrSiN	High speed machining of M2 die steel
nc-(Al1-xCrx)N/ Si_3N_4(nACRo)	Hobbing

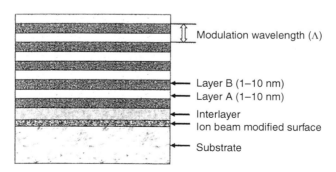

Figure 4.66 Layer structure of generic nanolaminate coating [297].

Table 4.15 Applications of transition metal nitride based nanolaminate coatings [297].

Coating System	Application
TiN/NbN	Machining stainless steel Rolling contact fatigue of M-50 steel
TiAlN/CrN	Protection of titanium aluminides used in aerospace industry Moulds for high temperature Cu semi-solid processing
CrN/NbN	Printing industry, leather industry and surgical blades Textile industry Cutlery industry
TiAlN/VN	Dry high speed machining of low alloyed and Ni based steels Dry high speed machining of Al alloys Machining of Al alloys for aerospace and automotive components
TiN/AlTiN	Machining of inconel superalloys High speed machining of stainless steel
$TiAl_xN/TiAl_yN$	Dry machining of tray and ductile cast iron
$TiAlN_x/TiAlN_y$	Machining of difficult to cut materials (CrNiMoTi, CrMo, etc.)
TiN/CrN	Moulds for high temperature Cu semi solid processing
TiN/TaN	Machining of stainless steel
TiN/SiN_x	Hard disk and tribological applications
TiN/CrN-Ti/ Cr-TiN/CrN	Periodontal instruments
TiAlCrN/ TiAlYN	Dry high speed machining of steels Coatings on moulds for glass industry Protection of titanium aluminides used in aerospace industry
TiAlCrN/NbN	Dry high speed machining of hardened H 13 steel
$CrN_x/a-Si_3N_4$	Water hydraulic components

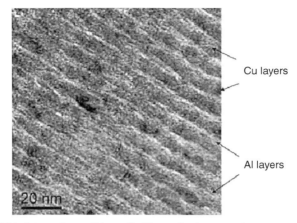

TEM picture of a multi layer Cu/Al nanolaminate coating

Figure 4.67 TEM picture of magnetron sputtered nanolaminate film deposited on PET [328].

hardness, this film could also be free standing [328]. Films with 5000 alternating Cu and Au layers with thicknesses of 2 and 3 nm, respectively, were deposited in a vacuum roll coater using Ar gas. The web substrate was continually rotated over Cu and Al targets to obtain the large number of layers. The nanolaminate had a hardness of 6.5 GPa, impressive for an all "soft" metal system. This type of structure is being developed for ultralight deployable mirrors in space. The nanolaminte could be mounted on a mandrel to fix its final figure and then coated with a highly reflective metal such as Au, Al, or Ag.

We have just scratched the surface on this topic. Hundreds of papers have been written on nanocomposites, nanolaminates, and superlattices, and mechanical properties depend on a complex series of effects. The search for new extrinsic superhard coatings continues and is being directed toward oxide/nitride nanostructures. Transition metal nitride based nanostructures have limitations for high temperatures while superhard coatings consisting of oxide materials are highly stable with very high melting temperatures (ZrO_2, Al_2O_3, and Y_2O_3 have melting temperatures = 2700, 2000, and 2400, respectively). Recall that yttria stabilized zirconia (YSZ) is used as thermal barrier coatings in jet and gas turbine blades.

References

1. Ludvik Martinu *et al.*, in *Handbook of Deposition Technologies for Films and Coatings*, Third Edition, Peter M Martin, Ed., Elsevier (2009) 417.
2. M. Ma, SVC 46*th* *Annual Technical Conference Proceedings* (2003) 535.
3. P. M. Martin *et al.*, *SVC 40th Annual Technical Conference Proceedings* (1997) 187.
4. J. M. Anton *et al.*, *SVC 48th Annual Technical Conference Proceedings* (2005) 599.
5. C. Metzner *et al.*, *SVC 51st Annual Technical Conference Proceedings* (2008) 370.
6. D. M. Devia *et al.*, *SVC 52nd Annual Technical Conference Proceedings* (2009) 596.
7. C. Misiano *et al.*, *SVC 53rd Annual Technical Conference Proceedings* (2010) 166.
8. M. Podob, *SVC 39th Annual Technical Conference Proceedings* (1996) 72.
9. A. Leonhardt *et al.*, VIII. *Int Pulvermetallurgische Tagung*, Bd. 3, Beitrag 36, Dresden (1985) 197.
10. A. Leonhardt *et al.*, 31. *Jahrg, Heft* 11 (1986) 423.
11. R Bunshah, *Handbook of Hard Coatings*, Noyes (2001).
12. Phani J. *Vac Sci Technol* A28(3) (2010) 341.
13. Kutzaki J. *Vac Sci Technol* A28(3) (2010) 341.
14. Huili Wang *et al.*, *Thin Solid Films* 516 (2008) 5419.
15. F. Arrondo *et al.*, *Surf Coat Techol* 68/69 (1994) 536.
16. O. Knotek *et al.*, *Surf Coat Techol* 61 (1991) 320.
17. M. Pancielejko *et al.*, *Vacuum* 53 (1999) 57.
18. N. J. Archer, *Thin Solid Films* 80 (1981) 221.
19. K. G. Stjernberg *et al.*, *Thin Solid Films* 40 (1977) 81.
20. Mitsunori Kabayashi and Yashihico Doi, *Thin Solid Films* 54 (1978) 67.
21. Y. H. Cheng *et al.*, *J Vac Sci Technol* A28(3) (2010) 431.
22. L. C. Agudelo *et al.*, *Phys Scr* T131 (2008) 014006.
23. L. F. Senna *et al.*, *Surf Coat Technol* 94–95 (1997) 390.
24. L. Karlsson *et al.*, *Thin Solid Films* 371 (2000) 167.
25. Y. Y. Guu and J. F. Lin, *Wear* 210 (1997) 245.
26. J. M. Lackner *et al.*, *Surf Coat Technol* 188– 189 (2004) 519.
27. Y. Cheng and Y. F. Zheng, *Surf Coat Technol* 201 (2007) 4909.
28. A. Larsson and S. Ruppi, *Thin Solid Films* 402 (2003) 203.
29. Jianguo Deng *et al.*, *J Vac Sci Technol* A12(3) (1993) 733.
30. M. Eerden *et al.*, *SVC 48th Annual Technical Conference Proceedings* (2005) 584.
31. M. Eizenberg *et al.*, *J Vac Sci Technol* A13(3) (1995) 590.
32. Y. H. Cheng *et al.*, *J Appl Phys* 104 (2008) 093502.
33. M. W. Barsoum *et al.*, *J. of Appl. Phys.*, 87(12) (2000) 8407, 2000.
34. T. Leyendecker *et al.*, *Surface and Coatings Technology*, 48 (1991) 175.

35. J. Vetter, *Surface and Coatings Technology* 76–77 (1995) 719.
36. Ali Erdemir and Andrey A. Voedvodin, in *Handbook of Deposition Technologies for Films and Coatings*, 3rd Ed., P M Martin, Ed., Elsevier (2009).
37. *Nanostructured Thin Films and Coatings: Mechanical Properties*, San Zhang, Ed, CRC Press (2010).
38. Donald Mattox, *SVC 50th Annual Technical Conference Proceedings* (2007) 5.
39. G. C. Vandross and HH Abu-Safe, *SVC 51st Annual Technical Conference Proceedings* (2008) 635.
40. M. F. Cano *et al.*, *SVC 52nd Annual Technical Conference Proceedings* (2009) 601.
41. W. D. Munz, *SVC 36th Annual Technical Conference Proceedings* (1993) 411.
42. W. D. Munz, *J Vac Sci Technol* A4(6) (1988) 2717.
43. W. D. Munz *et al.*, *J Vac Sci Technol* A11(5) (1993) 2583.
44. S. Veprek and M. Jilek, *Pure Appl Chem* 74(3) (2002) 1.
45. Andre Anders, in *Handbook of Deposition Technologies for Films and Coatings*, 3rd Ed., P. M. Martin, Ed., Elsevier (2009).
46. E. Le Bourhis *et al.*, *Surf Coat Technol* 203 (2009) 2961.
47. G. S. Kim *et al.*, *Surf Coat Technol* 201 (2006) 4361.
48. J. L. Endrino *et al.*, *Surf Coat Technol* 200 (2006) 6840.
49. R. Franz *et al.*, *Tribo Lett* 23 (2006) 101.
50. G. S. Kim and S.Y. Lee, *Surf Coat Technol*.
51. S. Lee, *Solid State Phenom* 124–126 (2007) 1609.
52. A. E. Rieter *et al.*, *Surf Coat Technol* 200 (2005) 2114.
53. O. Knotek *et al.*, *Surf Coat Technol* 42 (1990) 21.
54. O. Knotek *et al.*, *Surf Coat Technol* 43/44 (1990) 107.
55. C. Mitterer *et al.* *Surf Coat Technol* 41 (1990) 351.
56. Pi-Chuen Tsai *et al.*, *Thin Solid Films* 517 (2009) 5044.
57. M. Karnezos, Matl Sci forum 54 & 55, Trans Tech Publications (1990).
58. P. M. Martin *et al.*, *SVC 40th Annual Technical Conference Proceedings* (1997) 197.
59. I. Konyashin *et al.*, *Chem Vapor Dep* 3(5) (1997) 353.
60. V. N. Demin *et al.*, *J Vac Sci Technol* A18(1) (2000) 94.
61. S. Watanabe *et al.*, *Surface and Coatings Technol*, 49 (1991) 406.
62. M. Murakawa and S. Watanabe, *Application of Diamond Films and Related Materials* (Y. Tzeng, M. Yoshikawa, M. Murakawa and A. Feldman, eds., Elsevier (1991) 661.
63. I. Konyashin *et al.*, *Chem. Vap. Deposition* 3(5) (1995) 239.
64. J. L. Andujar *et al.*, *J Vac Sci Technol* A16(2) (1998) 578.
65. S. M. Gorbatkin *et al.*, *J Vac Sci Technol* A11(4) (1993) 1863.
66. Wilfrado Ortano-Ravera *et al.*, *J Vac Sci Technol* A16(3) (1998) 1331.
67. H. Klosterman *et al.*, *SVC 45th Annual Technical Conference Proceedings* (2002) 402.

68. Minoru Sueda *et al.*, *J Vac Sci Technol* A16(6) (1998) 3287.
69. M. Mieno and T. Yoshida, *Jpn J Appl Phys*, 27 (1990) L1175.
70. S. Watanabe *et al.*, *Surface and Coatings Technol*, 49 (1991) 406.
71. A. F. Jankowski *et al.*, *Thin Solid Films* 308–309 (1997) 94.
72. S. Miyake, *Thin Solid Films* 493(2005) 164.
73. C. W. Ong, *Thin Solid Films* 307 (1997) 152.
74. J. D. Ferguson *et al.*, *Thin Solid Films* 413 (2002) 16.
75. Y. K. Le and H. Oechsner, *Thin Solid Films* 437 (2003) 83.
76. X. Z. Ding *et al.*, *Thin Solid Films* 429 (2003) 22.
77. S. S. Dana, in *Synthesis and Properties of Boron Nitride, Materials* Science Forum 54 & 55, Tran Tech Publications (1990).
78. W. Cermingnani *et al.*, *SVC 41ˢᵗ Annual Technical Conference Proceedings* (1998) 66.
79. V. Cholet *et al.*, *Thin Solid Films* 188 (1990) 143.
80. M. Olsson *et al*, *Thin Solid Films* 172 (1989) 95.
81. Terry Hu *et al.*, *Thin Solid Films* 332 (1998) 80.
82. Shin-ichi Aoqui *et al.*, *Thin Solid Films* 407 (2002) 126.
83. S. Ulrich *et al*, *Thin Solid Films* 518 (2009) 1445.
84. V. Solozhenko *et al.*, *Phys Rev Lett* 102 (2009) 015506.
85. Jiaqian Qin *et al.*, *Advanced Materials* 20 (2008) 47
86. Dong Ho Kim *et al*, *Thin Solid Films* 447–448 (2004) 192.
87. Hidemitsu Aoki *et al.*, *Thin Solid Films* 518 (2010) 2102.
88. Fei Zhou *et al.*, *Thin Solid Films* 497 (2006) 210.
89. R. Gago *et al.*, *Thin Solid Films* 273(2000) 277.
90. Shojiro Miyake, *Thin Solid Films* 493(2005) 160.
91. P. M. Martin, W. D. Bennett and C. H. Henager, *SVC 50ᵗʰ Annual Technical Conference Proceedings* (2007) 43.
92. C. Morant *et al.*, *Surface and Coatings Technology* 180–181(Complete) (2004) 512.
93. R. G. Hoagland *et al.*, *Scripta Mater* 50(6) (2004) 775.
94. Allan Matthews, *J Vac Sci Technol* A21(5) (2003) S224.
95. M. Weis and B. Manty, *Proceedings of the 38ᵗʰ SVC Annual Technical Conference* (1995) 325.
96. William D. Sproul, Chapter 6: *50 Years of Vacuum Coating Technology and the Growth of the Society of Vacuum Coaters*, Donald M. Mattox and Vivienne Harwood Mattox, Ed., Society of Vacuum Coaters (2007).
97. R Bunshah, Ed., *Handbook of Deposition Technologies for Films and Coatings*, 1ˢᵗ Ed., Noyes (1982).
98. R Bunshah, Ed., *Handbook of Deposition Technologies for Films and Coatings*, 2ⁿᵈ Ed., Noyes (1994).
99. Milton Ohring, Engineering Materials Science, Academic Press (1995).
100. William D. Sproul, *J. Vac. Sci. Technol.*, A 21(5) (2003) S222.
101. A. Matthews *et al.*, *J. Vac. Sci. Technol.*, A13(3) (1995) 1202.
102. A. Matthews, *J. Vac. Sci. Technol.*, 214(5) (2003) S224.

103. J-E Sundgren and H. T. G. Hentzell, *J. Vac. Sci. Technol.*, A 4(5) (1986) 2259.
104. B. R. Anton, *Proceedings of the 45th SVC Annual Technical Conference* (2002) 407.
105. U. Kopacz and S. Schulz, *Proceedings of the 34th SVC Annual Technical Conference* (1991) 48.
106. Philippe Schmid, Masako Sato Sunaga, and Francis Le´vy, *J. Vac. Sci. Technol.*, A 15(5) (1998) 2870.
107. D. E .Siegfried *et al.*, *Proceedings of the 39th SVC Annual Technical Conference* (1996) 97.
108. R. Wei, J. Arps, and B. Lanning, *Proceedings of the 46th SVC Annual Technical Conference* (2003) 431.
109. Peter J. Clarke, *J. Vac. Sci. Technol.*, 14(1) (1977) 141.
110. L. Maya and C. E. Vallet, *J. Vac. Sci. Technol.*, 15(4) (1997) 2007.
111. P. Jedrzejowski *et al.*, *Proceedings of the 47th SVC Annual Technical Conference* (2004) 704.
112. P. Jedrzejowski *et al.*, *Proceedings of the 46th SVC Annual Technical Conference* (2003) 530.
113. Douglas L Schulz *et al.*, *J. Vac. Sci. Technol.*, 27(4) (2009) 962.
114. M. Holzherr *et al.*, *Proceedings of the 48th SVC Annual Technical Conference* (2005) 452.
115. "Products" Ion Fusion Surgical.
116. P. J. Kelly *et al.*, *Proceedings of the 45th SVC Annual Technical Conference* (2007) 596.
117. M. M. Lacerda *et al.*, *J. Vac. Sci. Technol.*, 17(5) (1999) 2915.
118. Ju-Wan Lim *et al.*, *J. Vac. Sci. Technol.*, 18(2) (2000) 524.
119. P. J. Martin *et al.*, *Proceedings of the 36th SVC Annual Technical Conference* (1993) 375.
120. Z. Yu *et al.*, *Proceedings of the 49th SVC Annual Technical Conference* (2006) 538.
121. B. R. Anton and P.C. Esquibel, *Proceedings of the 45th SVC Annual Technical Conference* (2002) 407.
122. Xiao-Ming He *et al.*, *J. Vac. Sci. Technol.*, 18(1) (2000) 30.
123. Sanchez-Lopez *et al.*, *J. Vac. Sci. Technol.*, 23(4) (2005) 681.
124. K. Kashiwagi *et al.*, *J. Vac. Sci. Technol.*, 4(2) (1986) 210.
125. J. Musil, *Surf. and Coat. Technol.*, 125 (2000) 322.
126. F. Capotondi *et al.*, *La Metallurgea Italiana* (2007) 15.
127. J. M. Anton *et al.*, *Proceedings of the 48th SVC Annual Technical Conference* (2005) 599.
128. Marie-Paule Delplancke-Ogletree and Othon R. Monteiro, *J. Vac Sci Technol*, 15(4) (1997) 1943.
129. Jiangua Deng *et al.*, *J Vac Sci Technol*, 16(4) (1998) 2073.
130. Huili Wang *et al.*, *Thin Solid Films*, 516 (2008) 5419.
131. Donald Mattox, *Handbook of Physical Vapor Deposition (PVD) Processing*, Noyes (1998).

132. Donald Mattox, *Handbook of Physical Vapor Deposition (PVD) Processing*, Second Ed., Elsevier (2010).
133. B. Krzan *et al. Tribology International*, 42 (2009) 229.
134. *Handbook of Nanostructured Films and Coatings*, Sam Zhang, Ed., CRC Press (2010).
135. H. O. Pierson, *Handbook of Carbon, Graphite, Diamond and Fullerines*, Noyes (1993).
136. N. Sasvvides and T. J. Bell, *Appl Phys Lett*, 46 (1985) 146.
137. J. Robertson, *Mater Sci Eng*, R37 (2002) 129.
138. P. K. Bachmann *et al.*, *Diamond and Related Materials*, 3 (1994) 799.
139. D. Neerinck *et al.*, *SVC 40th Annual Technical Conference Proceedings* (1997) 86.
140. A. Erdemir, *J. Vac. Sci. Technol.* A18(4) (2000) 1987.
141. A. Erdemir, *SVC 44th Annual Technical Conference Proceedings* (2001) 397.
142. B. Bhushan and S. Venkatesan, *J Appl Phys*, 74 (1993) 4174.
143. S. J. Bull *et al.*, *Surf Coat Techolo*, 68/69 (1994) 603.
144. A. K. Gangopadhyay and M. A. Tamor, *Wear*, 169 (1993) 221.
145. M. N. Gardos, "Tribology and Wear Behavior of Diamond", in *Synthetic Diamond: Emerging CVD Science and Technology*, K. E. Spear and J. P. Dismukes, eds., Wiley (1994).
146. M. N. Gardos and B. L. Soriano, *J Mater Res*, 5 (190) 2599.
147. B. K. Gupta *et al.*, *ASME J Tribology*, 116 (1994) 445.
148. I. P. Hayward, *Surf Coat Technol*, 49 (1991) 554.
149. I. P. Hayward *et al.*, *Wear*, 157 (1992) 215.
150. M. Kohzaki *et al*, *J Mater Res*, 7 (1992) 1769.
151. C. T. Kuo *et al.*, *J Mater Res*, 5 (1990) 2515.
152. E. Broitman *et al.*, *J Vac Sci Technol* A21(4), (2003) 851.
153. C. Y. Chan *et al.*, *J. Vac. Sci. Technol.* A 19(4) (2001) 1606.
154. B. J. Knapp *et al.*, *Proceedings of the 35th SVC Annual Technical Conference* (1992) 174.
155. L. Haubold *et al.*, *Proceedings of the 51st SVC Annual Technical Conference* (2008) 193.
156. Joel I. Gersten and Frederick W. Smith, *The Physics and Chemistry of Materials*, Wiley (2001).
157. See www.azom.com
158. J. J. Cuomo *et al.*, *J Vac Sci Technol* 16 (1979) 299.
159. Donald Mattox, Chapter 6, Ion Plating, *Handbook of Deposition Technologies for Films and Coatings*, 3rd Ed., Peter M. Martin, Ed., Elsevier (2009).
160. A. Y. Liu and M. L. Cohen, *Science* 245 (1989) 841.
161. L. Sun *et al.*, *J Vac Sci Technol* A28(6), Nov/Dec 2010, 1299.
162. A. Y. Liu and M. L. Cohen, *Science*, 245 (1989) 841.
163. T. Inoue *et al.*, *Appl Phys Lett*, 67 (1995) 353
164. Jorg Neidhardt and Lars Hulman, *J Vac Sci Technol* A25(4), Jul/Aug 2007, 633.

165. Masami Aono *et al., J Vac Sci Technol* A26(4), Jul/Aug 2008, 966.
166. Hyn S. Myung *et al., Thin Solid Films*, 506–507 (2006) 87.
167. Toshiyuki Hayashi *et al., Thin Solid Films*, 376 (2000) 172.
168. L. F. Johnson and M. B. Moran, *SVC 40th Annual Technical Conference Proceedings* (1997) 192.
169. E. F. Motta and I. Pereyra, *J Noncryst Solids*, 338–340 (2004) 545.
170. Z. Y. Chen *et al., J Vac Sci Technol* A20(5), Seot/Oct 2002, 1639.
171. Dongping Liu *et al., J Vac Sci Technol* A22(6), Nov/Dec 2004, 2329.
172. I. Alexandrou *et al., Phys Rev B*, 60 (1999) 10903.
173. M. Johansson *et al., Vacuum* 53 (1999) 451.
174. D. L. Strout, *J Phys Chem.* A, 104 (2000) 3364.
175. M. Chhowalla and G. A. J. Amaratunga, *Nature* (London), 407 (2000) 164.
176. S. Avivi *et al., J Am Chem Soc.*, 122 (2000) 4331.
177. G. Seifert *et al., Solid State Commun.*, 115 (2000) 635.
178. A. Rothschild *et al., J. Am. Chem.* Soc., 122 (2000) 5169.
179. R. Tenne, *J Mater Res.*, 21 (2006) 2726.
180. Å. Johansson, Doctoral thesis, Linkoping University, 2002.
181. W. W. Porterfield, *Inorganic Chemistry—A Unified Approach*, Academic, London (1993).
182. S. Stafstrom, *Appl Phys Lett*, 77 (2000) 3941.
183. L. G. Jaconsohn *et al., J Vac Sci Technol* A18(5) (2000), 2230.
184. T. E. Karis *et al., Vac Sci Technol* A 15 (1997) 2382.
185. C. Donnet *et al Surf Coat Technol* 94–95 (1997) 531.
186. D. Sarangi *et al., Thin Solid Films*, 447–448 (2004) 217.
187. S. K. Mishra *et al., Thin Solid Films*, 515 (2007) 4738.
188. H. L. Chang and C. T. Kuo, *Mater Chem Phys*, 72 (2001) 236.
189. H. L. Chang and C. T. Kuo, *Diamond Relat Mater*, 10 (2001) 1910.
190. Yongqing Fu *et al., Surf Coat Technol*, 160 (2002) 165.
191. A. Badzian *et al., Diamond Relat Mater*, 1 (1998) 80.
192. C. Uslu *et al., J Electron Mater*, 25 (1996) 23.
193. W. Cheng *et al., Mater Chem Phys*, 85 (2004) 370.
194. J. Vleck *et al., Surf Coat Technol*, 160 (2002) 74.
195. J. J. Wu *et al., Thin Solid Films* 355–356 (1999) 472.
196. S. Ulrich *et al., Diamond Relat Mater*, 6 (1997) 645.
197. N. Laidani *et al., Thin Solid Films* 223 (1993) 14.
198. J-P Riviere *et al., Sur. Coat Technol* 100/101 (1998) 243.
199. P. R. Chalker *et al., Diamond Relat Mater* 4 (1995) 632.
200. A. M. Haghiri-Gosnet *et al., J Vac Sci Technol* B8 (1990) 1565.
201. Y. Hirohata *et al., Thin Solid Films* 63 (1979) 237.
202. H. Sung *et al., Surf Coat* Technol 54/55 (1992) 541.
203. M. Zaytouni *et al., Thin Solid Films* 287 (1996) 1.
204. H. Yoshihara, H. Mori, M. Kiuchi, *Thin Solid Films* 76 (1981) 1.
205. A. K. Costa *et al., Thin Solid Films* 377–378 (2000) 243.
206. Yongwa Chris Cha *et al., Thin Solid Films* 253 (1994) 212.

207. A. Czyzniewski, *Thin Solid Films* 433 (2003) 180.
208. J. Esteve *et al.*, *Thin Solid Films* 373 (2000) 282.
209. Yongxin Wang *et al.*, *Surf Coatings Technol* 205 (2011) 2770.
210. Phillip D. Rack *et al.*, *J Vac Sci Technol* A 19(1), Jan/Feb 2001, 62.
211. Y. H. Cheng and B. K. Tay, *J Vac Sci Technol* A 21(2) Mar/Apr 2003 411
212. Jianhong He *et al.*, *Surf Coatings Technol* 157 (2002) 72–79.
213. H. Hogberg *et al.*, *Thin Solid Films* 272 (1996) 116.
214. Hugh O. Pierson, *Handbook of Chemical Vapor Deposition (CVD): Principles, Technology, and Applications*, William Andrew (1992).
215. Jianhong He and Julie M. Schoenung, *Surf Coatings Technol* 157 (2002) 72.
216. L. G. Yu *et al.*, *Surf Coatings Technol* 182 (2004) 308.
217. M. Watanabe *et al.*, *Surf Coatings Technol* 201 (2006) 619.
218. J. Nerz *et al.*, *J Thermal Spray Technol* 1 (2) (1992) 147.
219. M. R. Hilton and P. D. Fleischauer, *Surf Coat Technol*, 54/55 (1992) 435.
220. L. Cizaire, *Surface and Coatings Technology* 160 (2002) 282–287.
221. D. G. Teer *et al.*, *40th SVC Annual Technical Conference* Proceedings (1997) 70.
222. Z. Z. Xia *et al.*, *Surface and Coat Technology* 201 (2006) 1006–1011
223. E. W. Roberts, *Thin Solid Films* 181 (1989) 461.
224. Talivaldis Spalvins, *Thin Solid Films* 80 (1973) 291.
225. D. G. Teer *et al.*, *42nd SVC Annual Technical Conference Proceedings* (1999) 357.
226. M. R. Hilton and P. D. Fleischauer, Surf Coat Technol, 54/55 (1992) 435.
227. G. Jayaram *et al.*, *Surf Coat Technol*, 76–77 (1995) 393.
228. G. Weise *et al.*, *Surf Coatings Technol*, 76–77 (1995) 382.
229. G. Jayaram *et al.*, *Surf Coat Technol*, 68–69 (1994) 439.
230. A. Aubert *et al.*, *Surf Coat Technol* 41 (1990) 127.
231. M. S. Donley *et al.*, *Surf Coat Technol* 36 (1988) 329.
232. Da-YungWang *et al.*, *Surf Coat Technol* 120–121 (1999) 629.
233. Hsi-Hsin Chien *et al.*, *Thin Solid Films* 518 (2010) 7532.
234. Jeffrey R. Lince *et al.*, *Thin Solid Films* 517 (2009) 5516.
235. N. M. Renevier *et al.*, *Surf Coatings Technol* 123 (2000) 84.
236. G. Hoetzch *et al.*, *SVC 40th Annual Technical Conference Proceedings* (1997) 77.
237. J. D. Affinito *et al.*, *SVC 46th Annual Technical Conference Proceedings* (2003) 89.
238. Y. Siddiqui and D. Crawford, *SVC 49th Annual Technical Conference Proceedings* (2006) 697.
239. F. Sabary *et al.*, *SVC 50th Annual Technical Conference Proceedings* (2007) 489.
240. W. D. Sproul *et al.*, *SVC 47th Annual Technical Conference Proceedings* (2004) 96.
241. C. Cibert *et al.*, *Thin Solid Films* 516 (2008) 1290.
242. DAP Bulla and N. I. Moramoto, *Thin Solid Films* 334 (1998) 60.

243. F. Weist *et al.*, *Thin Solid Films* 496 (2006) 240.
244. A. Roy Chowdhuri and C. G. Takoudis, *Thin Solid Films* 446 (2004) 155.
245. S. Shiller *et al.*, *SVC 38th Annual Technical Conference Proceedings* (1995) 18.
246. S. Sneck, *SVC 51st Annual Technical Conference Proceedings* (2008) 413.
247. M. Aguilar Frutis *et al.*, *Thin Solid Films* 389 (2001) 200.
248. Z. W. Zhao *et al.*, *Thin Solid Films* 447–448 (2004) 14.
249. Zywitski and Hoetzsch, *Surf Coat Technol* 96 (1997) 262.
250. Kari Koskia, Jorma HoÈlsaÈ, Pierre Julieta, *Thin Solid Films* 339 (1999) 240.
251. M.P. Singh, S.A. Shivashankar, *Surf Coatings Technol* 161 (2002) 135.
252. A.F. Jankowski *et al.*, *Thin Solid Films* 420–421 (2002) 43.
253. Nigel Dansen *et al.*, ., *Thin Solid Films* 289 (1996) 99.
254. P. M. Martin, W. D. Bennett and C. H. Henager, *SVC 50th Annual Technical Conference Proceedings* (2007) 643.
255. R. B. Heil, , *SVC 38th Annual Technical Conference Proceedings* (1995) 33.
256. A. W. Smith, *SVC 45th Annual Technical Conference Proceedings* (2002) 525.
257. J. Madocks *et al.*, *SVC 50th Annual Technical Conference Proceedings* (2007) 233.
258. F. Benitez *et al.*, , SVC 45th Annual Technical Conference Proceedings (2002) 280.
259. H. Kakiuchi *et al.*, *Thin Solid Films* 519 (2010) 235.
260. Arto Pakkala and Matti Putkonen, Atomic Layer Deposition, in *Handbook of Deposition Technologies for Films and Coatings*, 3rd Ed., P. M. Martin, Ed, Elsevier (2009) 364.
261. Donald Mattox, Ion Plating, in *Handbook of Deposition Technologies for Films and Coatings*, 3rd Ed., P M Martin, Ed, Elsevier (2009) 297.
262. S. Guruvenket *et al.*, *Thin Solid Films* 478 (2005) 256.
263. Gang Xu *et al.*, *Thin Solid Films* 425 (2003) 196.
264. H. Hirohata *et al.*, *Thin Solid Films* 253 (1994) 425.
265. Bao-Shun Yau, Jow-Lay Huang, *Surf Coat Technol* 176 (2004) 290.
266. T. A. Brooks and D. W. Hess, *Thin Solid Films* 153 (1987) 521.
267. V. Verlaan *et al.*, *Surf Coat Technol* 201 (2007) 9285.
268. Jerrgen Ramm and Ralph E Pixley, *Surf Coat Technol* 52 (1992) 187.
269. Robert B. Heimann, *Surf Coat Technol* 205 (2010) 943.
270. H. Schlemm, *Surf Coat Technol* 174 –175 (2003) 208.
271. M. Vila *et al.*, *J Appl Phys* 94(12) (2003) 7868.
272. S. J. Holmes and V. W. Biricik, *SVC 36th Annual Technical Conference Proceedings* (1993) 146.
273. A. G. Evans and R. M. Cannon, *Acta Met* 34 (1986) 761.
274. P. J. Martin and R. P. Netterfield, *SVC 36th Annual Technical Conference Proceedings* (1993) 375.
275. D. A. Glocker *et al*, *SVC 37th Annual Technical Conference Proceedings* (2004) 183.

276. J. W. Moore and N. K. Tsujimoto, *SVC 37th Annual Technical Conference Proceedings* (1994) 110.
277. Corina Nistorica *et al.*, *J Vac Sci Techno.* A 23(4), Jul/Aug 2005 836.
278. G. Garcia *et al.*, *Thin Solid Films* 370 (2000) 173.
279. D. D. Hass and H. N. G. Wadley, *J Vac Sci Techno.* A 27(2) Mar/Apr 2009 404.
280. A. Erdemir, *Tribol Int,* 38 (2005) 249.
281. A. A. Voevodin and J. S. Zabinski, *Composite Sci Technol* 65 (2005) 741.
282. S. Veprek and M. J. G. Veprek-Heijman, *Surf Coat Technol* 202 (2008) 5063.
283. *Nanostructured Thin Films and Coatings*, Sam Zhang, Ed., CRC Press (2010).
284. P. M. Ajayan, L. S. Schadler, P. V. Braun, Nanocomposite science and technology, Wiley (2003).
285. Michael T. Taschuk, Matthew M. Hawkeye and Michael J. Brett, Glancing Angle Deposition, in *Handbook of Deposition Technologies for Films and Coatings*, Third ed., P. M. Martin, Ed, Elsevier (2009).
286. A. A. Voevodin and J. S. Zabinski, *Thin Solid Films* 370 (2000) 223.
287. W. D. Sproul, *SVC 50th Annual Technical Conference* (2007) 591.
288. Hsi-Hsin Chien *et al.*, *Thin Solid Films* 518 (2010) 7532.
289. N. M. Renevier *et al.*, *Surf Coatings Technol* 123 (2000) 84.TALIVALDIS
290. A. A. Voevodin, J. P. O'Neill, J. S. Zabinski, *Thin Solid Films* 342 (1999) 194.
291. Y. Liu *et al.*, *Thin Solid Films* 488 (2005) 140.
292. J. S. Yoon *et al.*, *Surface and Coatings Technology* 142–144 (2001) 596.
293. Riccardo Polini *et al.*, *Thin Solid Films* 519 (2010) 1629.
294. P. Eh Hovseian *et al*, *Surf Coat Technol* 133–134 (2000) 166.
295. R. G. Hoagland *et al*, *Scripta Mater* 50(60 (2004) 775.
296. Bernd Schultrich, *New Hard Coating by Nanotechnology*, Jahrbuch Oberflaechentechnik (2003) 59 128.
297. Harish C. Barshila, B. Deepthi and K. S. Rajam, Handbook of Nanostructured *Thin Films and Coatings*, San Zhang Ed, CRC Press (2010) 427.
298. A. Madan *et al.*, *Appl Phys Lett* 68 (1996) 2198.
299. X. Chu *et al.*, *J Mater Res* 14 (1999) 2500.
300. H. Barshilia and K. S. Rajam, *Surf Coat Technol* 183 (2004) 174.
301. H. C. Barshilia *et al.*, *Vacuum* 77 (2005) 169.
302. P. B. Mirkarimi *et al*, *J Mater Res* 9 (1994) 1456.
303. P. E. Hovsepian *et al.*, *Surf Coat Technol* 116–119 (1999) 727.
304. H. C. Barshilia *et al*, *Appl Surf Sci* 253 (2007) 5076.
305. H. C. Barshilia *et al*, *J Vac Sci Technol* A27 (2009) 29.
306. D. Glocker *et al.*, *SVC 48th Annual Technical Conference* (2005) 53.
307. Z. Zhong, *SVC 47th Annual Technical Conference* (2004) 493.
308. D. M. Devia *et al.*, *SVC 52nd Annual Technical Conference* (2009) 596.

309. Wei *et al.*, *J Vac Sci Technol* A17(6) (1999) 3412.
310. R. F. Huang *et al.*, *Diamond Rel Mat* 10 (2001) 1850.
311. S. Veprek, S. Reiprich, *Thin Solid Films* 268 (1995) 64.
312. S. Veprek, *Thin Solid Films* 317 (1998) 449.
313. J. Musil, Hruby, *Thin Solid Films* (2000) in press.
314. A. A. Voevodin, S. V. Prasad, J. S. Zabinski, *J. Appl. Phys.* 82 (1997) 855.
315. C. Mitterer, *et al.*, *Surf Coat Technol* 120–121 (1999) 405.
316. J. Musil, *Surf Coat Technol* 125 (2000) 322.
317. G. E. Dieter, *Mechanical Metallurgy*, 2nd ed., McGraw-Hill, 1976.
318. A. Czyzniewski, *Thin Solid Films* 433 (2003) 180.
319. Shou-Yi Chang *et al.*, Multicomponent $AlCrTaTiZr(N_xC_y)$ Nanocomposites, *Thin Solid Films*, In Press (2011).
320. Youn J. Kim *et al.*, *Thin Solid Films* 516 (2008) 3651.
321. C. C. Tseng, *Thin Solid Films* 516 (2008) 5424.
322. Sam Zhang *et al.*, *Thin Solid Films* 467 (2004) 261.
323. P. Jedrzejowski, J. E. Klemberg-Sapieha, L. Martinu, *Thin Solid Films* 466 (2004) 189.
324. M. Stueber *et al.*, *Surf CoatTechnol* 200 (2006) 6162.
325. J Musil, Surf Coat Technol 125 (2000) 322.
326. J. Musil *et al.*, *Surf Coat Technol* 142–144 (2001) 557.
327. P. J. Martin *et al.*, *Surf Coat Technol* 200 (2005) 2228.
328. P. M. Martin *et al.*, *SVC 50th Annual Technical Conference* (2007) 643.
329. P. Barna, M. Adamik, Formation and characterization of the structure of surface coatings, in: Y- Paleau, P.B. Barna, Eds., *Protective Coatings and Thin Films*, Kluwer Academic Publisher, 1997.
330. Harish C. Barshilia *et al.*, *Vacuum* 77 (2005) 169.
331. Harish C. Barshilia *et al.*, *Vacuum* 83 (2009) 427.

5

Optical Thin Films and Composites

Modification and enhancement of the optical properties of solid surfaces, including transparent, reflective, opaque or absorptive surfaces, is an important application for surface engineering. Except for reducing scattering, there is very little that can be done to modify the optical properties of a bulk surface without some external process (such as application of thin films, smoothing layers, etching and patterning). Because the optical constants are ultimately properties of the lattice and bonding in materials, stress and strain can also affect their behavior. Unlike many other properties, modification of the optical properties of a surface can totally change the optical properties and even the color of the substrate. However, in addition to modifying the optical properties, in many cases they serve two or more SE functions, such as improving abrasion and wear resistance, electromagnetic shielding, and as gas, chemical, and water permeation barriers. The optical properties of a surface can be modified by a simple single layer thin film coating or by complex multilayer coatings with hundreds of layers. Optical properties of a surface can thus be modified using:

- Thin films and multilayer thin films
- Patterned structures

283

- Etching
- Low dimensional structures
- Photonic crystals
- GLAD coatings

The major types of multilayer optical coatings are:

- Antireflection coatings used to reduce Bragg reflections
- High reflector coatings used to increase the reflectance from narrow to broad spectral ranges
- Filters used to transmit narrow to broad wavelength ranges
- Notch filters
- High and low pass filters
- Dichroic coatings
- Rugate filters
- Low-e coatings
- Absorptive/emissive coatings
- Metal mirrors
- Photocatalytic coatings for self-cleaning and surface oxidation
- Transparent gas and water vapor barriers

Although there is considerable overlap, Table 5.1 lists thin film optical materials and their approximate usable spectral ranges. Note that this table is not all inclusive. Deposition processes described in Chapter 2 are used to deposit single and multilayer optical thin films. Also note that many of these materials are used in a variety of applications, other than optical coatings.

Table 5.1 Useful spectral ranges of thin film optical materials.

Material	Optical Property	Spectral Range	Examples
Transition metal oxides	Transmission, photocatalyst	UV - MWIR	Ta_2O_5, Nb_2O_5, ZrO_2, WO_3, TiO_2, V_2O_3, VO_2
Semiconductor oxides	Transmission	UV – NIR	SiO_2, SiO, Ge_2O_3

Table 5.1 (cont.) Useful spectral ranges of thin film optical materials.

Material	Optical Property	Spectral Range	Examples
Other metal oxides	Transmission	UV – NIR	Al_2O_3, BeO, BO
Transparent conductive oxides		UV – VIS	ITO, In_2O_3, ZnO, ZnO:Al, SnO_2,
Transition metal nitrides	Reflection	VIS - LWIR	TiN, HfN, ZrN, TaN
Semiconductor nitrides	Transmission	UV - MWIR	Si_3N_4, Ge_3N_4
Other metal nitrides	Transmission and reflection	UV - MWIR	AlN, CrN
Oxynitrides	Transmission	UV - MWIR	SiO_xN_y, AlO_xN_y
Metal fluorides	Transmission	UV - LWIR	MgF_2, CaF_2, HfF_4
Carbides	Transmission	VIS - LWIR	SiC, B_4C, GeC
Semiconductors	Transmission and absorption	UV - LWIR	Si, Si:H, Ge, Ge:H, C, DLC, CdTe
Metal sulfides	Transmission and absorption	UV - LWIR	CdS, PbS, ZnS
Metals	Reflection and absorption	UV – mm wave	Ag, Au, Al, Cu, Ni, Ta, Cr, Pt

5.1   Optical Properties at an Interface

The basic optical properties (transmittance, reflectance, absorptance, scattering) of solid surfaces are introduced in Section 1.3. However, it is important for thin films that we understand refraction and polarization of light at an interface in more detail because more than one interface is generally involved. The Fresnel equations define the components of the reflected and transmitted waves with respect to the angle of incidence (AOI) of light, as shown in Figure 5.1 for two media with refractive indices n_1 and n_2, and $k_1 = k_2 = 0$. Polarization of light plays an important role in the interaction of light with matter, for example [1]:

- The amount of light reflected at a boundary between materials depends on the polarization of the incident wave.
- The amount of light absorbed by certain materials is polarization dependent
- Light scattering from matter (solids in our case) is generally polarization sensitive
- The refractive index of anisotropic materials depends on polarization
- Optically "active" materials (such as liquid crystals) have the natural ability to rotate the polarization plane of linearly polarized light.

Light can be considered a plane wave with frequency v (wavelength = $c/v = \lambda$) traveling in a specific direction with velocity c. Consider light traveling in the z direction. The electric field generate by the wave can be described by

$$E(z,t) = \mathrm{Re}\left\{\mathbf{A}\exp i\left(t - \frac{z}{c}\right)\right\} = E_x\mathbf{x} + E_y\mathbf{y} \text{ where} \quad (5.1)$$

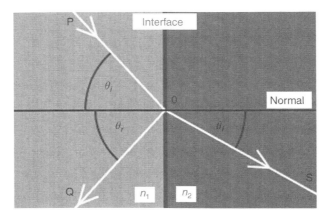

Figure 5.1 Reflection and refraction at the boundary of two media.

A is the complex amplitude envelope $= A_x\mathbf{x} + A_y\mathbf{y}$. As usual, **bold** characters are vector quantities. **A** can be expressed as functions of phase ϕ:

$$A_x = a_x \exp(i\phi x) \tag{5.2}$$

$$A_y = a_y \exp(i\phi y) \tag{5.3}$$

$$\text{Phase difference} = \phi = \phi_x - \phi_y. \tag{5.4}$$

Light is considered linearly polarized in the x or y direction if either a_x or $a_y = 0$ (but not both).

Light is circularly polarized if $\phi = \pm\pi/2$ and $a_x = a_y = a_0$. Right circular polarization occurs when $\phi = \pi/2$ and left circular polarization occurs when $\phi = -\pi/2$.

Thus, light traveling in the z direction has electric field components in the x and y directions, as shown in Figure 5.2. Figure 5.3 defines circular and elliptical polarizations.

Polarization plays an important role in reflection and refraction of a plane wave light. Figure 5.4 breaks Figure 5.1 into more detail and shows the reflection and refraction of the components of a plane wave at a boundary between two dielectric media.

The electric field vectors are shown for the incident and reflected waves. The plane wave is incident at the boundary with angle θ_1. The wave is reflected at θ_3 and refracted at θ_2. We know that $\theta_3 = \theta_1$ and, according to Snell's law, refraction is defined by

$$n_1 \sin\theta_1 = n_2 \sin\theta_2 \tag{5.5}$$

Reflected and refracted amplitudes are most conveniently represented by matrices, but we will just give final results here.

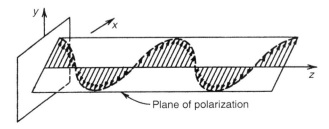

Figure 5.2 Linearly polarized light showing plane of polarization [1].

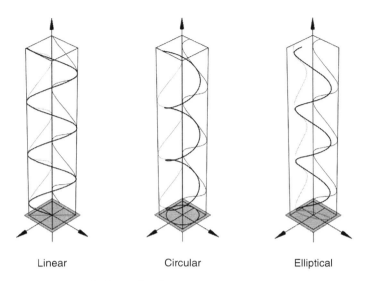

Linear   Circular   Elliptical

Figure 5.3 Circular and elliptical polarizations.

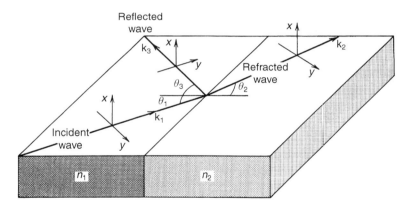

Figure 5.4 Reflection and refraction at the boundary between two dielectric media [1].

In addition to electric field components, the plane wave also has magnetic field components (H_x and H_y), similar to the electric field described above. Polarization in the x direction is called transverse electric (TE) polarization or orthogonal polarization since electric fields are perpendicular to the plane of incidence. Polarization in the y direction is called transverse magnetic (TM) polarization since the magnetic field is orthogonal to the plane of incidence. The TM mode, however, is perpendicular to the TE mode and thus parallel to the plane of incidence. These two modes are important when discussing transmission through waveguides. Convention for these modes is "s" polarization = TE mode and "p" polarization = TM mode. The Fresnel equations for reflection and transmission coefficients, which define reflection and refraction, can then be expressed as:

$$r_x = \frac{(n_1 \cos\theta_1 - n_2 \cos\theta_2)}{(n_1 \cos\theta_1 + n_2 \cos\theta_2)} - TE \tag{5.6}$$

$$t_x = 1 + r_x \tag{5.7}$$

$$r_y = \frac{(n_2 \cos\theta_1 - n_1 \cos\theta_2)}{(n_2 \cos\theta_1 + n_1 \cos\theta_2)} - TM \tag{5.8}$$

$$t_y = 1 - r_y \tag{5.9}$$

The reflectance R is the square of the reflectance coefficient:

$$R_S = \frac{(n_1 \cos\theta_1 - n_2 \cos\theta_2)^2}{(n_1 \cos\theta 1_i + n_2 \cos\theta_2)^2} \tag{5.10}$$

While the reflection coefficient of light polarized in the plane (p-polarization) is given by:

$$R_P = \frac{(n_1 \cos\theta_1 - n_2 \cos\theta_2)^2}{(n_1 \cos\theta_1 + n_2 \cos\theta_2)^2} \tag{5.11}$$

Referring to Figure 5.1, in this case $\theta_i = \theta_1$ and $\theta_t = \theta_2$ The average reflection coefficient is $R = (R_S + R_P)/2$.

Of course, all this assumes no or very little optical absorption. However, if there is significant optical absorption we must ultimately return to the equations discussed in Section 1.3:

$$T_S = 1 - R_S - A \tag{5.12}$$

$$T_P = 1 - R_P - A \tag{5.13}$$

Figure 5.5 shows s and p reflection for the cases of $n_1 < n_2$ and $n_1 > n_2$ against angle of incidence (AOI). For the first case, s polarization

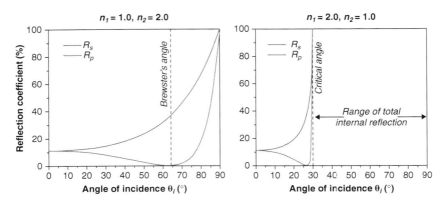

Figure 5.5 s and p reflection for $n_1 < n_2$ and $n_1 > n_2$.

reflectance continues to increase at all angles of incidence but p reflectance (r_y) approaches 0 at Brewster's angle $(\theta_B = \tan^{-1} n_2/n_1)$ and then increases rapidly with increased AOI. As demonstrated in the figure, total internal reflection is only possible when $n_1 > n_2$., and occurs at the critical angle

$$\theta_C = \sin^{-1}\left(\frac{n_2}{n_1}\right) \qquad (5.14)$$

Figure 5.6 demonstrates total internal reflection in a block of PMMA. Here $n_1 = 1$ and $n_2 = 1.5$. If $n_1 < n_2$, some light would be transmitted out of the material and some reflected back into it according to the Fresnel equations. These concepts are critical for waveguide operation.

For normal and near normal incidence the Fresnel equations simplify to [2]:

$$r = r_s = r_p = \frac{(n_2 - n_1)}{(n_1 + n_2)} \qquad (5.15)$$

$$t = t_s = t_p = \frac{2n_1}{(n_1 + n_2)} \qquad (5.16)$$

A word on absorptive media with respect to total internal reflection. There will be many cases in which the medium outside the dielectric boundary is a metal or is highly optically absorbing. In this case an evanescent wave field, frequently called a plasmon, is set up outside the dielectric medium, as shown in Figure 5.7 [3]. The amplitude of this wave decays exponentially with distance from the boundary. Energy is transmitted perpendicularly to the boundary. Thus, energy can be transferred to a dielectric medium via total internal reflection.

All surfaces scatter incident electromagnetic radiation at some level. Scattering causes attenuation of a light beam similar to absorption. The intensity of light due to scattering cross section σ_s [4] is

$$I(z) = I_0 \exp(-\sigma_s Nz) \text{ where} \qquad (5.17)$$

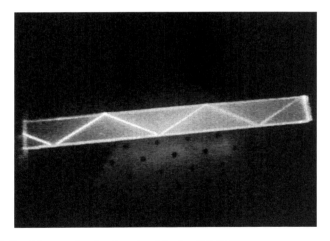

Figure 5.6 Total internal reflection in a block of PMMA.

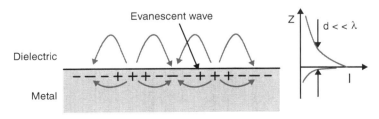

Figure 5.7 Formation of an evanescent wave at a metal/dielectric interface and the amplitude decay.

N is the density of scattering centers and z is the propagation direction through the solid. Note that scattering only gets worse at shorter wavelengths

$$\sigma_S \sim \frac{1}{\lambda^4} \qquad (5.18)$$

Thus when describing engineering of optical properties, it is important to specify which property is of interest and address all factors that affect this property. For example, maximizing transmittance requires that reflectance, absorptance, and scattering be minimized.

5.2   Single Layer Optical Coatings

Surface optical treatments can:

- Change a simple transparent ceramic window into a reflector, EMI shield, heat mirror, cold mirror, emissive display, laser shield, or switchable optical device
- Change the color of a surface, optic, or window
- Increase transmission of ceramic, semiconductor and semi-opaque substrates
- Modify the optical band structure of a surface
- Increase or decrease optical scattering of a surface
- Enable read-write capability on a substrate

The most widely used optical surface treatment is the metal mirror with the multilayer thin film optical coating as second. However, even a humble single layer coating can significantly change the optical properties of a surface or substrate. A simple metal thin film can change the optical properties of any non-scattering surface to a mirror or high reflector. Consider first the metallic reflector, which is essentially a broad band mirror. Figure 5.8 shows the reflectance of several metal coatings at 0° AOI. The color of each material depends on the wavelength range of high reflectance. Al and Ag have high reflectivity over the visible wavelength range and, essentially, have no color and appear "silvery". The reflectance of Au and Cu films only becomes high above 500 nm wavelength. We will address color in more detail later in this section. We see that Cu

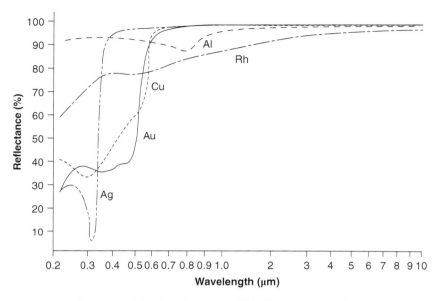

Figure 5.8 Reflectance of Al, Ag, Au, Cu and Rh films at 0° AOI [5].

reflectance contains a bit more blue - green. Rh appears more gray due to absorption at lower visible wavelengths.

Now consider a substrate or window with complex refractive index $n_S = n_S + i\kappa_S$ and a thin film with thickness t and complex refractive index $n_f = n_f + i\kappa_f$. Analysis even of this simple geometry is complex and requires summation of reflection and transmission over multiple passes through the film, as shown in Figure 5.9 [6]. Ignoring absorption and polarization, the reflectance and transmittance of a single layer on a substrate are given by the summations

$$R = \frac{\{(n_0^2 + n_f^2)(n_f^2 + n_S^2) - 4n_0 n_f^2 n_S + (n_0^2 - n_f^2)(n_f^2 - n_S^2)\cos 2\delta_f\}}{\{(n_0^2 + n_f^2)(n_f^2 + n_S^2) + 4n_0 n_f^2 n_S + (n_0^2 - n_f^2)(n_f^2 - n_S^2)\cos 2\delta_f\}}$$

(5.19)

$$T = \frac{8n_0 n_f^2 n_S}{\{(n_0^2 + n_f^2)(n_f^2 + n_S^2) + 4n_0 n_f^2 n_S + (n_0^2 - n_f^2)(n_f^2 - n_S^2)\cos 2\delta_f\}}$$

(5.20)

Here n_0 is the real part of the refractive index of the incident medium (usually air), δ_f is the phase change of the incident

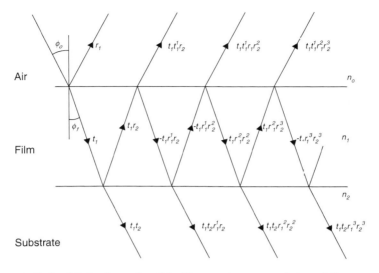

Figure 5.9 Path of light through a thin film coating on a substrate [6].

beam $= (2\pi/\lambda)n_f t\cos\theta$ passing through the thin film, and θ = angle of incidence (AOI). One can imagine the complexity of these terms if complex refractive indices and multilayers are used (a good reason to use optical design software). Also note that these expressions are polarization dependent, as described above.

A more compact expression for reflectance is given by [7]

$$R = \frac{\left(y_0 - \dfrac{y_f}{y_s}\right)^2}{\left(y_0 + \dfrac{y_f}{y_s}\right)^2} \qquad (5.21)$$

In this case, y_0, y_f and y_s are optical admittances of air, film, and substrate, which are related by

$$y_f = (y_0 y_s)^{\frac{1}{2}} \text{ and } n_f = (n_0 n_s)^{\frac{1}{2}}. \qquad (5.22)$$

With admittance is defined as $y = (n - ik)(\epsilon_0/\mu_0)^{1/2}$ [7].

As mentioned earlier, a single layer thin film is the simplest case of engineering the optical properties of a substrate. Figure 5.10

shows interference of transmitted and reflected waves at the interface of a coating with quarter wave ($n_f t = \lambda/4$) optical thickness. Such a film can reduce substrate reflectance (antireflection) or increase substrate reflectance (decrease transmittance) depending on the optical thickness ($n_f t$), as shown in Figure 5.11 for a thin glass film (n = 1.38) on TiO_2 (n = 2.35) at normal incidence. The reflectance of the surface of an untreated optical window depends primarily

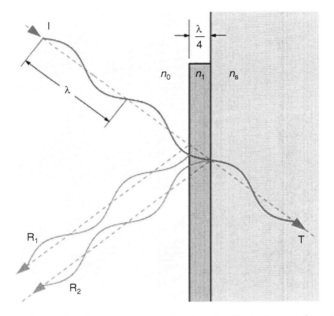

Figure 5.10 Relationship between transmitted and reflected wave for a thin film with quarter wave optical thickness [8].

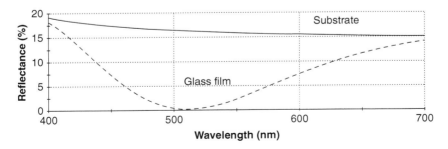

Figure 5.11 Reflectance of a thin glass film (n = 1.38) on TiO_2 (n = 2.35) at normal incidence at 550 nm compared to bare substrate (dashed line).

on its refractive index (assuming $\kappa \ll 1$ for optical windows) and can vary from 0.03 (reflects 3%) for magnesium fluoride (MgF_2) to 0.36 (reflects 36%) for germanium (Ge). Since an optical window reflects light from both its surfaces, optimal performance will be achieved if an antireflection (AR) coating is applied to both faces. We will see that high reflectance is usually required only on one side. The condition for a thin film to eliminate surface reflectance at a *single wavelength* is

$$n_f = (n_0 n_s)^{\frac{1}{2}} \qquad (5.23)$$

Assuming $n_0 = 1$ for air, then $n_f = n_s^{1/2}$ for the perfect AR coating at a single wavelength. For example, n (500 nm) for SiO_2 is ~ 1.45, which is close to a perfect AR coating for Ta_2O_5 (n = 2.15). Thin film AR coatings are commonly applied to ophthalmic (eyeglass) lenses, camera lenses, solar cells, sensor window, lenses in optical instruments (telescopes for example), fiber optics components, and flat panel displays. AR coatings are also used on more exotic applications, including increasing the transmission of aircraft windows, electro-optic (EO) sensors, missile guidance optical systems, LIDAR systems, and night vision apparatus.

5.3 Multilayer Thin Film Optical Coatings

Single layer thin film optical coatings are limited in how they can modify the optical properties of surface. The optical performance of a thin film can be enhanced and modified over wide ranges by adding alternating, or periodic, layers of high, low, and intermediate refractive index materials, as shown in Figure 5.12. Reflectance, transmittance, and absorptance can be enhanced or reduced over a specific wavelength band using multilayer coatings to create notch filters, low and high pass filters, bandpass filters, broadband high reflector or antireflection coatings, high and low emittance coatings, dichroic coatings, and many more designs.

The convention for describing the layer structure of a multilayer optical coating with n pairs of alternating layers is

Substrate (layer 1 optical thickness,
layer 2 optical thickness, ...)^n

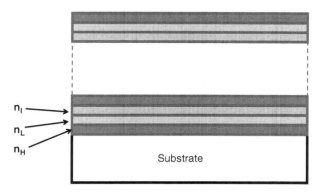

Figure 5.12 Schematic of multilayer optical coating.

If **H** = quarter wave optical thickness ($n_H t_H = \lambda/4$) of high index layer and **L** = quarter wave optical thickness ($n_L t_L = \lambda/4$) of low index layer, an all-quarter wave ten-layer high/low index coating is represented as S(HL)⁵. Similarly, if a metal layer (M) is included, the design would be S(HMH) for metal layer sandwiched between two quarter wave thickness high index layers. If layer thickness is not a single quarter wave, the thickness will be given in fractions or multiples of the quarter wave thickness, i.e., a half wave thickness = HH and an eight wave thickness = H/2.

5.3.1   Broad Band Antireflection Coatings

The problem with the single layer AR coating (see Figure 5.11) is that the above relation is good only at one wavelength. AR coatings, however, are required to operate over a narrow wavelength band to a very broad band of wavelengths from the ultraviolet (UV) to the long wavelength infrared (LWIR). They are even used to increase the transmission of semitransparent metal films in low-e coatings and induced transmission filters. They can have either discrete or graded layers with different refractive indices.

The width of the low reflectance region can be broadened by adding appropriate layers. A two layer quarter wave design, S(HL) offers a significant improvement over the single layer in both materials available and operating range. For example, quarter wave layers of TiO$_2$ (n ~ 2.4) and Al$_2$O$_3$ (n ~ 1.63) can be used to increase the transmission of a Ge window as shown in Figure 5.13. Note that this coating structure still has a region of higher reflectance, which is due to the high refractive index of the substrate. Real improvements in performance and increased operating wavelength band begin with three and more layers using combinations of low, intermediate, and high index layers, or just high and low index layers. Multilayer AR coatings on low index optics can have more than 10 layers. The bandwidth increases and reflectance generally decreases with increased number of layers, however, the improvements are less apparent as more layers are added. Figure 5.14 compares reflectance of 3, 5, and 9-layer AR designs [8]. From the figure we see that there is no free lunch in the AR world. It should be stressed that while adding layers can compensate for the fact that there will not be an optical material that meets all requirements for "perfect" antireflectance, no layer design will take reflectance to zero (as for the single layer case). Additionally, more complex layer structures are more complicated and costly to deposit.

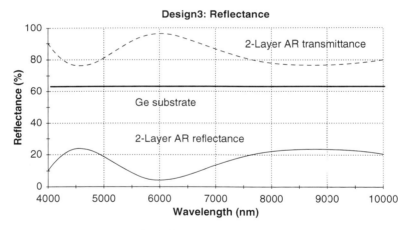

Figure 5.13 Transmittance of two-layer AR coating on Ge.

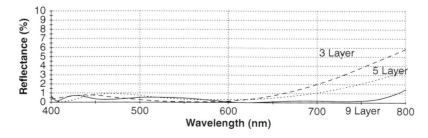

Figure 5.14 Reflectance of 3, 5 and 9 layer AR designs.

Another interesting case is the low emittance coating (heat mirror), which involves one or more semitransparent metal layers (typically Ag or Al) sandwiched between dielectric layers, which increases transmittance and lowers both reflectance and emittance of the metal layers. Typical designs are S(H′MH′), S(H′MH′MH′), ..., where M is the metal layer and H′ is a non-quarter wave optical thickness. Many current designs employ as many as five Ag layers. Thickness of the metal is ~ 1–10 nm. Figure 5.15 shows the transmittance and reflectance of a 5-layer Ag induced transmission filter (ITF) tuned at 550 nm with TiO_2 as the dielectric layer. Transmittance at visible wavelengths is ~ 90% while emittance is ~ and NIR – IR reflectance is > 90%. Thus this coating transmits visible light while reflecting heat. Below we address the inverse case of increasing emittance of a surface.

5.3.2 High Reflectance Multilayer Coatings

Let's now consider the opposite case: increasing reflectance of a surface. High reflection coatings have numerous applications, including telescope mirrors, laser mirrors, display optics, satellite optics, architectural glass, automobile mirrors and headlamps, defense optical systems, projectors, and much more. This family of coatings includes metal reflectors, protected metal reflectors, dielectric-enhanced metal reflectors, all dielectric multilayer reflectors, and rugate filters. Cousins to this group of coatings are dichroic coatings, high pass and low pass filters, beam splitters, dense wavelength division multiplexing (DWDM) filters, low-e coatings, and Fabry-Perot filters. Depending on the design and materials used, the high reflector can be broad band, covering a wide range of

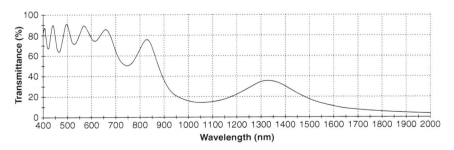

Figure 5.15 Transmittance and reflectance of a 5-layer Ag induced transmission filter (ITF) tuned at 550 nm with TiO_2 as the dielectric layer.

wavelengths, or selective, covering a specific wavelength band. An excellent summary of high reflector coatings can be found in *Optical Coating Technology* by Phil Baumeister [9].

There are six basic types of reflector coatings:

- First surface mirror
- Second surface mirror
- Overcoated metal mirror
- Enhanced metal high reflector
- Multilayer all-dielectric reflector (high reflector, low and high pass filter, dichroic filter, beam splitter, DWDM filter, etc.)
- Photonic band gap structure

As described previously, high reflectance can obviously be accomplished by adding a reflective metal layer. However, metal layers cannot be used if high reflectance is required in specific wavelength bands or for a high or low pass filter.

The majority of high reflectors are metal-based mirrors. A first surface mirror is a metal film deposited on an optically smooth (usually glass) substrate and has no protective overcoat. A second surface mirror is formed by depositing a metal film on the back of a transparent substrate, usually glass. Light is incident from the substrate side, and the metal film is protected by the glass plate. The metal film is typically protected on the back by paint. The reflectance of the metal layer can be degraded by as much as 4%–5% by Bragg reflections off the glass plate. The glass plate also absorbs a small amount of the incident light at visible wavelengths and much larger amounts at infrared wavelengths, degrading the reflectance. Instead of a metal or glass plate, the metal was protected from behind by electroless or electroplated Ni thin films [10]. The reflective metal was attacked through pinholes and by chemical degradation of the protective film.

The reflectance of a metal film can be improved over a limited spectral range by applying a multilayer H/L coating. Figure 5.16 shows an enhanced Ag design being considered for the Giant Segmented Mirror Telescope and Figure 5.17 shows an Al layer with a Ta_2O_5/SiO_2 H/L enhancement overcoat [11]. The coating design is $S/Ag/(LH)^2$, where H and L are quarter wave optical thicknesses of Ta_2O_5 and SiO_2 at the tuned wavelength of 510 nm in this case The reflectance is greater than that of bare Al between 425 and 700 nm, after which, it falls below Al reflectance. If we add more HL pairs, the high reflectance region (or stopband) will square up, as will be discussed in more detail for all dielectric high reflectors. The reflectance of even poor-reflecting metals such as Cr can be significantly improved by enhancement designs [12]. Figure 5.18 compares the reflectance of a chromium film enhanced with 700 Å of SiO_2/400 Å TiO_2 dielectric layers with a silver second surface mirror and first surface chromium mirror.

Consider the generic multilayer all-dielectric high reflector (HR) which has the basic design, $H(LH)^n$, and consists of alternating H/L layers with quarter wavelength thickness at a tuned wavelength λ_o. The reflectance spectrum has the following salient features:

- Only odd order reflectance peaks exist for purely quarter wave thicknesses
- Adding more HL pairs increases the reflectance and "squares" or sharpens the high reflectance region
- The width of the high reflectance regions is determined by the ratio n_H/n_L

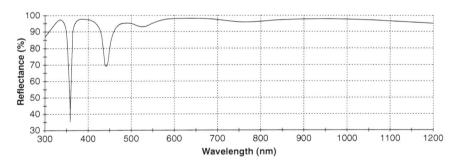

Figure 5.16 Design reflectance of giant segmented mirror telescope.

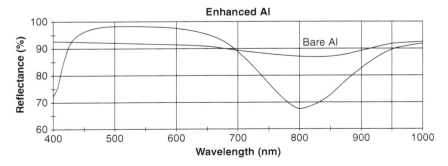

Figure 5.17 Reflectance of a dielectric enhanced Al coating tuned at 510 nm.

Figure 5.19 shows a reflectance spectrum of a quarter wave dielectric HR. All layers are a QWOT at the tuned wavelength. This design provides the optimum reflectance for the minimum number of layers. The first order peak is the high reflectance peak at the tuned wavelength λ_o. The third order peak is located at $\lambda_o/3$, the

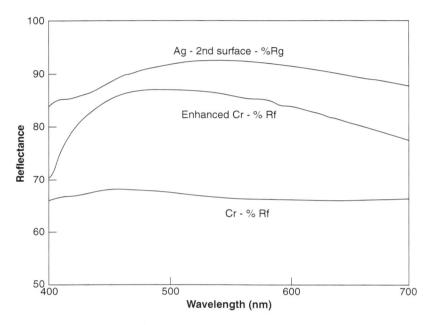

Figure 5.18 Comparison of reflectance of enhanced Cr mirror with first surface Cr and second surface Ag [12].

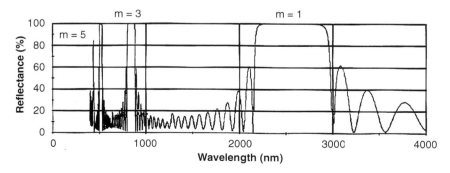

Figure 5.19 Reflectance spectrum of an H/L dielectric quarte rwave HR, showing higher order reflectance peaks.

fifth order peak is located at $\lambda_o/5$, and so on at $\lambda_o/(n+1)$. Here n = 0 and an even integer. Side bands are located in the high transmission regions between the high reflectance peaks. The amplitude of the interference pattern of the side bands can be minimized by altering the layer thicknesses or by employing a rugate design (address later). Note that the width of the stop bands get narrower at shorter wavelengths. Actually, the widths of the higher order stop bands are all the same if plotted against frequency ($\lambda = c/f$, where c is the speed of light and f is frequency), as shown in Figure 5.20.

The reflectance at the tuned wavelength can be expressed as:

$$R_o = \left\{ \frac{1 - \left(\frac{n_H}{n_L}\right)^{2n}\left(\frac{n_H^2}{n_s}\right)}{1 + \left(\frac{n_H}{n_L}\right)^{2n}\left(\frac{n_H^2}{n_s}\right)} \right\}^2 \quad [6],$$

(5.24)

where n_s is the index of the substrate.

The half width Δg depends on the ratio n_H/n_L by the relation:

$$\Delta g = \left(\frac{2}{\pi}\right)\sin - 1\left[\frac{n_H - n_L}{n_H + n_L)}\right].$$

(5.25)

Figures 5.21 and 5.22 summarize these features. Figure 5.21 shows how the reflectance increases with the increase in number of H/L pairs. It takes significantly fewer pairs to obtain high

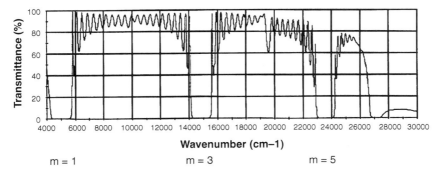

Figure 5.20 Reflectance spectrum of a H/L dielectric quarter wave HR as a function of frequency (i.e. wave number = 1/wavelength).

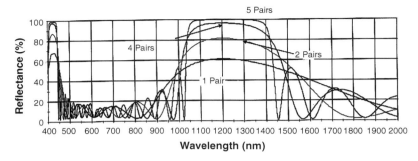

Figure 5.21 Increase in reflectance with number of n_H/n_L pairs.

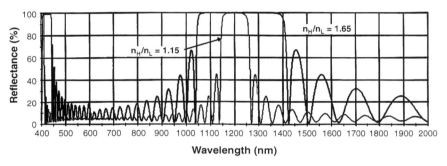

Figure 5.22 Reflectance for a QWOT HR with $n_H/n_L = 1.65$, compared to $n_H/n_L = 1.15$.

reflectance as n_H/n_L increases. For example, it take 5 HL pairs to achieve a reflectance of 0.99 for n_H/n_L (TiO_2/SiO_2) = 1.65, while 18 pairs are needed for n_H/n_L (Al_2O_3/SiO_2) = 1.15. Figure 5.22 shows the increase in bandwidth with increased n_H/n_L. Thus we need a large n_H/n_L to obtain broadband reflectors, while this ratio must be decreased for narrowband or notch filters. High reflectance bands can be located adjacent to each other to extend the high reflectance region. This, however, may not always be possible if the fundamental high reflectance bands overlap higher order bands.

The quarter wave dielectric stack is just the starting point for a number of functional multilayer dielectric coatings. Beam splitters, dichroics, highpass, and lowpass filters are all related to this basic design. The dichroic filter, shown in Figure 5.23, reflects a small band of wavelengths (or colors) and transmits all others. Modern fiber optics networks use a multitude of wavelengths that are closely spaced together. In order to pass a large number of wavelengths in a single fiber, filters are used to segregate the wavelengths and each filter must have pass bands on order of angstroms. Figure 5.24 shows the transmittance of a 146 layer Ta_2O_5/SiO_2 wavelength division wavelength multiplexing (WDWM) filter. The pass band of this filter is ~ 1 nm.

Many optical elements must either pass wavelengths below or above a certain wavelength. Highpass and lowpass filters, shown in Figure 5.25, are designed to transmit above and below a specific wavelength. The highpass filter blocks all wavelengths between 500 and 700 nm while transmitting wavelengths above 700 nm,

Figure 5.23 Picture of dichroic filters.

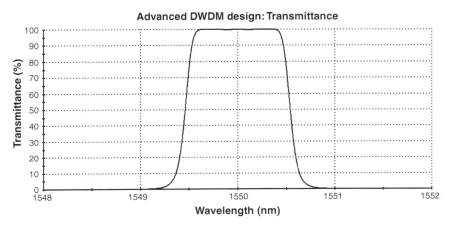

Figure 5.24 Transmission of a 146 layer Ta_2O_5/SiO_2 WDWM filter.

while the lowpass filter blocks wavelengths above 700 nm for both s and p polarizations and passes lower wavelengths. Both filters are composed of TiO_2/SiO_2 layers.

In many cases, a very narrow band of wavelengths must be reflected, for example, a laser reflector, while preserving good transmittance over all other wavelengths. The rugate filter optimizes the

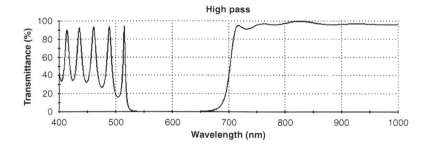

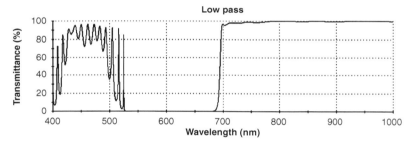

Figure 5.25 Transmission of high and lowpass filters.

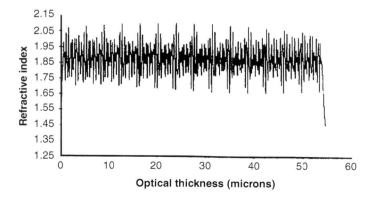

Figure 5.26 Refractive index variation of a rugate filter [13].

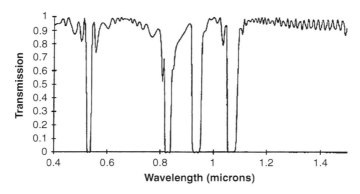

Figure 5.27 Transmittance spectrum of a rugate filter.

transmittance between reflectance bands and can have stopbands of various orders. Unlike the coatings described previously that had discrete layer thicknesses and interfaces, the refractive index of the rugate filter varies sinusoidally or with combinations of sine functions. The resulting coating has very narrow reflecting regions and broad transmitting regions. Figure 5.26 shows the refractive index variation of a rugate filter, and Figure 5.27 shows the transmittance spectrum of a rugate filter.

5.4   Color and Chromaticity in Thin Films

5.4.1   Color in Thin Films and Solid Surfaces

A wide variety of thin film and engineered decorative materials are concerned with color. The appearance of architectural glass, and a wide variety of coated objects depends on the "color" or "lack of color" of the thin film coating.

One of the primary applications of thin films is to change the color or reduce the color of a surface, window, optic, or object. A wide variety of thin film and engineered decorative materials are concerned with color. The appearance of architectural glass, and a wide variety of coated objects depends the "color" or "lack of color" of the thin film coating. Color is defined as the visual perceptual property corresponding in humans to the categories called red, green, blue, and others. We all "see" the color of an object differently. Color results from the visible light spectrum interacting in the eye with the spectral sensitivities of the light receptors located in the retina. Color categories and physical specifications of color associated with objects, materials, light sources, etc., are based on optical properties such as visible light absorption, transmission, reflection, or emission spectra. Several attempts have been made to quantify color by identifying numerically their coordinates in a color space model [14–21].

Because perception of color results from the varying sensitivity of different types of cone cells in the retina to different parts of the spectrum, colors are defined and quantified by the degree to which they stimulate these cells (color blindness occurs when some receptors are less sensitive to certain colors). We rarely visualize a pure color and the color of most objects is a mixture of primary colors and hues. So how are colors and hues mixed? Can we model color and color perception? Chromatics is the science of color, and includes the perception of color by the human eye and brain, the origin of color in materials, color theory in art, and the physics of visible electromagnetic radiation.

As mentioned above, a number of models have been presented to quantify color perception. Figure 5.28 summarizes color vision characteristics and associated models [22]. The figure diagrams color perception factors and how they are mapped into

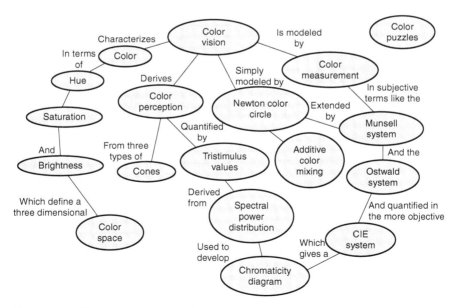

Figure 5.28 Diagram mapping color perception of human eye [18].

the chromaticity model. The entire picture ends up in the bottom chromaticity diagram and includes:

- Vision characteristics
- Color perception
- Color mixing
- Color measurement
- Spectral power distribution
- Chromaticity model

We start with tristimulus (three-color) values. Any color can be produced by the appropriate combination of primary colors:

$$C = BB + GG + RR \text{ (bold = unit value)}$$

Here **B**, **G** and **R** are unit values for blue, green and red and B, G and R are relative intensities of these primary colors. In this case, white is characterized by "1" unit:

$$W = 1B + 1G + 1R \text{ (in appropriate intensity units)}$$

Properties of color that are indistinguishable to the human eye are hue, saturation, and brightness. While spectral colors can be one-to-one correlated with light wavelength, the perception of light with multiple wavelengths is complex. Many different combinations of light wavelengths can produce the same perception of color. The most widely used model used to describe color perception is the C.E.I Chromaticity diagram (Commission Internationale de l'Eclairage) [16, 17]. A diagram mapping the color perception of the human eye is shown in Figure 5.28 and a typical C.I.E chromaticity diagram is shown in Figure 5.29. Applying this diagram involves inserting chromaticity matching functions (Figure 5.30) [23, 24]. For example, white can also be achieved with many different mixtures of light, e.g., with complementary colors. If you have two illuminating sources which appear to be equally white, they could be obtained by adding two distinctly different combinations of colors. This implies that if you used them to illuminate a colored object which selectively absorbs certain colors, that object might look very different when viewed with the two different "white" lights. For instance, an object will have a different color outdoors in direct sunlight than under fluorescent or incandescent illumination.

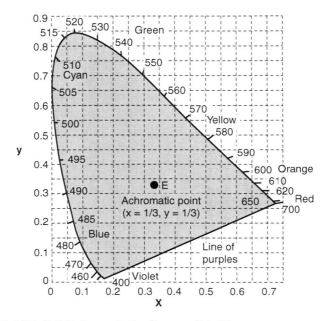

Figure 5.29 1931 C. I. E chromaticity diagram [16, 17].

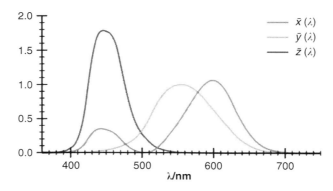

Figure 5.30 CIE chromaticity matching functions [23, 24].

Before we go into these diagrams in any detail, it will be instructive to review the assumptions and basic concepts of this model (note that this is not the only model to address chromaticity). Color is expressed in "x" and "y" coordinates. The rainbow spectrum of pure spectral colors falls along the outside curve of the chromaticity diagram. They can be described as fully saturated colors. This includes all perceived hues. The "line of purples" across the bottom represents colors that cannot be produced by any single wavelength of light. A point along the line of purples could be considered to represent a fully saturated color, but it requires more than one wavelength of light to produce it.

Color matching functions are used to quantify the chromatic response of the observer. Referring to Figure 5.30, the three CIE color matching functions are labeled $x(\lambda)$, $y(\lambda)$ and $z(\lambda)$. These functions can be correlated to the output of linear spectral detectors, which then match the output with the CIE tristimulus values X, Y, and Z [23, 24, 25]. These are not actual colors, but convenient mathematical constructs, and represent perceivable hue and various combinations of wavelengths that yield the same set of tristimulus values as described above:

$$\text{Color} = C = X\mathbf{X} + Y\mathbf{Y} + Z\mathbf{Z}.$$

The tristimulus values for a color with a spectral power distribution $I(\lambda)$ are given in terms of the standard observer by:

$$X = \int I(\lambda)x(\lambda)d\lambda \qquad (5.26)$$

$$Y = \int I(\lambda)y(\lambda)d\lambda \qquad (5.27)$$

$$Z = \int I(\lambda)z(\lambda)d\lambda \qquad (5.28)$$

Integrating from 0 to ∞ (for humans, the visible spectrum is generally adequate). We can now define the CIE x, y, and z coordinates (see Figure 5.29) as

$$X = \frac{X}{(X+Y+Z)} \qquad (5.29)$$

$$Y = \frac{Y}{(X+Y+Z)} \qquad (5.30)$$

$$Z = \frac{Z}{(X+Y+Z)} = 1 - x - y. \qquad (5.31)$$

Figure 5.31 shows more detail in the CIE diagram. Thus, each perceived color is defined by x and y values.

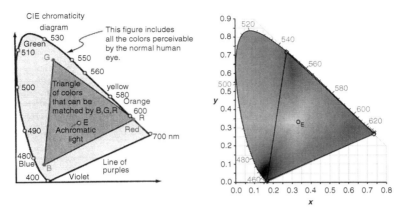

Figure 5.31 Detail in CIE chromaticity diagram.

CIE XYZ chromaticity diagrams have the following salient features:

- All chromaticities visible to the average person are represented, and are shown in color. This region is known as the gamut of human vision and, as shown in Figures 5.29 and 5.31, is tongue, or horseshoe shaped. The edge of the gamut is called the spectral locus and corresponds to monochromatic light. The straight edge on the bottom is known as the line of purples, which have no counterpart in monochromatic light. Less saturated colors are located in the interior near the white center region.
- All visible chromaticity values are > 0.
- All colors that lie in a straight line between two points on the diagram can be obtained by mixing the two colors. All colors formed by mixing three points are located inside the triangle defined by these colors.
- The distance between two colors does not generally correspond to the degree of difference between the colors.
- Three real sources cannot completely cover the gamut of human vision, i.e., a single triangle cannot include the entire gamut.
- Light with a flat power spectrum corresponds to $(x,y) = 1/3,1/3$.

Now, referring to Figure 5.31, and keeping the above factors in mind:

- The combination of wavelengths needed to produce a given perceived color is not unique. Lines between pairs CD, FG, and JH all intersect at the same point and thus, can each produce the color T (assuming the correct proportions).
- The outline encompasses all perceivable hues.
- E is the achromatic point; AB or any pair that connects through E forms a complementary color pair.
- If there are two illuminating sources that appear equally white, they can be obtinaed by adding two distinctly different color combinations.
- Colors along the outline are fully saturated.
- Colors along the line of purples are fully saturated but are not monochromatic.

One variation of the CIE model is the CIE RGB (red-green-blue) model [26, 27]. An RGB color space is any additive color space based on the RGB color model. A particular RGB color space is defined by the three chromaticities of the red, green, and blue additive primaries, and can produce any chromaticity that is the triangle defined by those primary colors (see above). The complete specification of an RGB color space also requires a white point chromaticity and a gamma correction curve (see Figure 5.31). CIE RGB matching functions ($r(\lambda)$, $g(\lambda)$, and $b(\lambda)$) are shown in Figure 5.32, and Figure 5.33 shows the gamut of RGB triangle. The area under each of the RGB color matching functions is normalized such that

$$\int r(\lambda)d\lambda = \int g(\lambda)d\lambda = \int b(\lambda)d\lambda \quad \text{(from } 0 - \infty) \qquad (5.32)$$

Thus, each primary gets equal weight and the resulting RGB tristimulus values are obtained by integrating

$$R = \int r(\lambda)I(\lambda)d\lambda \qquad (5.33)$$

$$G = \int g(\lambda)I(\lambda)d\lambda \qquad (5.34)$$

$$B = \int b(\lambda)I(\lambda)d\lambda \qquad (5.35)$$

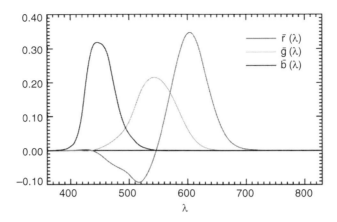

Figure 5.32 CIE RGB color matching functions [28].

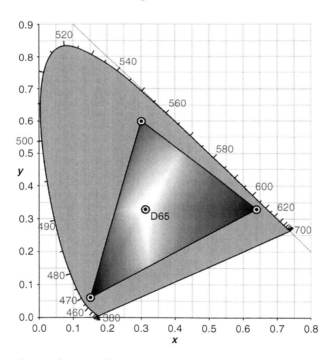

Figure 5.33 Gamut for the RGB triangle.

We finish with a discussion of the Lab Color Space system (CIELAB), which is a color opponent space with dimension L for lightness and a and b for the color-opponent dimensions, based on nonlinearly compressed CIE XYZ color space coordinates (see above). The coordinates of the Hunter 1948 L, a, b color space are L, a, and b [29]. However, the more widely used Lab is now used as a short cut abbreviation for the CIE 1976 (L*, a*, b*) color space (also called CIELAB, whose coordinates are actually L*, a*, and b*). Both spaces are offsprings of the CIE 1931 XYZ color space discussed above. Our focus will be on the CIELAB system.

The three coordinates of CIELAB represent:

- Lightness of the color (L^* = 0 black and L^* = 100 indicates diffuse white; specular white may be higher). This is the vertical axis.
- Position between red/magenta and green (a^*, negative values indicate green while positive values indicate magenta)
- Position between yellow and blue (b^*, negative values indicate blue and positive values indicate yellow).

The asterisk (*) after L, a, and b are part of the full name, since they represent L*, a* and b*, to distinguish them from Hunter's L, a and b [14, 15].

This is a three-dimensional model. Comparing to x, y and z coordinates where x and y form a plane, a* and b* form a chromaticity plane, similar to the CIE gamut. Brightness is fixed in each a*-b* plane. L is the z-axis and represents lightness. Figure 5.34 depicts these planes for 3 L values. One advantage of this model is that many of the "colors" within L*a*b* space fall outside the gamut of human vision (see above), and are therefore purely imaginary; these "colors" cannot be reproduced in the physical world but can be useful for computer graphics. CIELAB coordinates can be converted to CIE space using specific transformations [29].

We have briefly addressed only three of the many chromaticity models, albeit the most commonly used ones

5.4.2   Color in Thin Films: Reflectance

The color of a thin film, whether in reflection or transmission, is critical in many applications, including low-e windows, antireflection coatings, hardware, plumbing fixtures, high reflector coatings, jewelry, automotive parts (including paints), and architectural glass. The "color" of a thin film results from its optical properties: transmittance, reflectance, and absorption. The optical properties

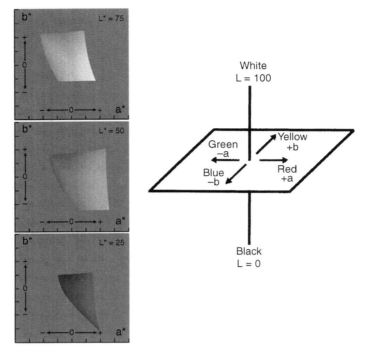

Figure 5.34 CIELAB color space.

of the substrate are often integral to the color of the thin film. When we look through a window we see both transmitted and reflected light. However, reflected and absorbed light are most important for coatings on structural components (plumbing fixtures, hardware, jewelry, etc.). Color can also be a function of angle of incidence (AOI) of the viewer.

Consider first the metallic reflector, which is essentially a broad band mirror. Recall that Figure 5.8 shows the reflectance of several metal coatings at 0° AOI. The color of each material depends on the wavelength range of high reflectance. Al and Ag have high reflectivity over the visible wavelength range and, essentially, have no color and appear "silvery". The reflectance of Au and Cu films only becomes high above 500 nm wavelength. Referring to the CIE 1931 Chromaticity diagram shown in Figure 5.35, we expect Au and Cu to have a considerable amount of yellow – red color. Gold is difficult to produce from CIE chromaticity and RGB chromaticity diagrams, but has CIE coordinates (x,y) ~ (0.55,0.46) [30]. We see that

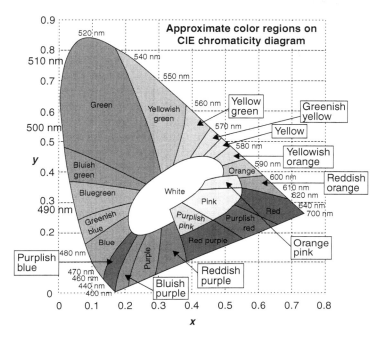

Figure 5.35 Color regions of CIE chromaticity diagram.

Cu reflectance contains a bit more blue - green. Rh appears more gray due to absorption at lower visible wavelengths.

Figures 5.8 and 5.35 are only part of the story. Reflectance has a strong dependence on angle of incidence of the viewer. It is possible to shift color by variation of AOI as will be described below. The Fresnel equations (Section 5.1) define the components of the reflected and transmitted waves.

Figure 5.36 shows s and p reflectance of Al for various AOI and recall that Figure 5.5 shows reflectance for the cases $n_1 = 1$ and $n_2 = 2$ and $n_1 = 2$ and $n_2 = 1$. The separation of reflectance of the two polarizations increases with increased AOI, yet note that the average reflectance is always the reflectance at $0°$ AOI. P polarization reflectance is always less than s polarization reflectance. Since our eyes see the average reflectance, the reflectance, and therefore the color, of metals is virtually the same at all AOI. Brewster's angle is defined as the AOI at which $R_p = 0$. Total internal reflection is possible when $n_2 < n_1$.

Color shift in multilayer optical coatings is much more interesting. Figure 5.37 shows the change in reflectance of a high reflector

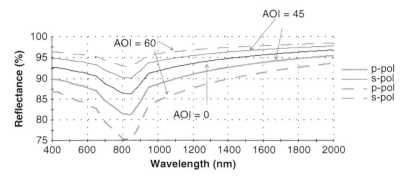

Figure 5.36 Reflectance of s and p polarizations for Al for AOI of 0°, 45° and 60°.

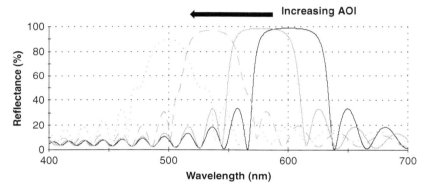

Figure 5.37 Shift in high reflectance peak of H(LH)^{20} (H = 1.63 (Al$_2$O$_3$) and L = 1.48 (SiO$_2$)) design with increased AOI.

with design H(LH)^{20} @ 510 nm with AOI. Here H = quarter wave optical thickness with n = 1.63 (Al$_2$O$_3$) and L = quarter wave optical thickness with n = 1.48 (SiO$_2$). The high reflectance peak shifts from 600 nm @ 0° AOI to 500 nm @ 60° AOI, essentially from red to blue-green. Figures 5.38a and 5.38b show the shift in CIE Chromaticity coordinates (x,y) for reflectance and transmittance (yellow X's in the diagrams). In reflectance the CIE coordinates shift from (0.58,0.41) @ 0° AOI to (0.16,0.28) @ 60°, from reddish orange to greenish blue. Transmitted chromaticity coordinates shift from (0.30,0.42) @ 0° to (0.42,0.42) @ 60°, or from blue green to yellow orange.

Color shift in a Fabry-Perot (FP) filter is the basis for color shift coatings, paints, and inks. Figure 5.39 shows the basic concept

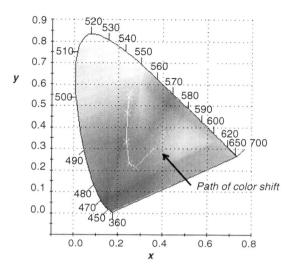

Figure 5.38a CIE chromaticity diagram for reflectance color shift of high reflector shown in Figure 5.37.

CIE 1931 chromaticity diagram

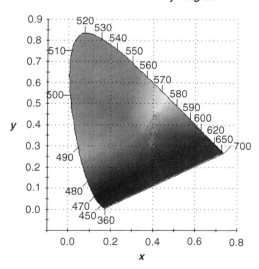

Figure 5.38b CIE chromaticity diagram for transmitted color shift of high reflector shown in Figure 5.37.

of the color shift coating. There are two basic types of FP filters, metal/dielectric (MD) and all-dielectric. We are concerned here with the MD type. The FP filter is a narrowband filter with basic design MDM, where M is a semitransparent metal layer, usually Al or Ag, and D is the dielectric cavity layer. While there are a number of more sophisticated designs, Figure 5.40 shows the reflectance of a $Ag/SiO_2/Ag$ structure and Figure 5.41 shows the chromaticity changes with AOI varying from $0°$ to $60°$ (these changes are small in comparison to what has actually been achieved). Figure 5.42 shows a picture of a car with color shift paint.

Table 5.2 shows the wide variation in colors achievable with ChromaFlair® color shift pigments, and Figure 5.43 shows the path of these shifts [31]. Chromaticity here is expressed in La*b* units (see previous section). Layer design is $Cr/MgF_2/Al/MgF_2/Cr$. The pigment is composed of micron-size flakes. Color shifts range

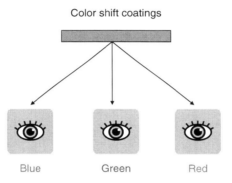

Figure 5.39 Color shift concept.

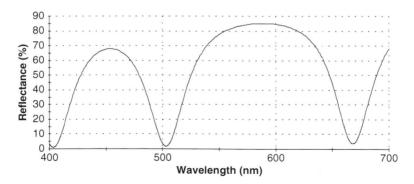

Figure 5.40 Reflectance of $Ag/SiO_2/Ag$ FP filter.

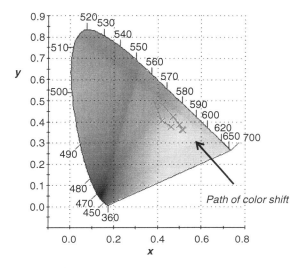

Figure 5.41 Chromaticity change with AOI of FP filter shown in Figure 5.40.

Figure 5.42 Car with color shift paint.

from gold/silver, magenta/gold, cyan/purple, to silver/green. In addition to auto paints, the many applications for these pigments are football helmets, fabrics, golf clubs, pens, skis, calculators, and electric guitars.

Table 5.2 Color properties of ChromaFlair® interference pigments [31].

Color Combination	Lightness	a*	b*	Chroma	Hue
Gold/Silver	177/219	26/−56	147/76	149/94	80/127
Red/Gold	92/173	87/71	−3.2/163	87/178	358/67
Magenta/ Gold	85/159	83/81	−35/148	90/169	337/61
Purple/ Orange	85/132	70/90	−95/77	118/118	306/41
Blue/Red	93/103	43/92	−117/−8	124/92	290/355
Cyan/Purple	116/110	−62/108	−41/−62	75/125	214/330
Green/Purple	118/122	−101/115	13/−50	102/126	173/336
Silver/Green	125/126	36/−92	8.9/22	37/95	14/166

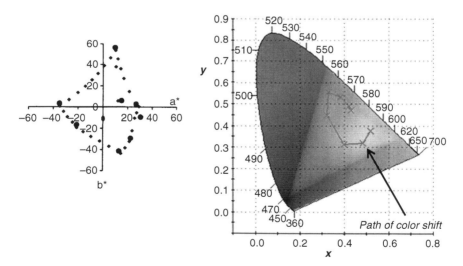

Figure 5.43 La*b* colors and CIE diagram based on blending pigments shown in Table 5.2 [31].

The last example of reflected colors is transition metal nitrides (TiN, HfN, ZrN, TaN) which are applied to a wide range of objects for decorated and protective applications. These materials can have a gold, brass, and silvery color. Applications include jewelry, hardware, cutting tools and medical implants. Figure 5.44 shows the reflectance spectrum of a TiN thin film. Figure 5.45 shows the CIE chromaticity diagram for this film compared to that of gold.

While metals and transition metal nitrides have virtually the same color with change in viewing angle, a wide range of colors and color shifts can be obtained from multilayer optical coatings.

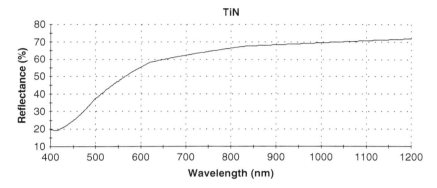

Figure 5.44 Reflectance spectrum of TiN thin film.

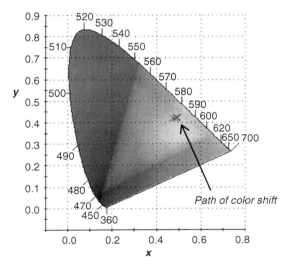

Figure 5.45 CIE chromaticity diagram for a TiN thin film.

5.4.3 Color in Thin Films: Transmission

Color is also extremely important in transparent coatings for two reasons:

- There is always some degree of reflected light, even in a transparent window
- It is desirable to have as neutral a transmitted color as possible

The window or glazing must also not reflect colors at off normal incidence [32]. Additionally, some architectural applications require colored glass. There should be no color variation within the window or optical element. As mentioned in an earlier Ed Guide, color variation is particularly problematic in large area applications. In addition to color properties, thin films must perform in harsh environments, heat, humidity and have a degree of wear resistance.

Applications include:

- Antireflection coatings for window glazings
- Low-e coatings
- Electrochromic coatings
- Automobile windshields and windows
- Solar control coatings
- Computer displays

We first examine color and color shift in antireflection (AR) coatings. AR coatings are the most widely used type of transmissive architectural coating. AR coating designs range from single layer narrow band to multilayer broad band. The simplest AR coating is a single layer and is only tuned at one wavelength. The condition for antireflection is

$$n_f = (n_S)^{\frac{1}{2}} \qquad (5.36)$$

Here n_f is the real part of the refractive index of the film and n_S is the real part of the refractive index of the substrate. The AR coating must also have a quarter wave optical thickness at the tuned wavelength. Figure 5.46 shows the reflectance of this design with $n_S = 2.5$ and $n_f = 1.58$ tuned at 550 nm. We see from the CIE chromaticity diagrams for reflectance and transmittance that reflectance

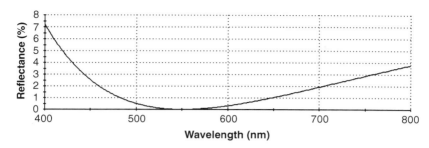

Figure 5.46 Reflectance of a single layer AR coating tuned at 550 nm with $n_s = 2.5$ and $n_f = 1.58$.

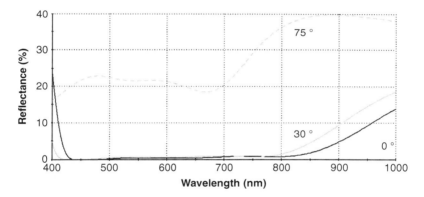

Figure 5.47 Reflectance spectrum of 9-layer AR coating at AOI of 0°, 30° and 75°.

color varies considerably from yellow-green to purple with a change in AOI from 0 to 75° while variation in transmittance color is insignificant. Thus, a coated window or optic will have significant off-normal reflected color while transmitting reddish-yellow at virtually all viewing angles.

While it is considerably more complicated and costly to deposit, the multilayer broadband AR offers significantly less off-normal color variation and wider AR performance. Designs can range from all-quarter wave to all-nonquarter wave layer optical thicknesses. Figure 5.47 shows the reflectance spectrum of a 9-layer Ta_2O_5/SiO_2 AR coating tuned at 550 nm at angles of 0°, 30°, and 75°. Figure 5.48a shows the CIE chromaticity diagram for reflectance with change in AOI from 0°–30° and Figure 5.48b shows *reflectance* change for AOI varying from 0°–75°. Figures 5.49a and 5.49b show the associated

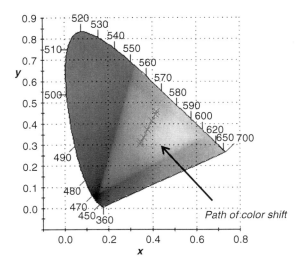

Figure 5.48a CIE chromaticity diagram for reflectance of 9-layer AR coating shown in Figure 5.47 for AOI from 0°–30°.

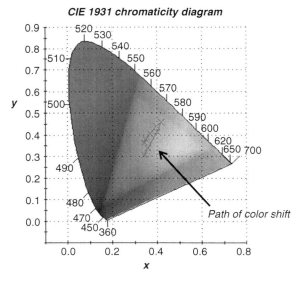

Figure 5.48b CIE chromaticity diagram for reflectance of 9-layer AR coating shown in Figure 5.47 for AOI from 0°–75°.

change in *transmittance* color. From Figure 5.50 we see that reflectance increases a small amount with AOI change from 0°–30° but the change is significant with AOI change from 0°–75°. While not

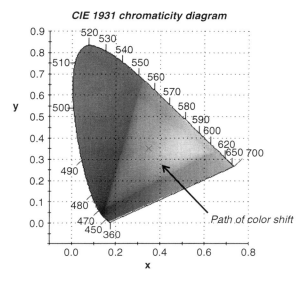

Figure 5.49a CIE chromaticity diagram for transmittance of 9-layer AR coating shown in Figure 5.47 for AOI from 0°–30°.

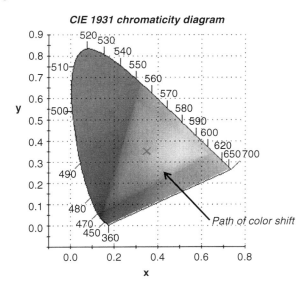

Figure 5.49b CIE chromaticity diagram for transmittance of 9-layer AR coating shown in Figure 5.47 for AOI from 0°–75°.

obvious from the reflectance spectrum, reflected color varies from essentially white to whitish-yellow. Transmitted color does not vary over both AOI ranges.

We see here then that the goal for designing a broadband AR coating for windows is to minimize color and reflectance variation over a specific range of AOI. While the change in color for the broadband AR is significant with AOI from 0°–30°, the actual change in visible reflectance is << 1%. Thus there would be no significant color change at least for viewing angles up to 30°.

The low-e window is now the most widely marketed and installed window treatment. The low-e coating is essentially an induced transmission filter, having one or more semitransparent Ag or Al layer sandwiched between transmissive dielectric layers, usually TiO_2, Ta_2O_5, SnO_2, or even a TCO such as ZnO. Figures 5.50 and 5.51

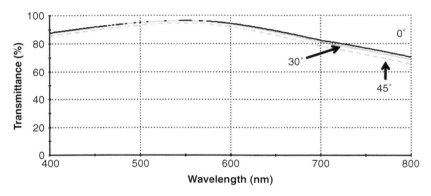

Figure 5.50 Transmission spectra of low-e coating for AOI of 0°, 30°, and 45°.

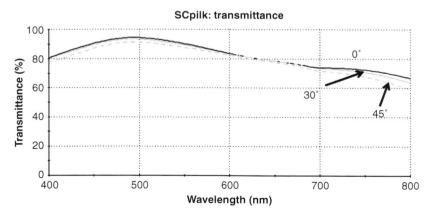

Figure 5.51 Transmission spectra of solar control coating for AOI of 0°, 30°, and 45°.

show transmission spectra of a low-e and solar control coatings marketed by Pilkington at AOI of 0°, 30° and 45°. The low-e coating has $SnO_2/ITO/Ag/ZnO/TiO_2$ layers and the solar control coating is a bit more complex with $SnO_2/ITO/Ag/ITO/Ag/ZnO/TiO_2$ layers. Figure 5.52 shows the change in color with AOI varying from

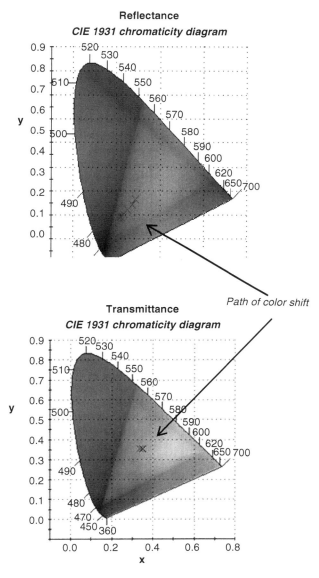

Figure 5.52 Change in color with AOI varying from 0°–45° for the low-e coating.

0°–45° for the low-e coating. We see that reflected color changes from white to blue-green with increased AOI. Transmission color did not vary significantly with this change in AOI. Again, because the reflectance does not increase significantly with increased AOI, this coating will keep essentially the same color with change in viewing angle.

Once we have the design that minimizes color variation, it is important that coating thickness be as uniform as possible to prevent color variation within the window or optic. Nothing is more troubling than seeing a "bulls eye" or color bands in the optic.

5.5   Decorative and Architectural Coatings

Decorative and architectural coatings were some of the first applications for vacuum metalizing and have evolved over the last five decades from essentially colored glass and glazings, chromium plating of automobile parts and hardware, and plating of jewelry with precious metals to sophisticated heat mirror, switchable windows, color shift coatings, and computer displays [33]. Coatings can be transmissive, reflective, and absorptive. A detailed treatment of optical properties of thin films and color can be found in the previous sections. Reflective coating applications include hardware, plumbing fixtures, high reflector coatings, jewelry, automotive parts (including paints). Decorative coatings should address the following requirements [34]:

- Reflect the appropriate specified color
- Scratch resistance
- Corrosion resistance
- Resistance against cleaning solvents
- Thermal shock resistance
- Excellent adhesion on the interlayer system.

A wide variety of materials must be coated, including glass, brass, stainless steel, copper, aluminum, zinc, steel, and plastics. Table 5.3 provides a list of some of the more widely used reflective decorative hard coatings and their colors [35–40]. Table 5.4 lists materials, substrates, and PVD processes used to coat glass, ceramics, and plastics [37]. Deposition processes for these materials are presented primarily in Chapter 4 on tribological coatings. However, in addition to hardness and wear resistance, color can depend on process parameters. Another important aspect of these coatings is that they are generally produced on an industrial scale which requires scale up of the deposition process for large area substrates and often three- dimensional structures.

Decorative coatings on plastics have become a particularly important and high volume industrial application for hard and wear resistant materials. In addition to process scale-up, adhesion of inorganic metal thin films to plastics is typically poor and exacerbated by the mismatch in thermal expansion coefficients

Table 5.3 Decorative hard coatings and colors [35–40].

Material	Colors
Gold	Gold
CrN_x	Silver → brown
TiN_x	Gold → brown- yellow → yellow
TaN_x	Blue → grey
NbN_x	
ZrN_x	Gold → green
TiAlN	Violet → Gold → black
TiC_xN_y	Bronze/gray
$TiZrN_x$	Gold
TiC_x	Silver → gray
CrC	Silver
TiC_xN_y	Red gold → violet
ZrC_xN_y	Silver → gold → violet
CrxOy	Copper → dark green
$(Ti,Al)O_xN_y$	Brown → dark blue
TiO_xN_y	Transparent → black

Table 5.4 Summary of decorative coating materials, substrates coated and deposition processes [37].

Substrate	Coating	PVD Technique
Tiles	TiN	Magnetron Sput-tering
	TiN	Cathodic arc
	Ti	Magnetron Sput-tering
	CrN	Magnetron Sput-tering
	CrCrN (sand blasted sub-strate)	Cathodic arc
	CrCrN	Cathodic arc
	AgNi(20min)	Cathodic arc
	AgNi(40min)	Cathodic arc
	Cu(15min)	Cathodic arc
	Cu(30min)	Cathodic arc
	TiN-Ti	Magnetron Sput-tering
Acrylonitrile buta-diene styrene (ABS)	Cr	Magnetron Sput-tering
Polypropylene (PP)	Cr	Magnetron Sput-tering
Polymethyl meth-acrylate (PMMA)	Cr	Magnetron Sput-tering
Polycarbonate (PC)	Cr	Magnetron Sput-tering
Polyamide (PA)	Cr	Magnetron Sput-tering
Glass	ZrN	Cathodic arc
	Zr oxide	Cathodic arc

(TCE), and must be improved by using a number of techniques [41, 42, 43]:

- Interstitial bonding or "glue" layer between thin film and substrate. Also known as a base coat.

- Layer with intermediate TCE placed between thin film and substrate.
- Stress reduction in thin film.
- Activation of the plastic surface by plasma or chemical treatment.
- Degassing and elimination of water in the plastic.
- Elimination of surface contaminants (grease, dust, adhesives, oxidation).

Additionally, the deposition process must not heat the plastic part above its softening temperature, which generally rules out CVD and related processes. PVD processes, and magnetron sputtering in particular, are thus well suited for deposition of hard decorative coatings on plastics [42]. PVD processes, reviewed in Chapter 2, have the following advantages for decorative coatings [34]:

- Wide range of substrate materials
- Widest range of color
- Scratch resistant with high hardness
- Chromium –free coatings
- Chemically and UV stable coatings
- High adhesion without delamination
- Environmentally friendly
- Low substrate temperatures

CVD and ALD processes, however, have the advantage of uniform deposition over three- dimensional structures, and plasma enhancement (PECVD) has permitted deposition at lower temperatures [43]. PVD processes have also been successfully used to coat 3-D parts [44]. Titanium (Ti), Cr, and Ni are often used as base coats to improved adhesion to all types of substrate materials.

Except for color shift, security, and decorative applications the reflective decorative coating should not noticeably change color when viewed at different angles. Transition metal nitrides (see above) are good examples of reflective coatings with virtually no color change.

References

1. Bahaa E. A. Saleh and Malvin Carl Teich, *Fundamentals of Photonics*, Wiley Interscience (1991).
2. F. Graham Smith and Terry A. King, *Optics and Photonics*, Wiley (2000).

3. Zum Gahr, Karl-Heinz, *Microstructure and Wear of Materials*, Tribology Series, 10. Elsevier (1987).
4. Robert W. Wood, *Physical Optics*, Optical Society of America (1988).
5. G. Haas, *J. Opt Soc*, 45 (1955) 945.
6. O. S. Heavens, *Optical Properties of Thin Solid Films*, Dover (1991).
7. H. Angus Macleod, *Thin-Film Optical Filters*, 3rd Ed, Institute of Physics (2001).
8. P. M. Martin, *Vacuum Technology and Coating*, December 2005, 6.
9. Phillip W. Baumeister, *Optical Coating Technology*, SPIE Press (2004). M. A. Lind, D. A. Chaudiere and T. L. Stewart, SPIE Vol. 324 Optical Coatings for Energy Efficiency and Solar Applications (1982) 78–85.
10. M. A. Lind, D. A. Chaudiere and T. L. Stewart, *SPIE* Vol. 324 Optical Coatings for Energy Efficiency and Solar Applications (1982) 78–85.
11. W. D. Bennett, P. M. Martin, A. Phillips, W. Brown, V. Wallace, J. Stilburn and J. Sebag, *49th Annual Technical Conference Proceedings of SVC* (2006) 16.
12. S. J. Nadel and T. Van Skike, *Proceedings of the 35th Annual Technical Conference of the Society of Vacuum Coaters* (1992) 365–369.
13. Rugate Technologies web site, www.rugate.com
14. Richard Sewall Hunter, *JOSA*, 38 (7) (1948) 661.
15. Richard Sewall Hunter, *JOSA*, 38 (12) (1948) 1094.
16. *CIE Commission internationale de l'Eclairage proceedings*, 1931. Cambridge University Press (1932).
17. Thomas Smith and John Guild, *Transactions of the Optical Society*, 33 (3) (1931–32). 73.
18. H. Albert Munsell, *The American Journal of Psychology*, 23 (2), University of Illinois Press (1912) 236.
19. W. Ostwald, *Die Farbenfibel*, Leipzig (1916).
20. W. Ostwald, *Der Farbatlas*, Leipzig (1917).
21. F. Birren, *The Principles of Color*, New York (1969).
22. See HyperPhysics: Color Vision.
23. Walter Stanley Stiles and Jennifer M. Birch, *Optica Acta*, **6** (1958). 1.
24. N. I. Speranskaya, *Optics and Spectroscopy*, **7** (1959). 424.
25. A. C. Harris and I. L. Weatherall, *Journal of the Royal Society of New Zealand*, 20 (3) (September 1990).
26. William David Wright,(1928). *Trans Opt Soc*, 30 (1928) 141.
27. John Guild, *Phil Trans Royal Soc London.*,Series A, A230: 149 (1931).
28. H. S. Fairman *et al.*, *Color Research and Application*, 22 (1) (1997) 11.
29. János Schanda , *Colorimetry*, Wiley-Interscience (2007)61.
30. Peter Signell, *Physnet*: MISN -0-227.
31. R. W. Phillips and M. Nafi, *Proceedings of the 42nd SVC Annual Technical Conference* (1999) 494.
32. F. Wallen *Proceedings of the 49th SVC Annual Technical Conference* (2006) 178.

33. Donald Mattox, in 50 Years of Vacuum Coating Technology and the Growth of the Society of Vacuum Coaters, Society of Vacuum Coaters (2007) 12.
34. W. Fleischer *et al.*, SVC 41st *Annual Technical Conference Proceedings* (1998) 33.
35. Rointan F. Bunshah, Ed *Handbook of Hard Coatings*, William Andrew (2001).
36. M. H. Bouix and C. P. Dumortier, SVC 43rd *Annual Technical Conference Proceedings* (2000) 52.
37. J. Esparza *et al*, SVC 52nd *Annual Technical Conference Proceedings* (2009) 592.
38. M. Podob, SVC 39th *Annual Technical Conference Proceedings* (1996) 72.
39. W. D. Munz SVC 36th *Annual Technical Conference Proceedings* (1993) 411.
40. P. Eh. Hovsepian *et al*, SVC 47th *Annual Technical Conference Proceedings* (2004) 528.
41. M. K. Shi *et al.*, SVC 42nd *Annual Technical Conference Proceedings* (1999) 307.
42. Donald Mattox, *Handbook of Physical Vapor Deposition (PVD) Processing*, Noyes (1998).
43. P. M. Martin, Ed., *Handbook of Deposition Technologies for Films and Coatings*, 3rd Ed., Elsevier (2009).
44. U. Kopacz *et al.*, SVC 36th *Annual Technical Conference Proceedings* (1993) 419.

6

Fabrication Processes for Electrical and Electro-Optical Thin Films

Elementary electrical and electro-optical (EO) properties of solid surfaces were introduced in Chapter 1. We will delve into more detail on electrical and EO properties in Chapter 8. In this chapter we address processing techniques that are used to engineer and modify the electrical and EO properties of solid surfaces. Thin films are used extensively, but not exclusively, to modify these properties. Deposition processes for thin films are addressed in Chapter 2. Processes discussed in this chapter are used to fabricate micro and nanostructures on surfaces to modify electrical, electromagnetic, and EO properties, and create microcircuitry and resonant structures. Methods used to modify the electrical and EO properties of a solid surface are:

- Etching (wet chemical, plasma, reactive ion, sputter, ion milling)
- Plasma treatment
- Ion implantation
- Thin film metallization

- Low and high k materials
- Thin film insulators
- Patterning and photolithography

It should be noted that electrical properties can also be modified by application of low dimensional structures and will be addressed in Chapter 7.

6.1 Plasma Processing: Introduction

Plasmas are by far the most common phase of matter in the universe, both by mass and by volume [1]. For example, all the stars are composed of plasma, and even the space between the stars is filled with a plasma, albeit a very sparse one. Virtually all supernova remnants, auroras and nebulae are plasma. For our purpose, however, plasmas are used in a wide variety of vacuum and atmospheric deposition and treatment processes. Plasmas can provide an energy selective reactivity to a surface without additional heating of the surface and essentially add another dimension to thin film deposition processes. Plasma processing is a plasma-based material processing technology that aims at modifying the chemical and physical properties of a surface. While plasma can take many forms, we are interested in the role they play in deposition and vacuum processing technologies. Processing can be performed in vacuum and atmospheric systems. Examples include:

- Plasma activation
- Plasma modification
- Plasma functionalization
- Plasma polymerization
- Plasma surface interactions
- Plasma electrolytic oxidation

Before we address specific plasma processing applications, it will be instructive to review plasma basics and how plasmas interact with surfaces. Plasmas are often called the fourth state of matter, as is broadly defined as a quasineutral gas that exhibits a collective behavior in the presence of applied electromagnetic fields [2]. Table 6.1 compares the properties of gases and plasmas. Plasmas

Table 6.1 Comparison of gas and plasma properties.

Property	Gas	Plasma
Electrical conductivity	Very low	Usually very high
Independently acting species	One, all the same	Two – three (electrons, ions, and neutrals)
Velocity distribution	Maxwellian	Often non-Maxwellian
Interactions	Binary: two particle collisions	Collective: waves or organized motion

are weakly ionized gases consisting of a collection of electrons, ions, and neutral atomic and molecular species [2]. Although it will not be discussed here, plasmas are often used to describe electrical and optical properties of solids. Figure 6.1 summarizes the range of plasmas and shows the various applications with respect to number of charged particles and plasma "temperature" [3]. As we see, plasmas are naturally occurring and can also be man-made. Ion densities of man-made high pressure plasmas can be as low as $10^7 cm^{-3}$ and as high as $10^{20} cm^{-3}$.

The basic plasma, or discharge, is created by a high DC or AC voltage applied between two metal electrodes located in a low pressure gas. The plasma consists of energetic ions with ionization potential V_i and critical breakdown potential

$$V_B = \frac{APd}{\{\ln(Pd) + B\}} \tag{6.1}$$

also known as Paschen's Law.

Here P is the pressure of the gas, d is the interelectrode spacing, and A and B are constants.

Figure 6.2 plots V_B against Pd for a variety of gases [4]. There are too few electron-ion collisions to sustain a discharge at low pressures. Conversely, there are too many electron-ion collisions at high pressures and electrons to not gain sufficient energy between

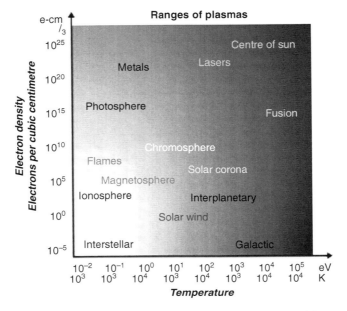

Figure 6.1 Range of naturally occurring and man-made plasmas [3].

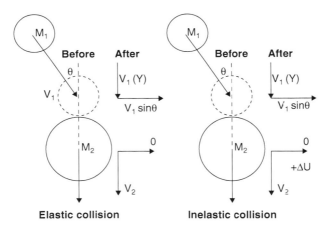

Figure 6.2 Models of elastic and inelastic collisions between moving and stationary particles [2].

collisions to ionize gas atoms. The "sweet spot" appears to be in the range of several hundred eV to achieve a sustainable discharge. The steps needed for a sustainable discharge and useable plasma using a high impedance power supply can be summarized as:

1. Small plasma current due to low number of charge carriers generated
2. Rapid increase of current but constant voltage limited by power supply impedance
3. Discharge becomes self-sustaining
4. Voltage drops and current rises significantly (normal glow)
5. Increased power supply voltage results in higher current and voltage
6. At this point sputtering and ion etching can occur

Plasma processing is concerned with utilizing the energy of the ions in the plasma to perform a variety of operations, which means understanding how plasmas interact with deposition processes and the nucleation of the thin film. Thus knowledge and control of plasma parameters for each specific deposition process and plasma process is essential. Ion sources are addressed in Section 2.1.2. The interior of a plasma consists of a partially ionized gas composed of densities of electrons (n_e), ions (n_i), and neutral species (n_0) [2]. The plasma has the following salient features:

- The velocity of electrons, while independent of other species, is significantly higher than that of ions.
- As stated above, averaged over all particles, the plasma is electrically neutral (all charges balance) and we thus must have $n_e = n_i = n_0$.
- While neutral gas atoms execute Brownian motion, an applied electromagnetic field can disrupt this motion
- Significant Coulomb interactions can occur between the charged particles, which determines many properties of the plasma

The degree of ionization is defined as

$$f_i = \frac{n_e}{(n_e + n_0)}$$

(6.2)

A typical value for f_i is $\sim 10^{-4}$ in a glow discharge. Refer to Figure 6.1 again for the range of densities.

Electron energies fall in the 1–10 eV range. The effective temperature of the plasma (see Figure 6.1) is given by

$$T_e = \frac{E}{k_B},$$

(6.3)

where k_B is the Boltzmann constant.

Thus the temperature of 2 eV electrons is 23,000 K! The neutral gas atoms have a much lower temperature ~ 293 K while ions have a temperature ~ 500 K. Various vibrational and translational modes of neutral gas atoms can become excited in the plasma. T_e for N_2 at 2 mT is $\sim 12,000$ K.

The trajectory of the electrons and ions depends on the type (s) of applied electromagnetic fields. The trajectory is linear for an applied electric field, helical for an applied magnetic field, helical for applied electric and magnetic fields, and cycloidal for crossed electric and magnetic fields. These types of trajectories create collisions between gas species, without which plasma reactions cannot be sustained. Both physical and chemical reactions can occur, depending on the nature of the gas species. Figure 6.2 shows models of elastic and elastic collisions [2]. Elementary physics tells us that the energy of colliding species is preserved in elastic collisions while it is not preserved in inelastic collisions. In an elastic collision, only kinetic energy is exchanged and momentum and translational kinetic energy are conserved. Regarding an inelastic collision, some of the kinetic energy of the incident particle is converted to potential energy changing the electronic structure of the particle (i.e., ionization). While total energy is conserved, some kinetic energy is lost after the collision. Atomic and molecular excitation can only occur in an inelastic collision.

In addition to energy considerations, the collision has to actually occur. The probability of collision is determined by the collision cross section:

$$n\sigma_C = \frac{1}{\lambda_{mfp}}$$

(6.4)

where λ_{mfp} is the mean free path of the colliding gas atoms and n is the number of collisions. Figure 6.3 shows the ionization cross

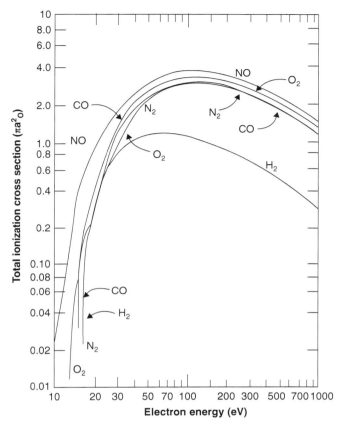

Figure 6.3 Total ionization cross section for various gases as a function of energy [5].

section for various gases as a function of energy [5]. Consider the ionization of a noble gas atom as a result of a collision with an electron. From Figure 6.3, we see that maximum ionization will occur at an electron energy ~ 100 eV. The minimum energy required to eject an electron is between 15 and 20 eV, known as the threshold energy E_{th}.

Collisions also lead to internal vibration and rotational excitations of gas molecules, which have different cross sections than σ_c [6, 7]. The total ionization cross section is the sum a number of individual cross sections

$$\sigma_T = \sigma_c + \sigma_V + \sigma_r + \sigma_a + \sigma_d + \ ......$$
(6.5)

where V = vibrational, r = rotational, a = attachment, and d = dissociation reactions.

In addition to transferring energy to the surface of a substrate by inert gas plasmas (includes sputtering, ion assist, etc.), chemical reactions can take place in multicomponent plasmas, which can be used for etching and a variety of chemical vapor deposition processes. In addition to electron collisions, ion-neutral, metastable-ion-excited atom and excited atom-neutral collisions can also occur. Table 6.2 shows the types of chemical reactions that occur in plasmas [2]. Electron collisions can totally change the chemical nature of the plasma; when an inert Ar gas atom loses an electron and becomes ionized, it now resembles Cl electronically and chemically, which changes with nature of the plasma. As a result of the very high electron temperature of the plasma, nonequilibrium and nonthermal processes can now take place. Plasma processes also include modified gas-solid heterogeneous reactions, which can modify surfaces and result in metastable and stable products.

There are two general categories of plasma chemical reactions [8]:

- Electron-atom (or molecule)
- Molecule-molecule

Consider the reaction rate for the bimolecular reaction A + B → P (products) [8]

$$\frac{dn_P}{dt} = k_{AB}(T)n_A n_B,\qquad(6.6)$$

where n denotes the concentration and $k_{AB}(T)$ is the rate constant, which is thermally activated:

$$k_{AB}(T) = k_o\ \exp-\left(\frac{E}{k_B T}\right).\qquad(6.7)$$

Here k_o and E are the initial reaction rate and E is the activation energy. Electron collision reactions have the general form

$$e^- + A \rightarrow P\ (products)\ and$$

the reaction rate strongly depends on electron energy. These reactions can be used for plasma etching (reactive plasma etching – REI) [9], addressed later in this chapter.

Table 6.2 Chemical Reactions in Plasmas [2].

Electron Collisions		
Type	**Generic Reaction**	**Example Reaction**
Ionization	$e^- + A \rightarrow A^+ + 2e^-$	$e^- + O \rightarrow O^+ + 2e^-$
	$e^- + A_2 \rightarrow A_2^+ + 2e^-$	$e^- + O_2 \rightarrow O_2^+ + 2e^-$
Recombination	$e^- + A^+ \rightarrow A$	$e^- + O^+ \rightarrow O$
Attachment	$e^- + A \rightarrow A^-$	$e^- + F \rightarrow F^-$
	$e^- + AB \rightarrow AB^-$	$e^- + SF_6 \rightarrow F_6^-$
Excitation	$e^- + A_2 \rightarrow A_2^* + e^-$	$e^- + O_2 \rightarrow O_2^* + e^-$
	$e^- + AB \rightarrow (AB)^* + e^-$	
Dissociation	$e^- + AB \rightarrow A^* + B^* + e^-$	$e^- + CF_4 \rightarrow CF_3^* + F^* + e^-$
Dissociative attachment	$e^- + A_2 \rightarrow A^+ + A^- + e^-$	$e^- + CF_4 \rightarrow CF_3^* + F^* + e^-$
Dissociative ionization	$e^- + AB \rightarrow A^+ + B^+ + 2e^-$	$e^- + N_2 \rightarrow N^+ + N + e^-$

Atom-Ion-Molecule Collisions	
Type	**Generic Reaction**
Symmetrical charge transfer	$A + A^+ \rightarrow A^+ + A$
Asymmetric charge transfer	$A + B^+ \rightarrow A^+ + B$
Metastable-neutral	$A^* + B \rightarrow B^+ + A + e^-$
Metastable-metastable ionization	$A^* + B^* \rightarrow B + A^+ + e^-$

We close with a brief description of plasma bombardment effects and ion-surface interactions. Plasma bombardment is used in a wide variety of deposition, cleaning, and etching processes. The types of plasma interactions that affect thin film deposition processes are shown in Figure 6.4 [10]. The reactions can be characterized as:

- Ions that bombard a surface.
- Incident ions can be reflected from, absorbed by, scattered from or attached (sticking) to a surface.
- Ion beam energy defines the type of reaction at the surface (see second bullet)
- Ion energies range from a few to several hundred eV; the lower range is used in surface treatment and etching during deposition while higher energies are used for ion implantation.
- Ion bombardment can either eject material from a surface (i.e, sputtering) or be used to modify the surface during thin film deposition
- Ion beams can be highly directional with a narrow band of energies.
- Ion bombardment of a surface can produce an assortment of charged particles, neutrals, and photons with a wide range of energies and abundances from the surface.

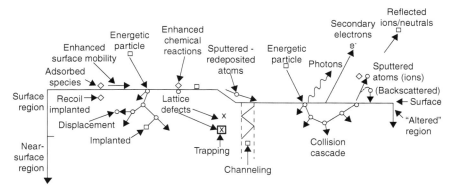

Figure 6.4 Types of plasma-film and surface interactions [10].

6.2   Etching Processes

In Chapter 2 we described deposition processes for thin films and low dimensional structures. Etching processes are essentially the inverse of deposition processes and are designed to remove material from a surface. Removal can be selective or nonselective. The goal of selective etching processes is to *exactly* transfer an image from a mask or projector to a thin film or substrate surface. Etching is used to define microelectronic devices, microciruitry, patterned thin films, metamaterials, frequency selective surfaces, MEMS devices, micro-resonant structures, and much more. Both plasma and chemical (wet) etch processes are used to define circuitry and structures in thin films and surfaces. Plasma etch processes can be categorized into five areas (see Figure 6.5) [2, 11, 12]:

- Sputter etching
- Chemical etching
- Anisotropic etching (accelerated ion assisted etching)
- Etching via an inhibitor ion-enhanced chemistry
- Reactive ion etching

Etch parameters are defined as [2]:

- Etch rate
- Etch rate uniformity
- Etch profile (isotropic etching and undercutting)
- Selectivity
- Etch bias

Etch rate is simply the rate of material removal during the etch process and equals thickness etched/etch time. High etch rates are generally desirable to allow higher production throughputs. Etch rate uniformity is a measure of the ability of the etch process to etch all parts of the surface at the same rate. Over and under etching can result from poor etch rate uniformity. It is important for an etch process to create a precise vertical edge profile, with no undercutting or rounding off of the etched pattern. An isotropic etch process removes material uniformly in all directions while an anisotropic etch process removes material exclusively in one direction (hopefully parallel to the edges of walls). Virtually all etch processes undercut pattern walls to some degree. Selectivity refers to the

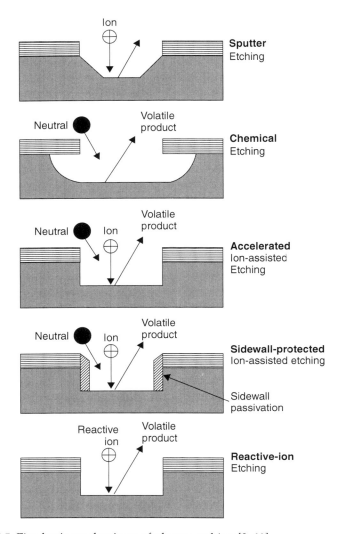

Figure 6.5 Five basic mechanisms of plasma etching [3, 11].

ability of the process only to etch the material and not the photoresist, mask or substrate. Etch bias is a measure of the change in line width or critical dimensions resulting from the etch process.

Sputter etching is probably the most purely physical etch process described here. Ion milling and ion etching are processes by which atoms of a surface material are removed by sputtering. The effectiveness of the etch process depends on a number of factors; ion energy and chamber pressure are the most important.

The type of gas used is also important. For an etch to be effective, it must be *selective* and *anisotropic*. Due to the controllable direction of ions, sputter etching and most ion etching processes are generally anisotropic (compared to wet etching, which is usually isotropic). As will be addressed later in this chapter, the above etch processes can be used with masks and photoresists, or can be used to clean the surface of a substrate.

Figure 6.6 compares wet (chemical) and plasma etching [13], and Figure 6.7 compares pattern transfer of isotropic and anisotropic etches. Here we see that severe undercutting results from an isotropic etch and pattern walls are sloped. Virtually no undercutting results from an anisotropic etch with significantly more vertical pattern walls.

The most important plasma etching process parameters are:

- Gas composition
- Reactor geometry
- Gas flow rates
- Power
- Frequency
- Temperature

These parameters influence etch rate, selectivity, anisotropy, and loading [13]. Sputter etching is essentially the same process as sputtering, except that some form of masking is used. Consequently, we will only address highlights of the process. The main advantage of sputter etching is that it is highly anisotropic and essentially line of sight for the incident ions. Operating pressures are generally the lowest of all etch processes (~ 1 mTorr). Ions striking the surface are sensitive to the magnitude of atomic bonding energies and structure. They, however, are not sensitive to the composition of the

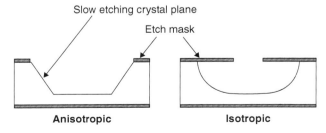

Figure 6.6 Comparison of wet and plasma etch profiles.

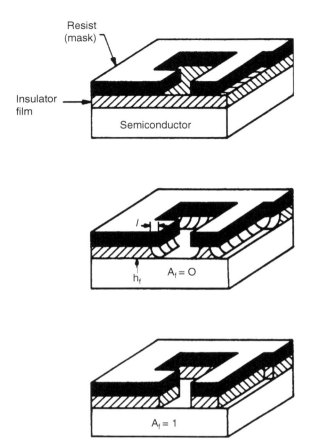

Figure 6.7 Pattern transfer for chemical and plasma etch processes.

surface, which is the main disadvantage, low selectivity. Another disadvantage is that the masking material (photoresist, metal mask, ceramic mask) is also etched, although in most processes the etch rate for the mask is much lower than that of the surface being etched. A good rule-of-thumb is that the mask should be at least 50% thicker than the etch depth.

Depth of etch then is determined by the power to the cathode element, duration of etch, inert gas used. The degree of anisotropy is defined as

$$A_f = \frac{1 - v_1}{v_v}, \tag{6.8}$$

Where v_l and v_v are the lateral and vertical etch rates. Note that $A_f = 0$ for isotropic etching with $v_l = v_v$. and $A_f = 1$ for $v_l = 0$. Sputter etching can locally heat the surface and has the slowest etch rate of all plasma etch processes. Also note that the intersection of the walls and bottom of the etched structure is generally not a clean right angle but concave in shape or an oblique angle. Another problem, if care is not taken, is re-deposition of the etched material in channels and on the substrate.

In chemical etching (not to be confused with *wet* chemical etching) the gas phase species reacts chemically with surface atoms to produce a volatile product, which is pumped out. The sole responsibility of the plasma is to carry the reactant species to the surface of the substrate. Plasma chemical etching is performed at higher chamber pressures (~ 1Torr) than sputter etching. As shown in Figure 6.6, etching is isotropic. However, etching is also highly selective. Primary etchant gases are F_2, NF_3 and CF_4. The etching of Si provides an excellent example of chemical etching. Figure 6.8 summarizes the

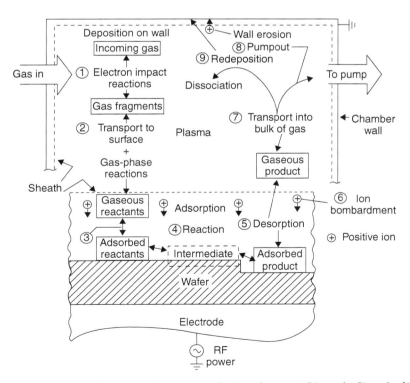

Figure 6.8 Microscopic process that occurs during plasma etching of a Si wafer [9].

process that transpires during plasma etching of a Si wafer. The etch process can be summarized with the following steps [9]:

- Generation of reactant etchant species in the plasma
    ○ $e^- + Cl \rightarrow Cl^+ + 2e^-$ for example
- Diffusion of reactant through dark space to substrate surface
- Adsorption of reactant species on the substrate surface
- Surface diffusion of active ions and radicals
- Chemical etching reaction with film atoms producing adsorbed byproducts
- Volatile compounds are formed and diffuse into the bulk gas, and are subsequently pumped out by the vacuum system

These steps somewhat simplify the process. The etch rates for the same material but with different crystal structures, microstructures, and impurity levels can vary widely. For example, the etch rate for n-type Si with Cl atoms is ~ 20 times higher than for undoped Si. Table 6.3 lists etchants for a number of electronic materials [6]. Etch rates can be expressed by an Arrhenius-like expression, for example for <100> Si in F [15].

$$\text{Re}(Si < 100 >) = 2.91 \times 10^{-12} T^{1/2} C_v \exp{-0.108eV / k_B T} \quad (6.9)$$

where C_v is the etchant concentration. The etch rate for SiO_2 is down by an order of magnitude:

$$\text{Re}(SiO_2) = 6.14 \times 10^{-13} T^{1/2} C_v \exp{-0.163eV / k_B T} \quad (6.10)$$

A simple parallel plate plasma etching configuration is shown in Figure 6.9 [16]. Plasma is confined between two closely spaced electrodes. Etchant gases are fed in through a gas ring. Etchant ions are attracted to the surface to be etched (wafer, thin film, etc.). Figure 6.10 shows the vertical and lateral etch rates for Si as a function of XeF_2 flow rate [16]. Recall that the etchant species in plasmas also bombards the surface with a range of energies. The lateral etch rate, however, does not involve the energetic bombardment of plasma species, but only how effectively Si is etched chemically.

Table 6.3 Etchants for selected electronic materials using Cl and F atoms [14].

Material Etched	Source Gas	Additive	Mechanism
Si	Cl_2	None	Chemical
	CCl_4 CF_4	O_2	Ion energetic
	$SiCl_4$	O_2	
	SF_6	O_2	Chemical
III-V	Cl_2	None	Chemical/ crystallographic
	CCl_4	O_2	
	$SiCl_4$	O_2	
GaN	Cl_2, H2, Ar		
Al	Cl_2	SiCl4, CCl4, BCl3	Ion Inhibitor
TiS_2, $TaSi_2$, $MoSi_2$, WSi_2	F_2	None	Ion energetic
	CF_4	O_2	
Ti, Ta, Mo, W, Nb	F_2	None	Ion energetic
	CF_4	O_2	
SiO_2/Si_3N_4	CF_4	O_2	Chemical

The vertical etch rate depends on the synergetic effect of neutral bombardment and chemical reactions. The degree of anisotropy (see above) can generally be increased by increasing the energy of the ions.

The etch rate and selectivity can further be modified by adding one or more gases. Hydrogen is often added to the etchant gas to

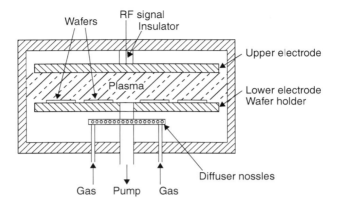

Figure 6.9 Parallel plate plasma etching system.

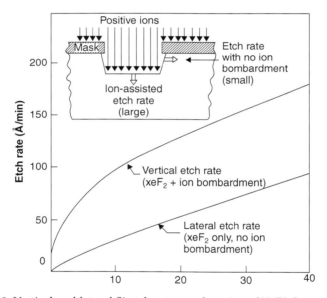

Figure 6.10 Vertical and lateral Si etch rate as a function of XeF2 flow rate [16].

increase selectivity. Figure 6.11 shows that the SiO_2 etch rate in CF_4 is approximately constant for H_2 additions up to 40%, but the rate for polysilicon drops exponentially for the same increase in H_2 [16]. SiO_2/ polySi selectivity is thus significantly enhanced, as shown by the bottom curve. The inverse is also true for mixtures of $Cl_2 + SF_6$, as shown in Table 6.4 [16]. The etch rate for Si can be increased by as much as a factor of 80 faster than that of SiO_2 by addition of SF_6. Oxygen can also

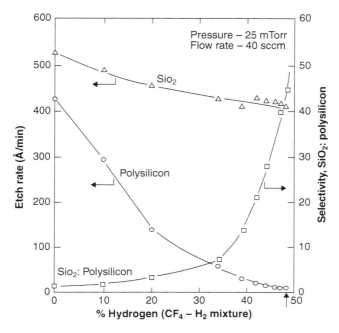

Figure 6.11 Etch rates for Si and SiO$_2$ and SiO$_2$/Si selectivity in mixtures of CF$_4$ + H$_2$ [16].

Table 6.4 Etch rates and selectivities for Si and SiO$_2$ [16].

Material (M)	Gas	Etch Rate (Å/min)	Selectivity		
			M/Resist	M/Si	M/SiO$_2$
Si	SF$_6$ + Cl$_2$	1000–4500	5	–	80
SiO$_2$	CF$_4$ + H$_2$	400–500	5	40	–
Al, Al-Si, Al-Cu	BCl$_3$ + Cl$_2$	500	5	5	25

influence etch rate in some cases. Mixtures of O$_2$ + CF$_4$ are also used to etch SiO$_2$, as shown in Figure 6.12 [17]. Figure 6.13 shows the result of a highly anisotropic etch with an aspect ratio of 31:1 into SiO$_2$ [18]. Width of the trench is only 25 nm with a depth of 788 nm.

The etch rate also depends on RF power to the plasma, pressure, gas flow rate, and temperature (see above Arrhenius relations) [9].

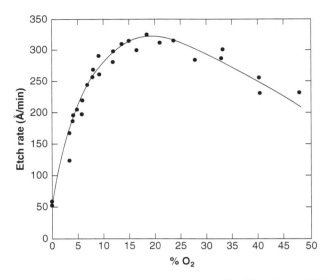

Figure 6.12 Dependence of the etch rate of SiO_2 on CF_4/O_2 mixture [5].

Figure 6.13 SEM of 25 nm wide trench etched into SiO_2 [9].

The power density affects the electron energy distribution, which in turn, determines generation rate of active species. Variations with pressure are independent of power. Figure 6.14 presents the dependence of etch rate of Si_3N_4, Si, and thermal SiO_2 with power. Etch rate generally increases monotonically with applied power. The dependence on pressure is less straightforward. Etch rate will generally increase with increased pressure, but due to eventual thermalization of the plasma ions, will then decrease due to lower energy collisions. Etch rate will also decrease at very low pressures due to reduced electron collisions.

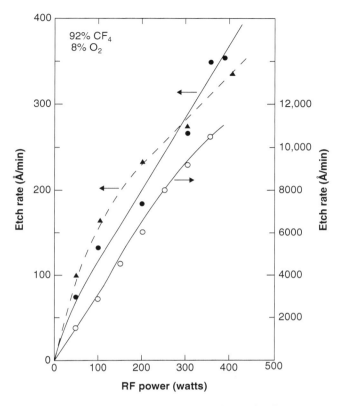

Figure 6.14 Variation of etch rate of Si_3N_4, Si and thermal SiO_2 in mixture of $0.92CF_4 + 0.08O_2$ with power [18].

There are a number of plasma etch reactor configurations, as shown in Figures 6.15 and 6.16 [18]. The operating frequency must be chosen carefully: too low a frequency will result in pulsating DC operation and too high a frequency may result in poor power transfer to the plasma. We discussed important plasma parameters in the previous section. Plasma decay time is critical in reactor operation [13].

$$\tau_p = \frac{qn_i\pi r^2 L}{(j_i[2\pi r^2 + 2\pi rL])} \tag{6.11}$$

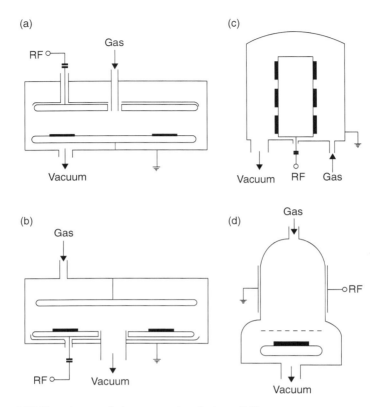

Figure 6.15 Four types of plasma reactor designs [18].

Figure 6.16 Pictures of industrial quartz barrel plasma systems (courtesy Anatech).

where r is the radius of the reactor, L is the spacing between electrodes, j_i is the current density, and n_i is ion density. This relation can be simplified to

$$\tau_p \sim L \left(\frac{m_i}{k_B T_e} \right)^{1/2} .$$
(6.12)

For L = 5 cm and $k_B T_e$ = 2 eV, τ_p ~ 20 μs, which gives an operating frequency of ~ 25 kHz. Operating frequencies can range as high as 2.45 GHz for industrial systems.

Referring to Figure 6.15, the most widely used reactor types are the anode-coupled and cathode-coupled barrel and single wafer reactors (6.15a and 6.15b). External electrodes are either capacitively or inductively connected to the power supply. Good etch rate and temperature uniformity are difficult to achieve in this type of reactor. The RF planar and hexode reactors have improved performance. The hexode type (c) has a RF-powered hexagonal tower on which substrates are mounted. In the RF planar configuration (d), gas is introduced radially over the electrodes. Wafers are placed on the smaller plat, which exposes them to a higher sheath potential and enhanced ion bombardment. Figure 6.16 shows a pictures of two- reactor and four-reactor quartz barrel etch systems (courtesy Anatech).

6.3   Wet Chemical Etching

Wet chemical etching was adopted in the 1950s for transistor and IC manufacture, and involves removal of material by immersing it in liquid reagents that attack the surfaces not protected by the photoresist or etch mask. All parameters relevant plasma etching are also involved with wet etching: etch rate, etch rate uniformity, etch profile, selectivity, and etch bias. In fact, most of these parameters are more difficult to control with wet etching. A wide variety of wet etch chemistries are available. For this process, selectivity becomes an important issue and is generally better than plasma processes. Feature size, however, is limited to 2 μm and larger [19].

In most wet etch processes, the material to be etched is not directly soluble in the etchant, but undergoes a (chemical) reaction with chemicals present in the etch solution. Reaction products can then be soluble in solution or be gaseous. One main advantage of wet etching is lower costs than plasma etching and the versatility of etching equipment. One of the disadvantages is that the photoresist or mask may lose adhesion to the underlying material when exposed to certain chemicals, such as hot acids. Table 6.5 lists etchants for several metals listed in the metallization section [20].

6.4 Metallization

Thin film metallization is used in semiconductor and molecular electronic devices to provide electrical contacts to the devices, interconnects, and bonding surfaces. Figure 6.17 shows single layer metal interconnection between two devices [21], and Figure 6.18 shows bond pads on a microelectronic chip. Insulation of the metal circuitry from the active areas of the device is accomplished using passivation layers. There are three major types of metallization schemes:

- Subtractive
- Fully-additive
- Semi-additive

Table 6.5 Common wet chemical etchants for metals [20].

Metal	Wet Etching Solution
Au	$KI + I_2 + H_2O$
Pb	$H_2O_2 + CH_3COOH$ $HNO_3 + H_2O$
Ti	$H_2SO_4 + HNO_3 + H_2O$ HF
AuZn	$KI + I2 + H_2O$
Ni	$HNO_3 + CH_3COOH + H_2SO_4$ HCl
Al	$H_3PO_4 + HNO_3 + CH_3COOH + H_2O$
Cu	HNO_3
Ag	HCl, H_2SO_4

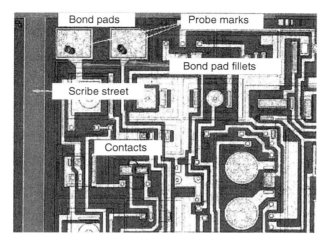

Figure 6.17 Single layer metal interconnection between two devices [21].

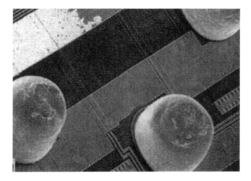

Figure 6.18 Bond pads on a semiconductor device (courtesy National Semiconductor).

Subtractive metallization involves the build-up of a blanket layer of metal on the substrate. This process uses a photoresist and metal etch to define traces. Fully-additive metallization involves the direct build-up or plating of metal traces on the substrate. Semi-additive metallization uses a blanket seed layer of metal. Traces are built-up using photoresist and plating. The seed layer is then removed.

Interconnect thickness is typically 1500 to 15000 Å. Metals most often used are Ag, Cu, Cr, Au, Al, W, and Ti. Table 6.6 lists bulk resistivities of these metals [2]. Each metal has a distinct application. For example, Ag has the highest conductivity and lowest resistive

Table 6.6 Resistivities of Some Bulk Materials.

Metal	Bulk Resistivity (20 °C, μΩ-cm)
Silver	1.55
Copper	1.7
Gold	2.4
Aluminum	2.8
Tungsten	5.5
Titanium	50
Platinum	10.5

losses and is used for circuitry while Au is extremely corrosion and oxidation resistant. Ag, however, is easily corroded. Al bonds well to semiconductor wafers and etches easily to form circuitry. Wires are easily bonded to Au films. Cu is a well known conductor and is also used in Cr/Cu/Au solder bond pads (see Figure 6.18). Al is very adherent and bonds well to oxidized surfaces. Pt and W are excellent diffusion barriers. TiN is also an excellent diffusion barrier. The resistivity of Pt is easily calibrated against temperature and used in resistance temperature detectors (RTD).

In order to combine the most desirable properties of each constituent material, metallization systems can also consist of multilayers. An example of this is the Cr/Cu/Au solder pad described above. Here the wire bonds readily to the Au layer while Cr is the bond layer to the Si wafer. Cu diffuses into both upper and lower layers to increase the mechanical integrity of the bond. Titanium/ Au is also an effective contact system. Again, Ti acts as the adhesion or glue layer. Note that these structures are also oxidation and corrosion resistant due to the Au layer.

Problems that can occur with metallization are electromigration, corrosion, oxidation, debonding, and degradation of conductivity. Electromigration is among the many phenomena that can degrade the performance of thin films. Prior to the advent of integrated ciucuits, electromigration was an interesting effect, but of no commercial interest. Electromigration, however, was found to significantly decrease the reliability of Al integrated circuit elements and any current carrying thin film, and could lead to eventual open and short circuits between metal films and semiconductors. Electromigration is the mass transport of atoms in a metal due to the momentum transfer between conducting electrons and diffusing metal atoms. While electromigration has been known for over 100 years, its deleterious effects on integrated circuits have made it a major concern. Electromigration, however, only exists whenever current flows through a metal wire or thin film, but the conditions necessary for this phenomenon to be a problem simply did not exist before the advent of microelectronics. Because of the high current densities and thin film involved, integrated circuits are most affected by electromigration. Because most of the heat generated by the current is conducted away into the chip, thin film conductors can withstand current densities at least two orders of magnitude greater than traditional bulk wires. This allows current densities of nearly 10^6 A/cm^2 with minimal joule heating, and this creates current densities high enough to cause significant electromigration.

With increasing miniaturization, the probability of failure due to electromigration increases in VLSI and ULSI circuits, because both power density and current density increase. Al was first used in microcircuitry, and immediately had significant electromigration problems. Figure 6.19 shows electromigration damage in an Al film [21]. Hillock growth (6.19a), whisker bridging (6.19b) and lateral mass growth areas (6.19c) are all manifestations of this phenomenon. As we shall see, these defects are caused by mass buildup and depletion due to flow of current. Because of its higher conductivity, Cu is now the interconnect material of choice. It is also intrinsically less susceptible to electromigration than Al. However, electromigration continues to be an ever present challenge to device fabrication, and as a result, research for copper interconnects is ongoing (though a relatively new field.)

Electromigration is influenced by several factors. Most important of these are:

- Structure effects
- Thermal effects
- Stress in the film

Anywhere there are different grain sizes and orientation distributions, local mass flux divergences can exist in a film-interconnect structure. Divergence also occurs at interfaces between interconnects, semiconductor contacts, and junctions between multilevel

(a) Hillocks   (b)   Bridging

(c) Lateral mass growth

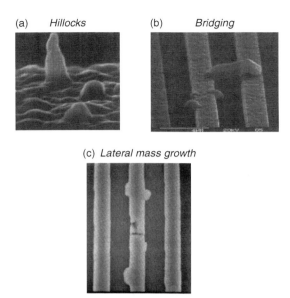

Figure 6.19 Electromigration damage in Al films in the form of hillock growth (a), whisker formation (b), and lateral mass growth (c) [1].

metallizations (vias, etc.). As a result, electromigration damage can occur at any of these sites. Essentially, more mass (atoms) enters these sites than exits them and growth occurs. Voids develop when more atoms exit the site than enter. Mass balance is given by

$$\frac{\partial C}{\partial t} + \text{Div. J} = 0 \qquad (6.13)$$

Here J is the atomic flux induced by electromigration and combines concentration, thermal, stress, and density fluxes. J_C is shown above.

Differences in temperature (temperature gradients) in the film can occur at regions of poor adhesion and interfacial regions. The flux due to nonuniform temperature distributions is

$$J_T = \frac{-CDQ(\text{DivT})}{kT^2} \qquad (6.14)$$

Where Q is the heat of thermal diffusion. Again, the diffusion coefficient plays a major role.

Mechanical stress also contributes to electromigration in a surprising way. Stress gradients can actually reduce the effects of electromigration. Mass tends to pile up at the anode side of the thin film strip and is depleted at the cathode side. The flux due to a stress gradient is

$$J_\sigma = \frac{CD\Omega(\text{DivH})}{kT} \qquad (6.15)$$

where H is the average stress along the diagonal of the stress tensor $(\sigma_{xx} + \sigma_{yy} + \sigma_{zz})/3$ and is Ω = atomic volume = $1/C$. There is a critical length of the conductor at which these two mass transport forces balance, which should reduce electromigration. The two forces balance when [2]

$$L_C = \frac{\Omega H_C}{Z^* q\rho jc} \qquad (6.16)$$

Electromigration can be completely halted when $L < L_C$, and thus we have a means to control this degradation mechanism (assuming these lengths are feasible). Additionally, shorter interconnects can withstand higher current densities. The diffusion coefficient can thus be expressed as

$$D = D_0 \exp\left\{\frac{(\Omega H - E_A)}{kT}\right\} \qquad (6.17)$$

where E_A is the effective activation energy for thermal diffusion. Here we see that D depends on stress and temperature.

Based on extensive electromigration data, we can express the mean time to failure due to electromigration and current density j [2]:

$$(\text{MTTF})^{-1} = K \exp\left(\frac{-E_C}{RT}\right) j^n \qquad (6.18)$$

For pure Al films, $n \sim 2$ and E_C, the activation energy for failure, ranges between 0.5 and 0.8 eV. The magnitude of activation energy is generally associated with grain boundary diffusion, while lattice diffusion activation energies are ~ 1.4 eV.

Electromigration is a particularly annoying problem for Ag films and circuitry. Ag is well known to be unstable at high temperatures, high humidity, and corrosive environments. Both temperature and humidity exacerbate electromigration and dendrite formation in this metal [22]. Electrolytic electromigration results when humidity is a factor. An example of dendrites and dendrite bridges is shown in Figures 6.20 and 6.21 [22]. The mechanism of electrolytic electromigration is water-dependent, and is enhanced when the insulator separating the conductors (as on PC boards, flexible circuitry, chip carriers, or IC ceramics) has adsorbsed sufficient moisture to allow electrolytic (ionic) conduction when an electrical potential is applied. Electromigration in Ag is generally moderate, even at elevated temperature, but humidity significantly accelerates this process [23]. Both colloidal "staining" and dendritic bridging result (see Figures 6.20 and 6.21). Deposits of colloidal silver (or copper) often appear as brownish stained regions which originate at the positively polarized conductor but do not necessarily remain in contact with it (see Figure 6.22). They are thought to result from reduction of the migrating ions, either by light or by chemical reducing agents on the insulator surface [24, 25].

The primary conditions that promote humid electromigration problems are:

- Moisture (i.e., high relative humidity)
- Contamination on the insulator surface
- Voltage difference between conductors
- Narrow spacing widths
- Elevated temperatures (at high relative humidity).

All other things being equal, higher temperatures enhance electromigration, particularly at very high relative humidity values. Particularly for Ag [26, 27]. This correlation has been observed for a variety of insulating materials with different silver migration susceptibilities [23–27]. Typical conditions include 85/85 T/RH and thermal cycling of ambient conditions.

Electrolytic electromigration in Ag can be curtailed by applying moisture barriers to the conductors and insulator surfaces [28]. The success of moisture barriers also depends on the operating

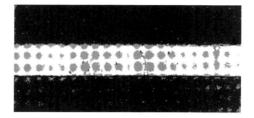

Figure 6.20 Dendrites and dendrite bridges in Ag [22].

Figure 6.21 Dendritic bridging across two Ag conductors on a printed circuit board [22].

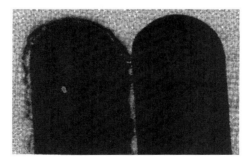

Figure 6.22 Staining due to electromigration [23].

condition of the circuitry; voltage, temperature and humidity all must be considered. Another method for mitigating electromigration is by reducing the thermodynamic "activity" or instability of the system. Techniques that have been tried with varying degrees of success are overcoating Ag with less "active" metals (Au, Pd, Sn) that act essentially as diffusion barriers [28, 29]. Alloying has also shown some success in suppressing electromigration. Alloys are significantly less prone to failure. For example, just a few percent Cu in Al will extend conductor lifetime by an order of magnitude.

Thus we see that electromigration is a real problem in thin film circuitry that is subjected to ultrahigh current densities, high temperatures, and humidity. Methods used to mitigate this destructive effect are choosing metals with reduced susceptibility, alloying, applying diffusion barriers, and keeping circuitry as short as possible.

6.5   Photolithography

The metal, however, must cover only certain areas of the surface to be useful. This is accomplished by lithographic patterning and masking processes. Photolithography (or "optical lithography") is a process used in microfabrication to selectively remove parts of a thin film or the bulk of a substrate or add thin films. It uses light to transfer a geometric pattern from a photo mask to a light sensitive photoresist applied to the substrate. Material may be added (lift off) or etched from the surface. A series of chemical treatments then transfers the exposure pattern into the material underneath the photoresist. In complex integrated circuits (ICs) a CMOS wafer will undergo up to 50 photolithographic steps.

The basic steps in the photolithography process are:

1. Substrate cleaning
2. Substrate preparation
3. Metallization or thin film deposition (for positive and negative photoresist)
4. Photoresist application
5. Prebake
6. Photoresist exposure
7. Exposure post bake
8. Developing
9. Development post bake
10. Metallization or thin film deposition (for negative photoresist)
11. Etching or thin film deposition (adding or subtracting material)
12. Photoresist removal

Figure 6.23 schematically defines the lithography process used to transfer patterns from a mask to a thin film on a wafer or to the wafer itself [19]. These patterns define metalized structures in ICs and other surfaces. Details of this process are presented elsewhere and literature is extensive. To this end we will present basic steps to develop a fundamental understanding of the patterning process. The process begins with substrate preparation and metallization of the substrate. A layer of photoresist is then applied by spin coating. Note that the substrate does not necessarily have to be a semiconductor wafer. Other applications, including micro-meshes, microfabrication, metamaterials, microlens arrays, and resonant optical structures (see Sections 7.3.3. and 8.4) also use photopatterning techniques.

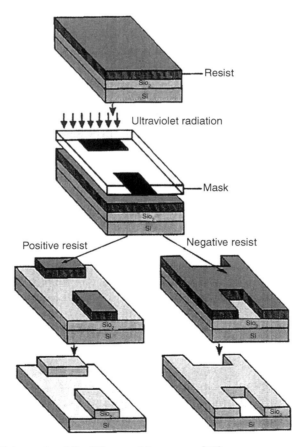

Figure 6.23 Schematic of the lithographic process [19].

Additionally, the substrate does not have to be planar, and can be hemispherical (outer and inner faces), cylindrical, and even a flexible material. An adhesion promoter such as hexamethyldisilizane (HMDS) is often used.

Photoresist (also called resists) is generally applied by a spin process but is also sprayed onto larger substrates. Thickness typically ranges from 0.5 μm to 2 μm, but can be up to 100 μm thick. Photoresists are classified into two groups (see Figure 6.23): positive resists and negative resists:

- A *positive resist* is a type of photoresist in which the portion of the photoresist that is exposed to light becomes soluble to the photoresist developer. The portion of the photoresist that is unexposed remains insoluble to the photoresist developer.
- A *negative resist* is a type of photoresist in which the portion of the photoresist that is exposed to light becomes insoluble to the photoresist developer. The unexposed portion of the photoresist is dissolved by the photoresist developer.

After application, the resist, being a monomer, is cured by baking. A wide range of photoresists is used. Composition of resists is a closely guarded secret; however, positive resists generally are a variation of diazonaphthoquinone. Common resists include Hoechst AZ 4620, Hoechst AZ 4562, Shipley 1400-17, Shipley 1400-27, Shipley 1400-37, and Shipley Microposit Developer. Microelectronic resists, presumably, utilize specialized products depending upon process objectives and design constraints. The general mechanism of exposure for these photoresists proceeds with the decomposition of diazoquinone, i.e., the evolution of nitrogen gas and the production of carbenes.

After prebaking, the photoresist is exposed to a pattern of intense collimated light. In some cases, a laser is used. Optical lithography typically uses ultraviolet (UV) light. Light may be transmitted through a mask or by projection of the pattern. Lithography systems have progressed from blue wavelengths (436 nm) to UV (365 nm) to deep-UV (248 nm) to today's mainstream high resolution wavelength of 193 nm. In the meantime, projection tool numerical apertures have risen from 0.16 for the first scanners to amazingly high 0.93 NA systems today producing features well under 100 nm in size. Even smaller features are obtained using electron and x-ray lithography. After exposure, the resist is baked to mitigate standing wave effects which can degrade the final pattern.

Once exposed and baked, the photoresist must be developed to generate the pattern, in analogy with photographic developers. Aqueous base developers are generally used. In particular, tetra-methyl ammonium hydroxide (TMAH) is used in concentrations of 0.2–0.26 N. This is one of the most critical steps in the photoresist process. The characteristics of the resist-developer interactions determine to a large extent the shape of the photoresist profile and, more importantly, linewidth control. Figure 6.24 shows a picture of a developed resist pattern.

The postbake is used to harden the final resist image so that it will withstand the harsh environments of implantation or etching. The process, performed at temperatures between 120 °C–150 °C, cross links the resist to form a polymer. Care must be taken not to overbake the resist or it will flow and degrade the pattern or be difficult or impossible to strip. In addition to cross-linking, the postbake can remove residual solvent, water, and gases and will usually improve adhesion of the resist to the substrate.

Other methods can be used to harden a resist image. Exposure to high intensity deep-UV light also crosslinks the resin at the surface of the resist forming a tough skin around the pattern. Deep-UV hardened photoresist can withstand temperatures in excess of

Figure 6.24 Developed photoresist pattern.

200 °C without dimensional deformation. Plasma treatments and electron beam bombardment have also been shown to effectively cross link and harden photoresist. Commercial deep-UV hardening systems are now available and are widely used.

Metallization, or other thin films, can then be etched in the areas not covered by the resist. Metallization can also be applied over a negative resist pattern and the photoresist then striped, or lifted off, to form the negative image of the mask or projected image. Etching can be accomplished using chemical etchants (acids or bases) or by ion and plasma bombardment in vacuum. Ions can also now be implanted.

After etching or metallization, the remaining photoresist must be removed using wet stripping using organic or inorganic solutions or dry (plasma) stripping. A simple example of an organic stripper is acetone. Although commonly used in laboratory environments, acetone tends to leave residues on the wafer (scumming) and is thus unacceptable for semiconductor processing. Most commercial organic strippers are phenol-based and are somewhat better at avoiding scum formation. However, the most common wet strippers for positive photoresists are inorganic acid-based systems used at elevated temperatures. Resists, preferably, can be oxidized and made volatile in an oxygen plasma. Understandably, this process will not work for metallization that readily forms oxides (Al, Cu, Ag, Ti, etc.) and the resist must be removed by wet chemical processes.

Optical techniques are effective in defining structures down to ~ 1 μm. X-ray and electron beam lithography are used to overcome the diffraction limits of optical lithography, reporting structure as small as 20 nm. The underlying system designs are basically the same for these techniques (beam focusing, masking, photoresist exposure, and developing), although these systems use very different beam focusing equipment and photoresists.

6.6   Deposition Processes for Piezoelectric and Ferroelectric Thin Films

Piezoelectric and ferroelectric thin films have numerous applications, including high frequency transducers, movable diaphragms, strain measuring devices, MEMS devices, computer memory, and robotics (controlled displacement and movement). Many piezoelectric (and semiconductor) materials are also photorefractive.

Ferroelectric materials exhibit an electric dipole moment even in the absence of an electric field. Deposition processes must achieve a specific microstructure, and to this end, we briefly review the piezoelectric and ferroelectric materials. The center of positive charge does not coincide with the center of negative charge, thus forming an electric dipole. A crystal that develops an electric polarization in response to an applied stress is called a piezoelectric crystal. Conversely, a mechanical stress, and subsequent deformation, can be introduced into a piezoelectric crystal through application of an electric field. The piezoelectric equations are [20]:

$$P = Zd + E\chi \qquad (6.19)$$

$$e = Zs + Ed \qquad (6.20)$$

Here P = polarization, Z = stress, d = piezoelectric strain constant, E = electric field, χ = dielectric susceptibility, e = elastic strain, and s = elastic compliance. Strain is a tensor quantity and is given by [20]

$$\varepsilon_{ij} = \Sigma d_{ijk} E_k \rightarrow = \Sigma d_{ku} E_k \qquad (6.21)$$

where i, j, k vary from 1–3 and

$$P_i = \Sigma\Sigma d_{ijk}\sigma_{jk} \quad P_i = \Sigma d_{i\mu}\sigma_\mu \, (i, \mu \text{ vary from } 1-6) \qquad (6.22)$$

(i, μ vary from 1–6). σ_{jk} is the piezoelectric strain tensor.

These effects are due to formation of polar crystals and, thus, the goal of deposition of piezoelectric and ferroelectric thin films is formation of polar crystals, or electric dipoles. To this end, columnar, polycrystalline, and highly textured microstructures generally are found to display piezoelectric and ferroelectric properties. A majority of ferroelectric and piezoelectric thin films are deposited by some form of sputtering [30–33], chemical vapor deposition [34, 35], although sol gel processing [36], pulsed laser deposition [37], and molecular beam epitaxy [38] are also used. Figure 6.25 shows the columnar structure of a 50 μm thick magnetron sputtered ZnO film used in a piezoelectric transducer [31]. Piezoelectric and ferroelectric films display two general features: a high degree of crystallinity, as shown in Figure 6.26 [20] and columnar structure as shown in Figure 6.27. Deposition conditions can be chosen based on the structure zone models presented in Chapter 2 to obtain highly columnar

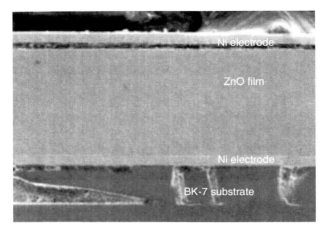

Figure 6.25 Columnar structure of ZnO film used in piezoelectric transducer [32].

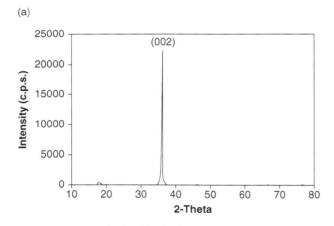

Figure 6.26 XRD spectrum of AlN film [20].

and textured morphologies. The crystal structures of these materials often are conducive for forming columnar microstructure. Twenty of the 32 crystal classes exhibit piezoelectric properties. The c-axis orientation of hexagonal crystal structure is often preferred, as shown in Figure 6.27 for AlN. In this case, the (002) orientation results in better piezoelectric properties. Examples of ferroelectric materials are $LiTaO_3$ (lithium tantalite), $LiNbO_3$ (lithium niobate), $BaTiO_3$

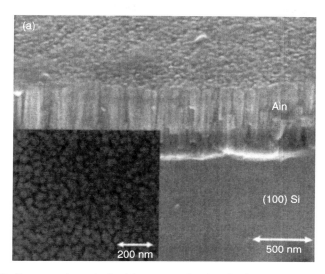

Figure 6.27 Cross section of a highly c-axis AlN film [32].

(BT–barium titanate), $PbTiO_3$ (PT–lead titanate), $PbZr_{0.95}Ti_{0.05}O_3$ (PZT), PLZT (lead-lanthanum-zirconium titanate), KH_2PO_4 (potassium di-hydrogen phosphate), and KD_2PO_4. Examples of piezoelectric materials are AlN, SiO_2 (quartz), ZnO, SiC, and the aforementioned ferroelectric materials. Polyvinylidene fluoride (PVDF) is an organic piezoelectric material. ZnS, PLZT, $BaTiO_3$, $LiNbO_3$ and $PbTiO_3$ additionally display photorefractive properties.

As mentioned earlier, highly oriented crystalline and columnar microstructures are required for good piezoelectric performance. Highly oriented c-axis AlN and ZnO films show strong piezoelectric effects [31, 32]. Referring to the structure zone models (SZM) shown in Chapter 3 for sputtering, the columnar structure such as that shown in Figures 6.25 and 6.27, is attributed to zone T [32]. Morphology typical of this zone is achieved using deposition conditions to achieve:

Zone T: ($0.1 < T_S/T_M < 0.4$ @ 0.15 Pa, T_S/T_M between 0.4–0.5 @ 4 Pa) – fibrous grains, dense grain boundary arrays.

Zone T is essentially just a transition region between zones 1 and 2 characterized by relatively low – moderate substrate temperatures, low – moderate chamber pressures and low deposition rates. We see for sputtering that inert gas pressure influences film structure through indirect mechanisms such as thermalization of deposited atoms with increased pressure and increased component of atom flux due to gas scattering [38]. Reduction of gas pressure results in more energetic particle bombardment which densifies the film.

It is, therefore, important that the SZM for each deposition process be consulted as a starting point for deposition of ferroelectric and piezoelectric films. Substrate morphology and crystal orientation are also important.

6.7   Deposition Processes for Semiconductor Thin Films

Semiconductor thin films are deposited by virtually every deposition process described in Chapter 2. The primary goal for these processes is to achieve electrical and optical properties as close to bulk as possible by:

- Minimizing structural and compositional defects
- Minimizing impurities
- Preserving stoichiometry
- Introducing dopants
- Minimizing diffusion between layers
- Obtaining high thickness uniformities
- High deposition rates
- Reducing production costs

However, one distinct advantage of thin film processes, because they are nonequilibrium in nature, is that they are able to synthesize compositions not possible with bulk processes. The following semiconductors are deposited in thin film and low dimensional forms: a-Si, a-Si:H, nanocrystalline Si, a-Ge, a-Ge:H, CdTe, CdS, GaAs, GaAs:Al, $CuInSe_2$, $CuInGaSe_2$, $Cu(In,Ga)Se_2$, Cu(InGa)SSe, ZnS, ZnSe, SiC, a-SiC:H, a-GeC, a-GeC:H, HgCdTe, PbSnTe, InP, InAs, PbS, ZnTe, $Al_xGa_yIn_{1-x-y}P$, $Al_xGa_{1-x}As$, AlInAs, and p- and n-doped compositions of these materials. Materials such as GaSa, GaP, ZnS, and ZnSe also display photorefractive properties. Description of deposition of all these materials is not possible. Instead, deposition processes for a-Si:H, CIGS, GaAS, and CdTe thin films will be addressed as typical for thin film semiconductors.

6.7.1 Amorphous Silicon

High quality a-Si:H films are deposited primarily by PVD processes, silane decomposition, CVD processes (including PECVD), and ALD. Applications include gate electrodes for thin film transistors (TFT), microphotonic IC, MEMS, sensors, detectors, microphotinic waveguides, as well as multiple junctions photovoltaic solar cells, large area electronics and liquid crystal displays. This material works well in thin film solar cells as a result of its broad wavelength absorption range; a-Si:H absorbs sunlight very efficiently and only very thin films (<500 nm) are needed for PV applications. PVD processes, described in Chapter 2, include reactive RF diode sputtering, reactive DC and RF magnetron sputtering, ion beam sputtering, and electron beam deposition. Hydrogen must be incorporated into a-Si to passivate dangling bonds to clean up the band gap and result in useful semiconductor properties. To this end, all sputtering processes generally use mixtures of Ar + H_2 as the sputtering gas [39, 40, 41]. Because it removes defects from the band gap, the optical properties of a-Si:H depend to a large degree on hydrogen content and bonding. Films with the best electrical properties have H bonded in the lattice as Si-H configuration, as opposed to the SiH_2 and SiH_3 configurations, which is true for all deposition processes.

Highest quality a-Si:H, used primarily in solar cells, is produced by PECVD and CVD processes [42–48]. Films are deposited in parallel plate and shower head reactors [48] onto 1 in – 12 in wafers as well as polymer webs. Precursor gases include mixtures of SiH_4 + Ar and typical substrate temperatures are around 300 °C. Dopant gases included PH_3 for n-type and B_2H_6 for p-type films. Typical electrical properties are:

- Undoped dard conductivity 10^{-10}–10^{-12} S/cm
- Undoped illuminated conductivity 10^{-5} S/cm
- Doped conductivity 10^{-3}–10^{-2} S/cm

As a result of precursor gases used, all films contain H.

Plasma decomposition (or glow discharge decomposition) of SiH_4 (silane) and Si_2H_6 (disilane) was the first technique used to deposit a-Si:H [49, 50, 51]. Gas mixtures are typically SiH_4 + Ar and Si_2H_6 + Ar, with ~ 10% Ar. The plasma excites and decomposes gas and generates radicals and ions. The main plasma reactions can be characterized as [51].

Electron-molecular: $SiH_4 + e \rightarrow SiH_n + (4-n)H + e$
Neutral-neutral: $H + SiH_4 \rightarrow SiH_3 + H_2$
Ion-molecule: $SiH_n^+ + SiH_4 \rightarrow Si_2H_m^+ + (4+n-m)H$

As with PECVD processes, doping is accomplished by adding PH_3 for n-type and B_2H_6 for p-type films. The glow discharge deposition system is typically a parallel plate capacitive reactor, shown in Figure 6.28, operating at an RF frequency of 13.56 MHz. Films can be deposited either on the cathode or anode. Substrate temperatures are in the neighborhood 200–300 °C with discharge pressure between 0.15–0.4 Torr. Dopant concentration is determined by amount of PH_3 or B_2H_6 in the discharge gas. Higher amounts of H are incorporated at lower substrate temperatures, which increases the band gap [14]. Less H is incorporated at higher substrate temperatures and, as a result, the band gap is reduced. This is a general rule of thumb for all hydrogenated semiconductors (and most thin films deposited using hycrocarbons) [39, 52]. Also note that fluorinated amorphous silicon (a-Si:F) is deposited by the above reactions using SiF_4 [51].

Typically, amorphous silicon thin-film cells are built in a p-i-n structure. Typical panel structure includes front side glass, transparent conductive oxide (TCO – see Chapter 7), thin film silicon, back contact, polyvinyl butyral (PVB), and back side glass. Backing material can also be flexible. Amorphous Si solar cells have efficiencies in the range 8–13%, compared to the average efficiency of 18% for single crystal cells. Advantages of a-Si:H cells are:

- The technology is relatively simple and inexpensive
- For a given layer thickness, a-Si:H absorbs 2.5 times more energy than c-Si cells

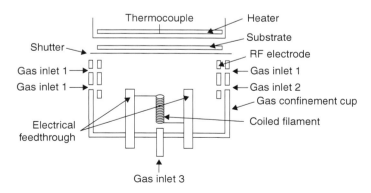

Figure 6.28 Parallel plate discharge reactor used to deposit a-Si:H.

- Much less material is required for a-Si:H cells, thus reducing costs and weight
- Cells can be deposited on a wide range of substrates, including flexible plastic, curved and rolled types
- Average efficiency is ~ 10%, less than c-Si cells but continually improving

6.7.2   Cadmium Telluride (CdTe) Thin Films

Cadmium telluride thin films have been used primarily in thin film solar cells as early as the 1960s [53–57]. This material is also used extensively in infrared optical applications. CdTe and doped CdTe are generally deposited by closed space sublimation. Figure 6.29 shows the progression of PV cell efficiency with time [58]. Note that all thin film semiconductor materials discussed here are included in this chart. CdTe sublimes in vacuum at temperatures near 650 °C, well below its melting point of 1092 °C. CdTe sources are radiantly heated (infrared heat lamps) in graphite boats at chamber pressures ~ 10^{-5} Torr [59]. Substrate temperatures in the range 400–600 °C are required to crystallize the deposited material [60]. Because the material sublimes, deposition rates are very high, ~ 0.1 μm/min, and increase with increased temperature but appear to peak at 520 °C [61].

Figure 6.30 shows a picture of CdTe deposited at temperatures of ambient (no heating), 400 °C, 500 °C, and 550 °C [61]. Note that the texture and associated grain size increases with increased substrate temperature.

Best CdTe solar cell efficiency has plateaued at 16.5% since 2001 [58]. Since CdTe has the optimal band gap for single-junction devices, it may be expected that efficiencies close to exceeding 20% (such as already shown in CIS alloys) should be achievable in practical CdTe cells. Modules of 15% would then be possible [58].

6.7.3   Gallium Arsenide (GaAs) and Aluminum Gallium Arsenide (AlGaAs) Thin Films

GaAs and AlGaAs are generally deposited by some form of epitaxial process: molecular beam epitaxy (MBE), metal organic chemical vapor deposition (MOCVD) epitaxy [62, 63]. However, hydrogenated GaAs (GaAs:H) and microcrystalline GaAs can also be deposited by

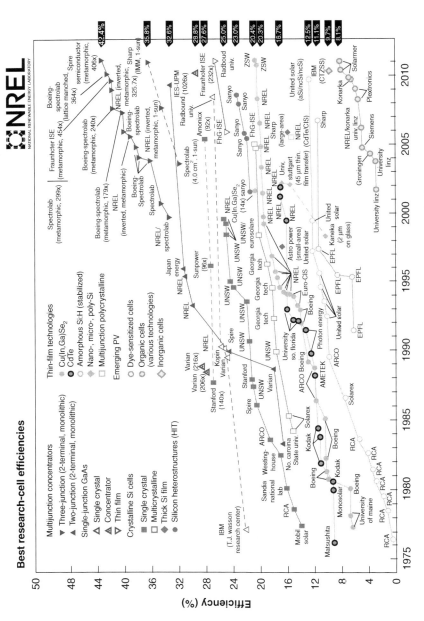

Figure 6.29 Tim line for best solar cell efficiency achieved for semiconductor materials [58].

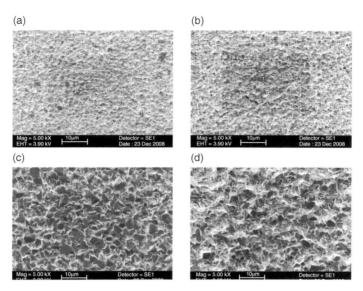

Figure 6.30 SEM image of CdTe showing the increase in grain size with increase in substrate temperatures: a. no heating, b. 400 °C, c. 500 °C, d. 550 °C [61].

reactive sputtering [64, 65]. Primary applications for these materials are thin film solar cells, multiquantum well structures (superlattices), and quantum cascade lasers. Source gases for the MOCVD process are trimethylgallium ($Ga(CH_3)_3$) and arsine (AsH_3) and GaAs is produced by the reaction [48]

$$Ga(CH_3)_3 + AsH_3 \rightarrow GaAs + 3\ CH_4$$

Another CVD route uses arsenic trichloride:

$$2\ Ga + 2\ AsCl_3 \rightarrow 2\ GaAs + 3\ Cl_2$$

GaAs films are grown by MBE using Ga and As sources. AlGaAs is always deposited by MBE processes and uses an additional AlAs source [66]. For MBE processing, a heated substrate wafer is exposed to gas-phase atoms of gallium and arsenic that condense on the wafer on contact and grow the thin GaAs film. In both MOCVD and MBE techniques, single-crystal GaAs layers grow epitaxially, i.e., new atoms deposited on the substrate continue the same crystal lattice structure as the substrate, with few disturbances in the atomic ordering. This controlled growth results in a high degree of crystallinity and in high solar cell efficiency.

Substrate temperature significantly affects surface morphology and can range from 500 to 850 °C. Films have smooth surfaces for substrate temperatures <600 °C. Figure 6.31 shows the layer structure of a GaAs/AlGaAs heterojunction structure [66].

One of the greatest advantages of GaAs and its alloys as PV cell materials is the wide range of design options possible. A cell with a GaAs base can have several layers of slightly different compositions that allow a cell designer to control the generation and collection of electrons and holes. (To accomplish the same thing, silicon cells have been limited to variations in the level of doping.) This degree of control allows cell designers to push efficiencies closer and closer to theoretical levels. For example, GaAs single junction solar cells have demonstrated efficiencies near 25% with a possible peak efficiency >40%. Figure 6.32 shows the layer structure of an advanced GaAs based multijunction thin film solar cell. Cell structure consists of top/middle/bottom cells connected by tunnel junctions. The layer structure includes GaInP, AlInP, and GaInAs layers as n/p junctions. The thin dimensions of the various layers allow charge carriers to be created close to the electric field at the junction.

Sputtering processes use GaAs and Si-doped GaAs targets in Ar or mixtures of Ar + H_2. There are several similarities between sputtered and MBE deposited films, mainly increased smoothness with increased substrate temperature. Sputtered films, however, have many of the same issues as other sputtered semiconductors,

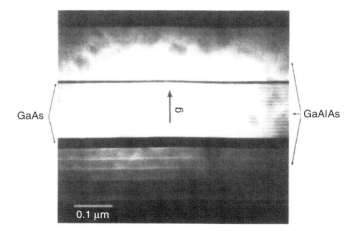

Figure 6.31 Layer structure of a GaAs/AlGaAs heterojunction structure [66].

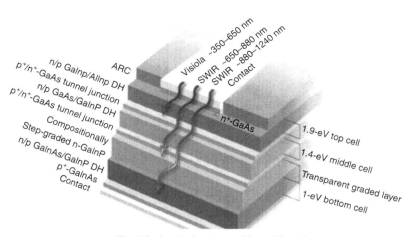

Figure 6.32 Layer structure of advanced GaAs based multijunction thin film solar cell (NREL).

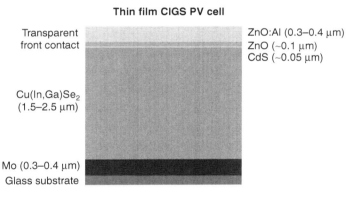

Figure 6.33 Layer diagram of CIGS solar cell.

namely, amorphous structures, dangling bonds that must be passivated by H bonding, low charge carrier mobilities, and poorer photoconductivities. Deposition rates are significantly higher than MBE films.

6.7.4   CIS and CIGS Thin Films

$CuInSe_2$ and $CuIn_xGa_{1-x}Se_2$ are two of the fastest growing thin film solar cell materials technologies. CIGS cells have demonstrated a record 19.9% efficiency and are deposited onto metal, glass, flexible metal and glass, flexible plastic and even glass tubes [67]. Figure 6.33 shows a layer diagram of a CIGS cell. The structure begins with a magnetron sputtered Mo layer followed by a 1.5–2.5 μm thick CIGS layer. The CIGS layer is deposited by co-evaporating or co-sputtering Cu, In, Ga and annealing in a Se vapor or alternately co-evaporating all constituents. Substrate temperature is 350 °C–400 °C. Top layers are magnetron sputtered CdS/ZnO/ZnO:Al(AZO). Sometimes an evaporated, sputtered, or electroplated CdS layer is placed between the CIGS and ZnO layers. The CIGS layer can also be deposited by electroplating.

References

1. D. A. Gurnett and A. Bhattacharjee, *Introduction to Plasma Physics: With Space and Laboratory Applications* (2005).
2. M .Ohring, *Materials Science of Thin Films*, Elsevier (2002).
3. See www.plasmas.org.
4. A. von Engel, *Ionized Gases*, Oxford University Press (1965).
5. S. C. Brown, *Basic Data of Plasma Physics*, 2nd Ed., MIT Press (1967).
6. J. L. Vossen, *Thin Film Processes II*, Academic Press (1991).
7. R. A. Powell and S. M. Rossnagel, PVD for Microelectronics: Sputter Deposition Appied to Semiconductor Manufacturing (Vol. 26, *Physics of Thin Films*), Academic Press (1999).
8. A. Grill, *Cold Plasma in Materials Fabrication*, IEEE Press (1994).
9. D. L. Flamm and G. K. Herb in *Plasma Etching- An Introduction*, ed. D. M. Manos and D. I. Plamm, Academic Press (1989).
10. D. M. Mattox, *J. Vac. Sci. Technol.* A7(3) (1989) 1105.
11. H. C. Casey *et al.*, *Appl. Phys. Lett.*, 24 (1974) 63.
12. J. W. Mayer and S. S. Lau, *Electronic Materials Science: For Integrated Circuits in Si and GaAs*, Macmillan (1990).
13. H. W. Lehmann, in *Thin Film Processes II*, J. L. Vossen and W. Kern eds., Academic Press (1991).
14. D. A. Flamm *et al.*, *J. Appl. Phys.*, 52 (1981) 3633.
15. Wen J. Li, www.is.city.edu.hk
16. C. M. Mellar-Smith and C. J. Mogab, in *Thin Film Processes*, John l Vossen and Werner Kern, eds., Academic Press (1978).
17. A. E. Braun, *Semiconductor International*, 24(2) (2001) 89.
18. A. J. van Roosmalen, A. G. Baggerman and S. J. Brader, *Dry Etching for VLSI*, Plenum (1991).
19. Stanley Wolf, *Microchip Manufacturing*, Lattice Press (2004).
20. Manijeh Razeghi, *Fundamentals of Solid State Engineering*, Kluwer Academic Publishers (2002).

21. M. Ohring and R. Rosenberg, *J. Appl. Phys.*, 42 (1971) 5671.
22. Simeon J. Krumbien, *Metalic Electromigration Phenomena*, AMP Corp. (1989).
23. G.T. Kohman *et al.*, *Bell Syst. Tech. J*, 34 (1955) 1115.
24. S. W. Chaikin *et al.*, *Indust. Eng. Chem.*, 51 (1959) 299.
25. S. J. Krumbein and A. H. Reed, *Proc. 9th ht. Conf. on Electric Contact Phenomena* (1978) 145.
26. E. Tsunashima, *IEEE Trans. Comp., Hybrids, Manuf. Technol.*, CHMT-1 (1978) 182.
27. J N. Lahti, R. H. Delaney, and J. N. Hines, *Proc. 17th Annu. Reliability Physics Symp.* (1979) 39.
28. J. F. Graves, *Proc. Int. Microelectronics Symp.* (1977) 155.
29. H. M. Naguib and B. K. MacLaurin, *IEEE Trans. Comp., Hybrids, Manuf. Technol.*, CHMT-2 (1979) 196.
30. Charles Ziman, *Introduction to Solid State Physics* (Eighth Ed.), Wiley (2005).
31. P. M. Martin *et al.*, *Thin Solid Films*, 379 (2000) 253.
32. A. Ababneh *et al.*, *Mat Sci Eng*, B172 (2010) 253.
33. S. Onishi *et al.*, *Appl Phys Lett*, 39 (1997) 643.
34. Tian-Ling Ren *et al.*, *Mat Sci Eng*, B99 (2003) 159.
35. N. D. Patel *et al.*, *Rev Prog Quant Nondestruct Eval*, 9 (1990) 823.
36. Zhihong Wang *et al.*, *J Ceram Soc.*, 27 (2007) 3759.
37. M. Benetti *et al.*, *Superlattices & Microstructures*, 39 (2006) 366.
38. J. Greene, in *Multicomponent and Multilayered Films for Advanced Microtechnologies: Techniques, Fundamentals and Devices*, O. Auciello and J. Engemann, eds., Kluwer (1993).
39. P. M. Martin and W. T. Pawlewicz, *Solar Energy Materials*, 2 (1979/ 1980) 143.
40. William Paul and David A. Anderson, *Solar Energy Materials*, 5 (1981) 229.
41. T. D. Moustakas, *Solar Energy Materials*, 13 (1986) 373.
42. Hae-Yeol Kim *et al.*, *J Vac Sci Technol* A 17(6) (1999) 3240.
43. Easwar Srinivasan *et al.*, *J Vac Sci Technol* A 15(1) (1997) 77.
44. Toshihiro Kamei and Akihisa Matsuda, *J Vac Sci Technol* A 17(1) (1999) 113.
45. M. Janai *et al.*, *Solar Energy Materials*, 1 (1979) 11.
46. G. N. Parsons *et al.*, *J. Vac. Sci. Technol.* A 7 (3) (1989) 1124.
47. Henry S. Povolny and Xunming Deng, *Thin Solid Films*, 430 (2003) 125.
48. *Handbook of Deposition Technologies for Films and Coatings*, 3rd Ed, Peter M. Martin, Ed., Elsevier (2009).
49. I. Sakata *et al.*, *Solar Energy Materials*, 10 (1984) 121.
50. J. L. Andujar *et al.*, *J. Vac. Sci. Technol.* A 9 (4) (1991) 2216.
51. F. Capezzuto and G. Bruno, *Pure Appl Chem*, 60(5) (1988) 633.
52. P. M. Martin and W. T. Pawlewicz, *J Non-Cryst Solids*, 45 (1981) 15.
53. D. A. Jenny and R. H. Bube, *Phys. Rev.* 96 (1954) 1190.
54. R. H. Bube, *Proceedings of the IRE 43* (12) (1955) 183.
55. D. A. Cusano, *Solid State Electronics* 6 (1963) 217.

56. B. Goldstein, *Phys. Rev* 109 (1958) 601.
57. Y. A. Vodakov *et al.*, *Soviet Physics, Solid State* 2 (1) (1960) 1.
58. X .Wu *et al.*, NREL/CP-520-31025 (2001).
59. N. A. Shaw *et al.*, *J Non Cryst Solids*, 355 (2009) 1474.
60. B. T. Ahn *et al.*, *Solar Energy Materials and Solar Cells*, 15 (1998) 155.
61. Benjamin R. Wakeling, Ph.D thesis "Close space sublimation of CdTe for solarcells and the effect of underlying layers", Cranfield University (2009).
62. U. Manmontri *et al.*, *Solar Energy Materials and Solar Cells*, 50 (1998) 265.
63. H. Liu *et al.*, *Thin Solid Films*, 231 (1993) 243.
64. Roberto Muri *et al.*, *J Non Cryst* Solids, 151 (1992) 253.
65. H. Reuter, *Thin Solid Films*, 254 (1995) 96.
66. Narish Chand, *Thin Solid Films*, 231 (1993) 143.
67. I. Repins *et al.*, Progress in Photovoltaics: Research and applications, 16 (2008) 235.

7

Functionally Engineered Materials

Engineered materials go one step beyond thin films to add additional functionality to either a thin film or a solid surface. As with many thin film materials, it is possible to engineer properties into thin films and surfaces that are not possible with bulk materials. We will find that properties can also be engineered into thin films and low dimensional structures that cannot be found in bulk solids. Thin films can have a wider variety and range of microstructures, compositions, bonding and short range structure, crystalline phase compositions and defects (structural and compositional) which makes them more sensitive to variations in these factors. Their thickness or "thinness" and large ratio of surface area to volume also contributes to a wide variation in properties. Properties of thin films can be attributed to:

- Energy band structure
- Lattice structure:
  - ○ Short range order or lack of short range order
  - ○ Crystalline phase composition
- Low dimensional structures
- Size effects

- Wide variation in density
- Void effects are more pronounced
- Higher mechanical stresses. Effects of stress are more pronounced.
- Surface and interface effects
- Quasi-two dimensional structures
- Quantum effects more likely

Examples of engineered materials are:

- Carbon nanotubes
- Superlattices and nanolaminates
- Photonic band gap
- Functional composites
- Sculpted thin films
- Thin film solar cells
- Dye sensitized solar cells

For our purposes, materials engineering will encompass design, synthesis, and application of man-made structures on micro and nano scales. At this scale, essentially in the range of atomic diameters to the wavelength of visible light, it is often necessary to use quantum mechanics to define and design physical properties and explain performance (as opposed to classical theory). Using many of the processes presented in the previous chapter, we can create:

- Artificial energy band structures
- Artificial lattices with extraordinary mechanical and physical properties
- Optical properties not possible in bulk solids or even thin films
- Nanocomposites and nanolaminates (see Section 4.2)
- Metamaterials
- Nanoplasmonic structures

7.1   Energy Band Structure of Solids

In order to fully understand how electrical and optical properties are engineered, it is necessary to review basic energy band structures.

Virtually every electrical and optical property of a solid can be linked to the energy band structure. Energy band structure, in its simplest form, consists of a valence band and conduction band that can be separated by a gap of forbidden energies, shown in Figure 7.1. Electrical conduction depends on whether or not there are electrons in the conduction band (or holes in the valence band). Figure 7.2 compares generic energy band structures of metals,

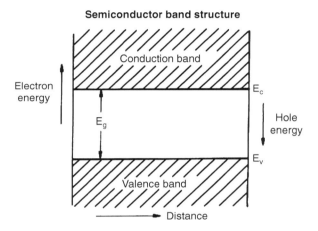

Figure 7.1 Generic energy band diagram.

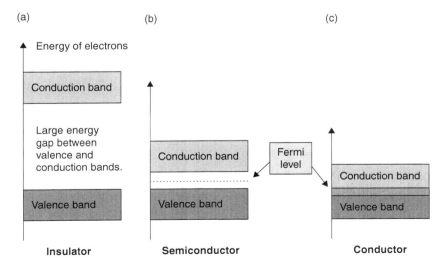

Figure 7.2 Energy band structure of insulators, semiconductors, and metals [2].

semiconductors, and insulators. In insulators, the electrons in the valence band are separated by a large gap from the conduction band, in conductors like metals the valence band overlaps the conduction band, and in semiconductors there is a small enough gap between the valence and conduction bands that thermal or other excitations can bridge the gap. With such a small gap, the presence of a small percentage of a dopant material can increase conductivity dramatically.

An important parameter in the band theory is the Fermi level, shown in Figures 7.3a and 7.3b, which is the top of the available electron energy levels at low temperatures. The position of the Fermi level with the relation to the conduction band is a crucial factor in determining electrical properties. Figure 7.3a plots the Fermi-Dirac energy distribution function at different temperatures [1]. At absolute zero (0 K), the distribution is a step function, which defines the Fermi energy E_F. This distribution "smears" out with increased temperature, indicating that electrons can have energies > E_F for T > 0. Since the number of energy states is constant, the same number will have energies < E_F. Figure 7.3b shows the position of E_F in the band gap. Electrons can thus be ejected into the conduction band by increasing their energy, using heat, electromagnetic (photons), or electrical energy (voltage).

Based on their energy band structure, metals are unique as good conductors of electricity. This can be seen to be a result of their

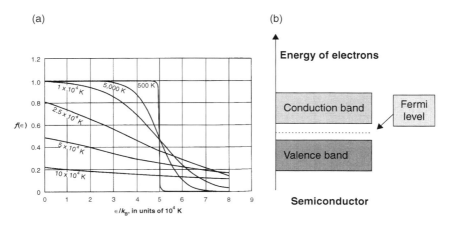

Figure 7.3 (a) Fermi-Dirac distribution function at different temperatures [1]. (b) Location of the Fermi level [3, 4].

valence electrons being essentially free. In the band theory, this is depicted as an overlap of the valence band and the conduction band so that at least a fraction of the valence electrons can move through the material.

It should be noted that Figures 7.1–7.3 show extremely simplified pictures of energy bands. Energy band structure depends on direction taken in the solid, impurities and defects in the lattice. As discussed below, energy band structure depends on the location and periodic structure of atoms in the crystal lattice, which depends on the direction in the lattice. Jumping ahead to Figure 7.23, we see that placement of atoms (and their resulting potential energy) depends on lattice structure and direction in the lattice. Conduction is often not a simple direct path of the electron from valence band to conduction band. Sometimes the electron needs a little help from a phonon (quantized lattice vibration). Figure 7.4 illustrates the two types of electron transitions in the energy bands. A direct transition is defined as the direct transfer of the electron from valence to conduction bands when valence band maximum matches conduction band minimum. An indirect transition occurs when valence band maximum does not match conduction minimum, and a phonon is needed to move the electron into the conduction band.

We can thus relate conductivity and other physical properties to energy band structure. This is complicated in thin films by the fact that defects, lack of long range order, compositional variations

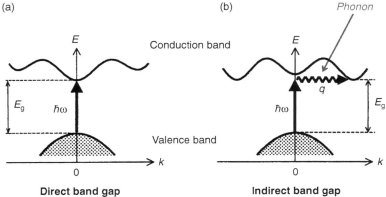

Figure 7.4 Direct and indirect electron transitions.

can affect and degrade their band structure [5]. These factors can totally smear out band structure in thin films, not to mention size effects. Quantum theory also predicts magnetic properties, magnetoresistance, thermal conductivity, thermoelectric effects, and optical phenomena.

7.2   Low Dimensional Structures

7.2.1   Quantum Wells

Carbon nanotubes, superlattices, nanowires, quantum dots, and photonic band gap materials are all low dimensional structures based on the quantum well and the multiquantum well. All reflect what is known as "man-made energy bands" and are typically based on semiconductors and optical materials. In order to understand low dimensional structures, it is important to understand the quantum well concept. The quantum well, shown in Figure 7.5, is constructed by sandwiching a thin (nm scale) semiconductor layer/film of one material between two layers/films of another semiconductor with a larger band gap. Figure 7.6 shows a quantum well formed by sandwiching a 1 nm thick GaAs (1.424 eV bandgap) layer between two AlGaAs (1.771 eV bandgap) layers [2]. The depth of the well is the difference in band gaps and equals 0.347 eV, and the width of the well is 1 nm (= 10 Å).

The purpose of the quantum well (QW) is to confine the electron's wave function and only allow certain energy levels. The electron can move in two directions with respect to the QW: (1) in the plane

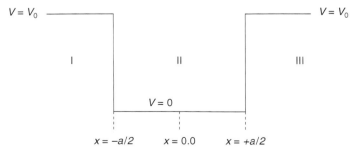

Figure 7.5 Potential energy profile of a quantum well [3].

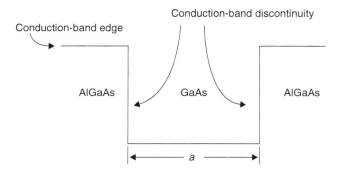

Figure 7.6 GaAs/AlGaAs quantum well [6].

of the well or (2) perpendicular to the plane of the well. Motion in the plane is similar to that of a free particle. The more interesting case is motion perpendicular to the plane of the well. We will not get involved with the math but just provide a pictorial and phenomenological tour through this concept. Figure 7.7 presents the concept of quantum confinement. In this example, we see that there are only two allowed confined energy states and that states with higher energy are not confined. Note that the wave functions of these states decay rapidly outside the well.

The density of electronic states (or just density of states) g(E) is the number of allowed electron energy states per unit energy interval around an energy E. Density of states is intimately connected to the energy band structure and essentially determines all transport properties. Engineering g(E) is accomplished by considering low dimensional structures and QWs. Figure 7.8 compares the density of state for bulk and 2D structures. g(E) for bulk varies as the square of the energy while the 2D g(E) has discrete steps at energies corresponding to quantized energy levels in the quantum well.

The density of states becomes even more discrete as we reduce dimensions from 2D to 1D and from 1D to 0D, as shown in Figures 7.9a and 7.9b. g(E) for the 2D case becomes highly peaked with energy compared to the steps for the 2D case. The quantum dot is the 0D case and is capable of confining electrons in all three dimensions and thus is the 0D case shown in Figure 7.9b. The density of states for the 0D case is essentially a series of delta functions, highly localized in energy. Thus we see that by confining the electron to smaller dimensions, we can achieve highly discrete energy states.

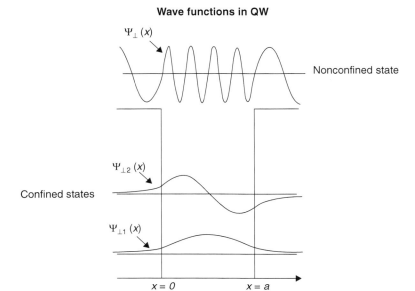

Figure 7.7 Shapes of wave functions for allowed energy levels of a quantum well [3].

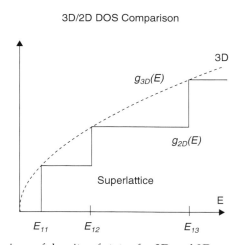

Figure 7.8 Comparison of density of states for 3D and 2D cases [3].

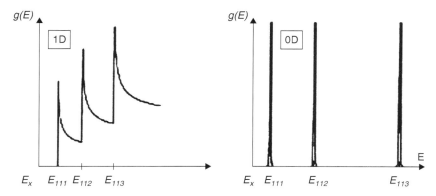

Figure 7.9 Comparison of density of states for 1D and 0D cases [3].

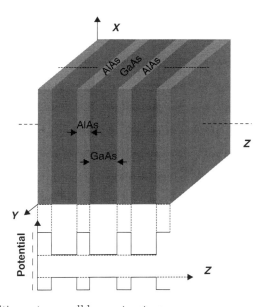

Figure 7.10 Multiquantum well layer structure.

7.2.2 Superlattices

Superlattices are multiquantum well structures, consisting of several to thousands of alternating narrow and wide band gap materials (usually semiconductors). Figure 7.10 shows the relationship between layer structure and quantum well structure. The band structure of superlattices can be engineered by

- Varying the width of the quantum well
- Varying materials with different band gaps (height of the well)
- Varying the number of layers/quantum well

Quantum well films consist of hundreds to thousands of nano-scale layers (1–20 nm) with alternating band gaps, one with a small band gap and one with a larger band gap. The layer with the larger band gap is used to confine charge carriers in the small band gap layer. The electical properties of these films can be engineered by varying layer thickness and period, and can be separated from those related to phonon propagation (thermal conductivity).

Minibands, such as those shown in Figure 7.11, can be formed in semiconductor devices using multiquantum wells such as quantum wires and superlattices. Figure 7.11 shows that as the number of energy bands confined in the quantum well increases, the number of minibands also increases. The miniband peaks in the absorption

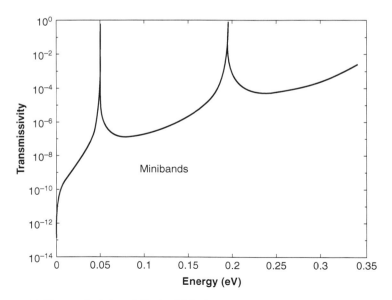

Figure 7.11 Transmissivity of GaAs/AlGaAs quantum wells vs incident electron energy showing miniband formation [6].

coefficient of GaAs/AlGaAs quantum wells (see Figure 7.11) correspond to the mini-subband n and well width = 30 Å. Each peak in the transmissivity of the quantum well corresponds to a confined energy subband [6]: 1,2 and 3. Here

$$n = 1: E_1 = 0.0273 \text{ eV}$$

$$n = 2: E_2 = 0.109 \text{ eV}$$

$$n = 3: E_3 = 0.236 \text{ eV}$$

Thus the optical properties, which depend on energy band structure, of the GaAs/AlGaAs structure can be engineered by building quantum wells.

Superlattices can have significantly better mechanical and tribological properties than those of the individual layers. Laminating different materials to achieve a combination of their various properties is a common technique for designing and engineering the mechanical and tribological properties of materials. Furthermore, when the laminate layers become very thin, the properties of the materials in individual layers often improve, relative to their macroscopic properties. Figure 7.12 shows how hardness increases with decreasing layer thickness for several bimetal systems [7, 8]. At layer thicknesses on the order of 10 nm, where the hardness approaches maximum, the mechanical properties of superlattices (also called nanolaminates) are strongly influenced by the nature of the mismatch of their crystal lattices at the interfaces as well as the very high ratio of interface volume to total material volume.

7.2.3 Quantum Wires

Quantum wires and carbon nanotubes (CNT), such as those shown in Figure 7.13, are one dimensional structures that possess unique optical, physical and mechanical properties that make them candidates for advanced technological applications, such as microelectronics, photovoltaics, thermoelectric and structural components. They are used in electrical and optical devices, structural composites, thermal devices and high frequency applications. CNT are high aspect ratio hollow cylinders with diameters ranging from 1–10 nm, and lengths on the order of centimeters. They are one of the many solid state forms of carbon (others are diamond,

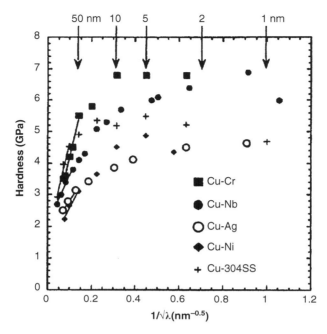

Figure 7.12 Results of hardness measurements showing the effect of layer thickness. The Cu–Cr, Cu–Nb, Cu–Ag, and Cu–Ni results are from [7, 8].

graphite, buckyballs). The three major types of CNT are single-walled (SWNT), multi-walled (MWNT) and Fullerite. Figure 7.13 shows the various types of single wall nanotubes (SWNT), including zigzag, armchair, and chiral [9]. Nanotubes are discussed in more detail later in this chapter (see Figure 7.58).

7.2.4　Quantum Dots

The quantum dot (QD) is the 0D structure in which charge carriers are confined in all three dimensions. A picture of QDs is shown in Figure 7.14. An exciton is an electron hole pair created by an external energetic source (such as photons), with the electron excited into the conduction band and the hole excited in the valence band of the semiconductor. As seen above, electrons in quantum wells have a range of discrete energies. Excitons have an average physical separation between the electron and hole, referred to as the exciton Bohr radius (EBR), and this physical distance is different for each material. In bulk, the dimensions of the semiconductor crystal

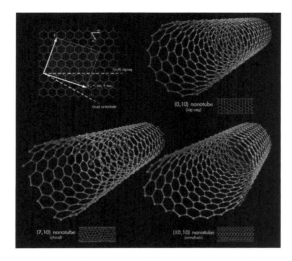

Figure 7.13 Various types of SWNT, including zigzag, armchair and chiral [9].

Bi₂Te₃/Sb₂Te₃ quantum dots

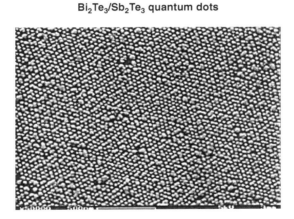

Figure 7.14 Picture of quantum dots.

are much larger than the EBR, allowing the exciton to extend to its natural limit. However, if the size of a semiconductor crystal becomes small enough that it approaches the size of the material's EBR, we have discrete energy levels formed in the quantum well, and the exciton is confined.

Because the QD's electron energy levels are discrete rather than continuous, the addition or subtraction of just a few atoms to the QD has the effect of altering the boundaries of the band gap. Changing the geometry of the surface of the quantum dot also changes the band gap energy, owing again to the small size of the dot, and the effects of quantum confinement. QDs can emit or absorb photons. The band gap in a quantum dot will always be larger than that of the bulk semiconductor and, therefore, emitted radiation will always be blue shifted (shorter wavelength). The minibands formed in the quantum well control the electrical and optical properties of the QD. Each peak in the miniband, such as those shown in Figure 7.11, corresponds to an energy transition between discrete electron-hole (exciton) energy levels. The quantum dots will not absorb light that has a wavelength longer than that of the first exciton peak, also referred to as the absorption onset. Like all other optical and electronic properties, the wavelength of the first exciton peak (and all subsequent peaks) is a function of the composition and size of the QD. Smaller QDs result in a first exciton peak at shorter wavelengths.

7.3   Energy Band Engineering

Knowledge of the energy band structure and how to modify it is important because band structure determines the electrical, optical, electro-optical, and magnetic properties of a thin film. One of the most important results of quantum theory is the description of available energies for electrons in the materials, i.e., energy bands. The energy band structure depends primarily on the lattice structure and electron configuration of a solid and is derived from Bloch's theorem [3]. Once we know the important parameters that determine energy band structure, we can begin engineering it. Bloch's theorem states that the eigenfunction for a particle moving in a periodic potential, such as a solid crystal, can be expressed as the product of a plane wave envelope function and a periodic function (Bloch function). Bloch wave functions can be expressed as a combination of a plane wave modulated by a periodic potential function:

$$\psi(k,r) = \exp(ik.r)u(k,r) \tag{7.1}$$

Here all vectors are in bold, $\mathbf{k}$ is the wavenumber vector, $\mathbf{r}$ is the position, and $u(\mathbf{k},\mathbf{r})$ is the space-dependent amplitude. The periodicity of the lattice $\mathbf{R}$ requires that

$$u(k, r + R) = u(k, r) \text{ and thus} \qquad (7.2)$$

$$\psi(k, r + R) = \exp(ik.R)u(k, r) \text{ and} \qquad (7.3)$$

$$k = 2\pi n / Nd$$

where n is an integer between $-N/2$ and $N/2$ and d is the period of the square well.

Figure 7.15 shows the behavior of the plane wave modulated by a periodic function. More generally, a Bloch-wave description applies to any wave-like phenomenon in a periodic medium. For example, a periodic dielectric in electromagnetism leads to photonic crystals, and a periodic acoustic medium leads also to photonic crystals. It is generally treated in the various forms of the dynamic theory of diffraction.

Every band structure model involves solving for the energy eigenvalues of a periodic lattice and the resulting periodic potential. It provides a simple understanding for the wide variation in

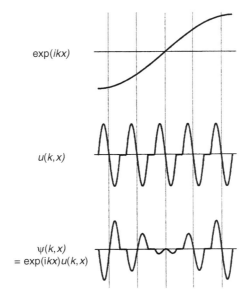

Figure 7.15 One- dimensional Block wavefunction: a plane wave (top) is modulated by a periodic potential (middle) to form the Bloch wavefunction (bottom) [3].

conductivity from insulators to metals, at least for crystalline solids. Every model involves an approximation. The fundamental energy band models are:

- Kronig-Penny model
- Nearly free electron approximation
- Tight binding approximation

We will not get involved with derivation or calculation of energy bands, which have been presented extensively in other publications [12, 13, 14, 15], but highlight approximations that are made for each model. Also note that the potential energy in a crystal lattice is extremely directional dependent. This should be apparent from the fact that the location of atoms depends on crystal structure, which is not uniform in every direction. This fact is very important for energy band structure. The Kronig-Penny (KP) model assumes a square wave potential, shown in Figure 7.16. Energy band structures are generally exhibited as plots of energy versus wavevector in the first Brillouin zone. Working through the math for this model results in allowed and forbidden energy regions. Figure 7.17 shows energy band regions as a function of parameter ζ (or energy) according to the KP model. We see that only certain energy bands

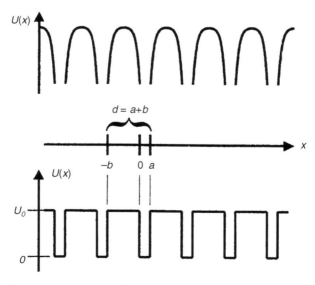

Figure 7.16 Square wave potential used in KP model [3].

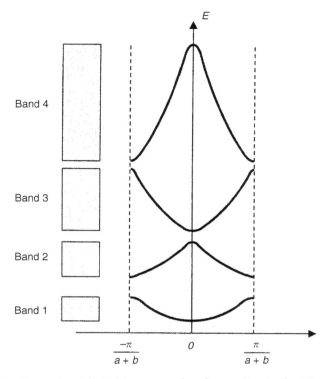

Figure 7.17 Allowed and forbidden energy band according to the KP model [3].

are allowed. The relationship between energy and wavevector is parabolic for a classical free electron (E ~ k²). However, if the KP model is applied to this relationship, we end up with the electron energy structure shown in Figure 7.18 [13]. Compare this to the energy of a free electron (dotted curve). Band gaps are apparent at Brillouin zone boundaries. This type of structure results from the other two models.

The nearly free electron approximation assumes that the periodic potential introduces a small perturbation on the free electron state. The perturbation term is added to the potential energy in the Schrodinger equation ($\hbar^2\nabla^2\psi/2m + [E - U]\ \psi = 0$). Discontinuities in the energy spectrum result from reflections of electron waves at zone boundaries (−K/2 and K/2), as shown in Figure 7.19. The resulting band structure is quite similar to the KP model.

The tight binding approximation employs atomic wavefunctions as the basis set for the construction of real electron wavefunctions.

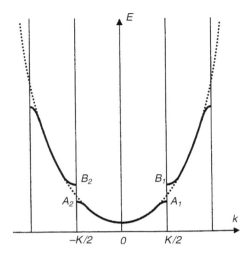

Figure 7.18 Electron energy in a periodic lattice for a classic free electron [13].

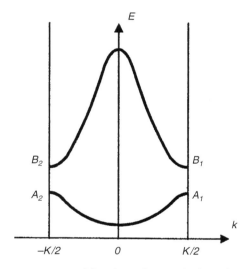

Figure 7.19 Electron energy resulting from the nearly free electron approximation [3].

The model brings isolated atoms with discrete energy levels together and arranges them in a lattice. As a result, the potential of each atom overlaps nearest neighbors and thus electrons in one atom have a non zero probability of being found in another atom (or lattice site). The electron thus does not live at a certain atomic energy indefinitely but travels from lattice site to lattice site, which is the formation of allowed energy bands. Electrons in this model are tightly bound to the atom to which they belong and have limited interaction with states and potentials on surrounding atoms in the lattice. As a result the wavefunction of the electron will be similar to the atomic orbital of the free atom. The energy of the electron will also be close to the energy of the electron in the free atom or ion because the interaction with potentials and states on neighboring atoms is limited. When the atom is placed in a crystal, this atomic wavefunction overlaps adjacent atomic sites, and so are not true eigenfunctions of the crystal Hamiltonian. The overlap is less when electrons are tightly bound. Any corrections to the atomic potential ΔU required to obtain the true Hamiltonian H of the system are assumed small. The Hamiltonian can thus be expressed as (we will go no further here due to the complexity of the mathematics):

$$H(r) = \Sigma H_{at}(r - R_n) + \Delta U(r) \quad \text{(summation over n)} \quad (7.4)$$

The resulting wavefunction can be written as a linear combination of atomic orbitals ($\phi_m(\mathbf{r} - \mathbf{R}_n)$):

$$\psi(r) = \Sigma \exp - ik(r - R_n)(\phi_m(r - R_n) \quad \text{(summation over n)}$$
$$(7.5)$$

Where m is the mth atomic energy level and $\mathbf{R}_n$ is the location of the nth atomic site in the lattice. Essentially, a number of atomic energy levels are thus consolidated into discrete energy bands for the electron, as shown in Figure 7.20.

With this information under our belt, how do we engineer band structure in a thin film or other type of structure? We see that band structure depends on lattice potential (resulting from lattice structure), composition, bond structure, and atomic energy level structure. Actual energy band theory is extremely complicated and involves sophisticated quantum mechanics [16]. It will first be instructive to review the types of energy band structures found in solids and then address how modifications in the aforementioned parameters affect band structure. The structure of thin films runs the gamut from crystalline to amorphous. Amorphous and disordered solids can also be included, but with increased complexity. Referring to Figures 7.1 and 7.2 which show a generic diagram of

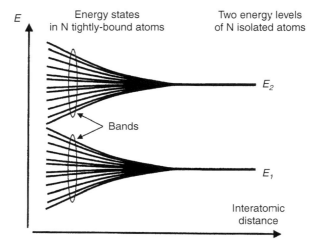

Figure 7.20 Consolidation of discrete atomic energy levels into allowed electron energy bands [3].

energy bands and provide a simple summary of the energy band structure of insulators, semiconductors, and metals, energy band structure, in its simplest form, consists of a valence band and conduction band that can be separated by a gap of forbidden energies (called the bandgap) [17]. Electrical conduction depends on whether or not there are electrons in the conduction band (or holes in the valence band). In insulators, the electrons in the valence band are separated by a large gap from the conduction band, in conductors like metals the valence band overlaps the conduction band, and in semiconductors there is a small enough gap between the valence and conduction bands that thermal or other excitations can bridge the gap. With such a small gap, the presence of a small percentage of a doping material can increase conductivity dramatically.

As a simple example of how lattice structure affects energy band structure, we compare the energy bands of Si and Ge semiconductors. Both of these semiconductors have the diamond, or the face center cubic (fcc) crystal structure. They differ, however, in their spacing between atoms in the lattice and electron configurations: Si spacing is 5.43 Å with $3s^23p^2$ configuration, while Ge lattice spacing is 5.64 Å with $3d^{10}4s^24p^2$ configuration [18]. The energy band structures for Si and Ge are completely different, as shown in Figure 7.21. In general, energy bands with negative energy correspond to the

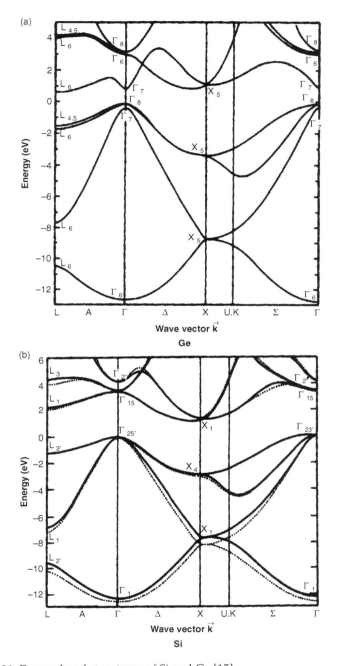

Figure 7.21 Energy band structures of Si and Ge [15].

valance band and those with positive energy correspond to the conduction band. Additionally, Si has an indirect band gap while Ge has a direct gap.

It should be noted that Figures 7.1, 7.2, and 7.21 show extremely simplified pictures of energy bands. Energy band structure depends on direction taken in the solid, impurities and defects in the lattice. Conduction is often not a simple direct path of the electron from valence band to conduction band. Sometimes the electron needs a little help from a phonon (quantized lattice vibration). Figure 7.4 illustrates the two types of electron transitions in the energy bands. A direct transition is defined as the direct transfer of the electron from valence to conduction bands when valence band maximum matches conduction band minimum. An indirect transition occurs when valence band maximum does not match conduction minimum and a phonon is needed to move the electron into the conduction band.

Engineering of energy bands is thus accomplished by modification of the crystal lattice, introduction of compositional and structural defects, reducing the dimensions of the solid to the nanoscale, and building artificial crystal lattices. As with crystal symmetry, energy band structure also depends on the direction in the lattice, as demonstrated in Figure 7.22. Examples of these actions are:

- Photonic crystals, compound semiconductors
- Substitutional doping of semiconductors
- Nanowires, quantum dots, and nanotubes
- Superlattices and nanolaminates

An example of how composition can alter band structure is GaAs and AlGaAs. Figure 7.23 compares crystal structures for these two

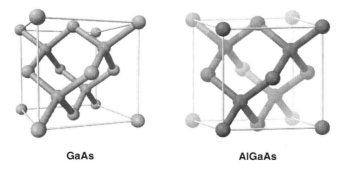

GaAs   AlGaAs

Figure 7.22 Crystal structures of GaAs and AlGaAs.

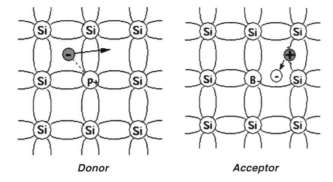

Figure 7.23 Donor (n-type) and acceptor (p-type) doping of silicon.

very similar materials. AlGaAs (actually $Al_xGa_{1-x}As$) has nearly the same lattice constant as GaAs but a larger band gap: 1.42 eV (GaAs) and 2.16 eV (AlGaAs). Band gap structure depends on Al content (x) and is direct for x < 0.4.

Substitutional doping is the easiest method used to modify the density of states and the band gap, and can significantly increase the electrical conductivity of the semiconductor film. The dominant charge carrier type (electron or hole) can be achieved by doping. The addition of a small amount of the right type of foreign atom in the crystal lattice can change electrical properties, producing n-type or p-type semiconductors (see Figure 7.23). Doping introduces donor states just below the conduction band or acceptor state just above the valence band, as shown in Figures 7.24 and 7.25 [19]. For example, in Si, impurity atoms with 5 valence electrons produce n-type semiconductors by contributing extra electrons (donors) and impurity atoms with valence 3 produce p-type semiconductors by producing holes, or electron deficiency (acceptors). Donor atoms for Si are Sb, P, and As while acceptor atoms are B, Al, and Ga.

Band structures can also be engineered artificially to achieve electrical and optical properties not possible with conventional thin film or bulk materials, which is addressed in the following sections.

7.3.1   Carbon Nanotubes

Carbon nanotubes (CNT) are low dimensional structures that possess unique optical, physical, and mechanical properties that make them candidates for advanced technological applications, such as microelectronics, photovoltaics, thermoelectric and structural

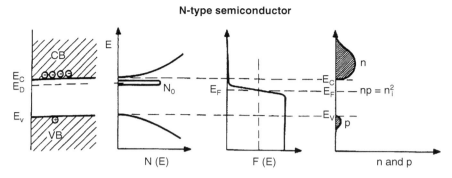

Figure 7.24 Donor states and density of states for an n-type semiconductor [19].

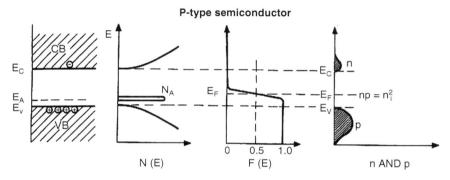

Figure 7.25 Acceptor states and density of state for a p-type semiconductor [19].

components. They are used in electrical and optical devices, structural composites, thermal devices, and high frequency applications. Properties of carbon nanotubes (CNT) that make them so useful are:

- Metallic or semiconductor properties possible
- Direct band gap that allows transitions between valence and conduction bands without the help of a phonon
- Multiple bands can participate in conduction
- Low defect density
- Metallic interconnects possible
- Very high thermal conductivity
- High structural integrity; they are the world's strongest fibers

Additionally, nanotubes conduct heat as well as diamond at room temperature, are very sharp, and as a result can be used as probe tips for scanning-probe microscopes, and field-emission electron sources for lamps and displays.

CNTs are high aspect ratio hollow cylinders with diameters ranging from 1–10 nm, and lengths on the order of centimeters. They are one of the many solid state forms of carbon (others are diamond, graphite, buckyballs). The three major types of CNT are single-walled (SWNT), multi-walled (MWNT), and Fullerite. Figure 7.13 in Section 7.1.3 shows the various types of single wall nanotubes (SWNT), including zigzag, armchair, and chiral [20].

The structure of a SWNT can be conceptualized by wrapping a one-atom-thick layer of graphite called graphene into a seamless cylinder. Figure 7.26 shows the atomic structure of a SWNT [21]. The nanotube can be envisioned as a strip of graphene rolled up in a closed cylinder. The way the graphene strip, shown in Figure 7.27, is wrapped is represented by a pair of indices (n, m) called the chiral vector. Basis vectors are $a_1 = a(\sqrt{3},0)$ and $a_2 = a(\sqrt{3}/2, 3/2)$, with the circumferential vector defined as $C = na_1 + ma_2$, corresponding to the edge of the graphene strip [20]. Integers n and m denote the number of unit vectors along two directions in the honeycomb crystal lattice of graphene. The types of SWNT are defined as:

- $m = 0$ for zigzag
- $n = m$ for armchair
- All others, chiral

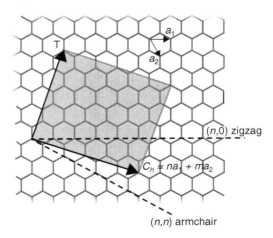

Figure 7.26 Atomic structure of a SWNT [21].

Figure 7.27 Graphene strip [21].

The radius of the NT is

$$R = \frac{C}{2\pi} = \left(\frac{\sqrt{3}}{2\pi}\right)\alpha\sqrt{\{n^2 + m^2 + nm\}} \qquad (7.6)$$

SWNTs are very important because they exhibit important electrical and optical properties that are not achieved by MWNT structures. The energy band structure and density of states of NT gives them unique optical properties not present in graphite. Optical and transport properties are derived using quantum mechanics and depend on the orientation of the NT. Figure 7.28 shows the density of states for a metallic and semiconducting SWNT [21]. Properties are metallic when $n - m = 3p$ ($p = 0$ or positive integer) and semiconducting for all other combinations [22]. In this case, the band structure is described much like that of a polymer: by the highest occupied molecular orbital (HOMO) and lowest unoccupied molecular orbital (LUMO). The fundamental energy gap (HOMO - LUMO) depends on the diameter (2R) of the NT:

$$E_g = 5.4a/2R \text{ (metallic SWNT) [22].}$$

E_g is thus in the range 0.4–0.7 eV for metallic SWNT. This small energy gap reflects the conductive nature of the metallic nanotubes.

The density of states for semiconducting SWNT is very different, as shown in Figure 7.28b, and is derived from the same relation, however, the resultant band structure is very different. Figure 7.29 shows the optical band gaps for light polarized parallel and perpendicular to

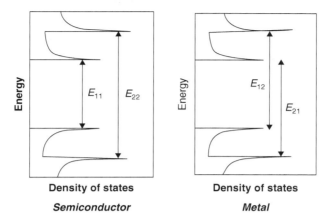

Figure 7.28 Density of states for (a) metallic and (b) semiconducting SWNT [22].

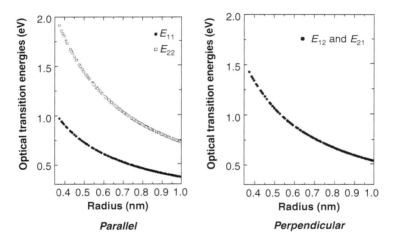

Figure 7.29 Optical band of SWNT with light polarized parallel and perpendicular to the axis of the tube [20].

the axis of the nanotube [20]. Band gaps range from 1.7 to 2.0 eV, and decrease with increased tube radius [22].

As with many low dimensional quantum structures (OLEDs for example), excitons play an important role in the optical and electrical properties [20]. Excitons are bound electron-hole pairs. The Coulomb attraction between the electron and hole leads to new quantized energy levels that affect the physical properties. To demonstrate this, Figure 7.30 shows the optical absorption spectra with and without excitons [23]. Exciton interactions create significant

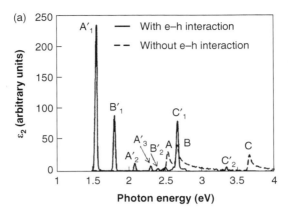

Figure 7.30 Comparison of semiconductor SWNT absorption spectra with and without exciton interactions [23].

absorption peaks near 1.5 eV, 1.8 eV, and 2.7 eV, not present if they were ignored. Excitons play an important role even in metallic SWNT, as shown in Figure 7.31 [24].

Similar to the optical properties, electrical conduction of a SWNT is quantized, and a nanotube acts as a ballistic conductor. Nanotubes also have a constant resistivity, and a tolerance for very high current density. They are ready made for nanoscale electronic devices. Figure 7.32 shows the energy band structure for semiconducting (a) and metallic (b) zigzag SWNTs [20]. The Fermi level is at zero energy. Note that the metallic SWNT has no band gap at ka = 0 and 2. Similar to the optical band gap, the electronic band gap decreases with increased radius, as shown in Figure 7.33 [20]. Resistivity is ~10^{-4} Ω.cm at 300 K [26]. High current densities ~10^7 A/cm² have been achieved and even higher values ~10^{13} A/cm² are considered possible [26, 27].

One would expect the thermal properties of CNTs to be as unique as the optical and electrical properties. Thermal conductivity of SWNT is dependent on the temperature and the large phonon mean free paths in the lattices. Recall that thermal conductivity has contributions from charge carriers and lattice vibrations (phonons). It has been shown to depend on the length of the nanotube as well, and level off at 28 W/cmK at a length ~10 nm [28, 29]. Figure 7.34 shows the thermal conductivity as a function of temperature [28]. Note that the thermal conductivity is high ~100 W/cm.K at 300 K.

SWNTs are promising candidates for miniaturizing electronics beyond the micro electromechanical scale that is currently the basis

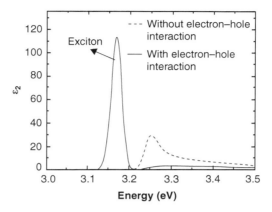

Figure 7.31 Comparison of metallic SWNT absorption spectra with and without exciton interactions [24].

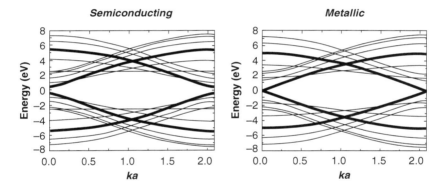

Figure 7.32 Band structure of semiconducting and metallic zigzag SWNT [20].

of modern electronics. The most basic building block of these systems is the electric wire, and as we have seen, SWNTs can be excellent conductors as well as semiconductors. One useful application of SWNTs is in the development of the first intramolecular field effect transistors (FETs). Other applications include thermoelectric and optoelectronic devices and structural components.

7.3.2   Optoelectronic Properties of Carbon Nanotubes

Before we address nanotube optoelectronic devices, it will be instructive to go into more detail on the optical properties of SWNTs. As

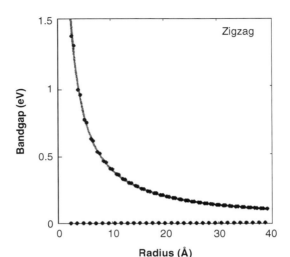

Figure 7.33 Electronic band gap of zigzag WSNTs [25].

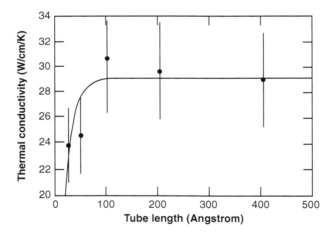

Figure 7.34 Thermal conductivity of SWNT [28].

with bulk and thin film semiconductors, the optical properties of CNTs are governed by selection rules, which place us in the realm of quantum mechanics. Common to all low dimensional structures (superlattices, quantum dots, nanowires, etc.) new states are introduced into the band gap and these states modify physical and optical properties. As shown in Figure 7.35, the nanotube can be illuminated with light polarized parallel or perpendicular to its axis [30].

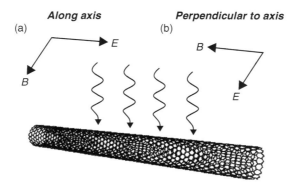

Figure 7.35 Illumination of CNT with light polarized parallel or perpendicular to its axis [30].

Optical properties of SWNTs result from electronic transitions within one-dimensional density of states (recall that nanotubes are one-dimensional structures). Figure 7.36 shows the band to band transitions for light polarized parallel to the nanotube axis for metal and semiconductor nanotubes, and Figure 7.37 shows these transitions for light polarized perpendicular to the axis [30, 31]. One major difference between nanotubes and bulk materials is the presence of sharp singularities, known as *Van Hove* singularities, in the density of states. Referring to these figures, the following optical properties result from these singularities:

- Optical transitions occur between the v_1-c_1, v_2-c_2, etc., states of semiconducting or metallic nanotubes and are traditionally labeled as S_{11}, S_{22}, M_{11}, etc. (S – semiconductor, M – metal). If the conduction properties are not known, the states are labeled as E_{11}, E_{22}, etc. Crossover transitions c_1-v_2, c_2-v_1, etc., are dipole-forbidden and are extremely weak [31].
- Energies (E_{11}, E_{22}, etc.) between Van Hove singularities depend on the nanotube structure, and as a result, the optoelectronic properties can be tuned by varying this structure [32].
- Optical transitions are sharp ($\sim$10 meV) and strong. As a result, it is relatively easy to selectively excite nanotubes having certain (n, m) indexes, as well as to detect optical signals from individual nanotubes.

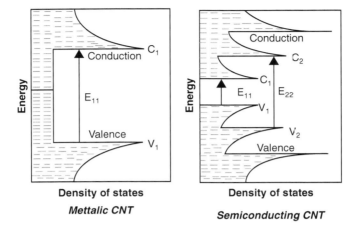

Figure 7.36 Optical transitions between valence and conduction bands along the axis of SWNT [2].

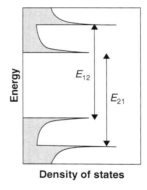

Figure 7.37 Optical transitions between valence and conduction bands perpendicular to the axis of SWNT [30, 31].

Figure 7.38 shows that the energy of these electronic transitions decreases with increasing tube diameter [33]. Note that this appears to be true for all transitions. The dependence on direction in the nanotube appears to be significant also. Each branch displays an oscillatory behavior, which reflects the dependence on the indices (n, m) of the direction in the nanotube (recall that nanotubes can be uniquely identified by these indices). For example, (10,0) and (8,3) tubes have almost the same diameter, but very different properties; the former is a metal, but the latter is semiconductor.

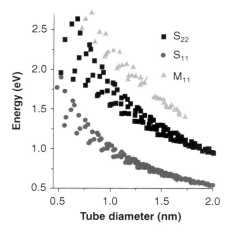

Figure 7.38 Relationship between tube diameter and optical transition energies [33].

The dependence on direction and optical transition is reflected in the absorption spectrum of a nanotube, as shown in Figure 7.39 [34, 36]. Absorptions due to S_{11}, S_{22}, and M_{11} transitions are labeled. The π peak is due to carbon. The transitions are relatively sharp and can be used to identify nanotube types; however, many nanotubes have very similar E_{22} or E_{11} energies, and significant overlap can occur in absorption spectra. The sharpness of the peaks degrades with increasing energy. Also, lines tend to get washed out if a number of different nanotubes are present.

The optical properties of bulk materials can generally be described using a single particle model. This is not the case for SWNTs, in which many-body effects dominate the optical properties [30]. Excitons rule in the world of nanotubes. Recall that an exciton is an excitation of an electron-hole pair bound by the Coulomb interaction. Excitons have quantized energy levels and the difference between these levels and the free electron energy is referred to as the exciton binding energy. Figure 7.40 shows a metallic NT absorption spectrum with and without exciton contributions [37]. Absorption peaks are either enhanced, shifted, or new ones added.

Excitons are also responsible for photoluminescence, which can be exited by absorption of light via the S_{22} transition, thus creating an exciton. With the help of a phonon, the electron and hole relax from c_2 to c_1 and from v_2 to v_1 states (see Figure 7.41), respectively [38]. Then they recombine through a $c_1 - v_1$ transition resulting in light emission (photoluminescence). Metallic NTs cannot create photoluminescence.

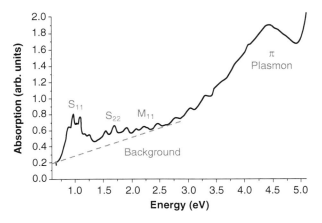

Figure 7.39 Optical absorption spectrum of SWNTs [34, 35, 36].

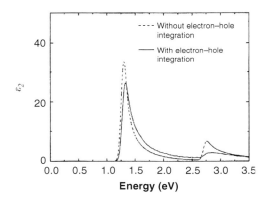

Figure 7.40 Optical absorption spectra of a metallic CNT with and without exciton contribution [38].

A photoluminescence map from single-wall carbon nanotubes is shown in Figure 7.41 with (n, m) indexes identifying specific semi-conducting nanotubes [39]. Thus photoluminescence is one technique to identify the type of nanotube.

The photoluminescent properties of SWCNTs can be summarized as:

- Photoluminescence (PL) from SWCNT, as well as optical absorption and Raman scattering, is linearly polarized along the tube axis. This allows monitoring of the SWCNTs orientation without direct microscopic observation.
- PL is quick: relaxation typically occurs within 100 ps [38].

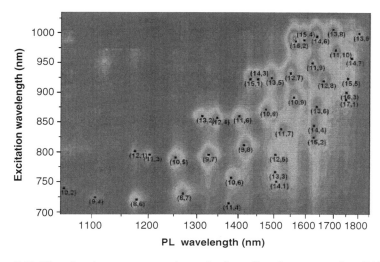

Figure 7.41 Photoluminescence map from single-wall carbon nanotubes [39].

- PL efficiency is usually low (~0.01%) [10], however, it can be enhanced by improving the structural quality of the nanotubes and clever nanotube isolation strategies [40].
- The spectral range of PL is rather wide. Emission wavelength can vary between 0.8 and 2.1 μm depending on nanotube structure [25, 26].
- As discussed above, interaction between nanotubes or between nanotube and another material (e.g., substrate) quenches PL [41]. Detachment of the tubes from the substrate drastically increases PL.
- The environment of the CNT can affect transition peaks. Again, the shift depends on the (*n, m*) indices (see above).
- If electrical contacts are attached to a nanotube electron-hole pairs (excitons) can be generated by injecting electrons and holes from the contacts 42].

There are a number of modes by which a CNT can emit radiation. These can be summarized as:

- Radial breathing mode (RBM): radial expansion/ contraction of the NT and frequency depends on NT diameter, and in fact, can be used to determine NT diameter [43].

- Bundling mode: collective vibrations in a bundle of RBM mode NTs [44].
- G (graphite) mode: planar vibrations of carbon atoms (~6 μm wavelength)
- D mode: due to structural defects in carbon atoms and used to determine the structural integrity of CNTs
- G' mode: second harmonic of D mode

Finally, the photoconductivity of CNTs must be addressed before optoelectronic properties can be thoroughly described. Photoconductivity is essential for operation of solar cells and virtually all optoelectronic devices. Simply put, photoconductivity is the onset of electric current when a material is illuminated. However, there must be an electric field in order for electric current to flow. This is accomplished by simple p-n junctions and Schottky barriers in solar cells and optoelectronic devices. We have seen that CNTs can be used to make these devices. Figure 7.42 shows the process of photon absorption by a nanotube p-n junction [45]. Note the similarities to bulk semiconductor photoconduction. In process #1, an electron is excited from the valence to the conduction band by absorption of a photon with energy ħω larger than the band gap. Processes 2 and 3 are indirect with the absorption of a photon with energy less than the band gap are assisted by a phonon. Because there are a number of sub-bands and optical transitions allowed in CNTs (see Figure 7.40), the photoresponse of a CNT p-n junction can have a number of peaks, as shown in Figure 7.43 [30]. Each peak corresponds to a different CNT band transition. This transition is allowed

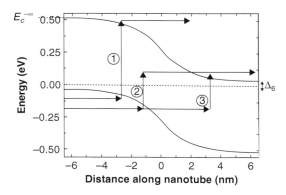

Figure 7.42 Photon absorption by a nanotube p-n junction [45].

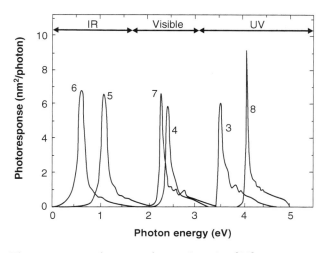

Figure 7.43 Photoresponse of a nanotube p-n junction [30].

if $\hbar\omega$ is greater than the band gap plus the potential step across the junction. The CNT p-n junction has wide photoresponse across a broad range of wavelengths, including UV - Visible - IR. Band J8 shows a UV response while band J6 moves to IR wavelengths.

Because the nanotube is illuminated from the side, its photoresponse also depends on length of the tube. As expected, photoresponse increases with increased tube length l [45]. For example, for $\hbar\omega = 0.612$ eV, photoresponse increases from 0 nm²/photon at $l = 0$ to ~30 nm²/photon at $l = 100$ nm. However, if phonon assist is required, photoresponse saturates with length. The dependence with length is attributed to increased band-to-band transitions, and scales with length.

As with bulk semiconductor p-n junctions, the current-voltage (I-V) characteristics of the nanotube p-n junction change under illumination. Dark I-V response can be expressed as

$$I_{dark} = I_S(e^{eV/kT} - 1) \qquad (7.8)$$

where I_S is given by

$$I_S = I_{0S}e^{-eVs/kT} \quad (V_S = E_g - 2\Delta)$$

and Δ is the difference between the asymptotic conduction band edge in the n-region and the Fermi level [30].

When the nanotube is illuminated, the device produces photo-current in the opposite direction of the dark current, which shifts the I-V curve, as shown in Figure 7.44 [46]. Thus, total current

$$I = I_{dark} - I_{ph}. \tag{7.8}$$

This brief introduction will lead us to CNT and NT optoelectronic devices, including solar cells, dye sensitized solar cells, bolometers, and electroluminescent devices.

7.3.3 Nanotube-based Electronic Devices and Photovoltaic Cells

Thin film solar cells, which include p-n and p-i-n junction semi-conductor cells, thin film semiconductor cells, dye sensitized cells, Gratzel cells, and organic cells, are addressed in detail in the next chapter. We will also include titanium dioxide (TiO_2) nanotubes since they are becoming very important in solar and energy applications. The anatase phase of TiO_2 is the premier photocatalytic material used in a wide range of applications, including Gratzel cells, self cleaning surfaces, hydrogen production, reduction of organics, oxygen production, and biomedical applications. TiO_2 is a semiconductor with a band gap ~3.2 eV, and the anatase phase of this has just the right energy band gap to produce electron hole pairs using ultraviolet radiation. Figure 7.45 shows the various photocatalytic

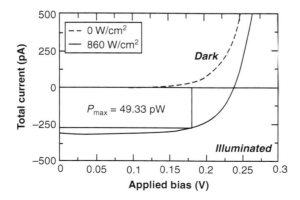

Figure 7.44 I-V plot of dark and illuminated nanotube p-n junction [49].

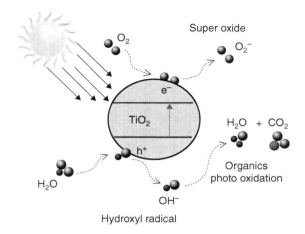

Figure 7.45 Photocatalytic processes occurring in TiO_2.

processes taking place in TiO_2. A UV photon excites an electron into the conduction band, forming an electron-hole pair. To review:

- Electrons can be used to form the photocurrent of a solar cell
  - ○ $TiO_2 + h\nu \rightarrow e + h$
- Holes can perform hydrolysis (photolysis)
  - ○ $2H_2O + h\nu \rightarrow^{(TiO2)}$ Active oxygen (AO) $+ 2H^+ + 2e^-$
- Holes can oxidize organics by
  - ○ $2H_2O + h\nu \rightarrow^{(TiO2)\text{-hole}} + OH^-$
  - ○ $OH^- + organic \rightarrow CO_2 + H_2O$

We are specifically interested in the first bullet and will see how nanotubes can replace TiO_2 thin films and nanoparticles [47] in these cells. A TiO_2-based cell functions as follows (see Figure 7.46) [48, 49]:

- A monolayer of dye is bonded to the TiO_2. Ideally, the dye should be black, i.e., absorb completely in the UVA, visible, and NIR.
- Sunlight enters the cell through the top transparent conducting (TCO) contact, striking the dye on the surface of the TiO_2. More photons will be absorbed with increased porosity or surface area (keep this in mind for nanotubes).

- Photons striking the dye with enough energy to be absorbed will create an excited state in the dye, from which an electron can be injected directly into the conduction band of the TiO_2. We will see how modification of the band structure of nanotubes can tailor photon absorption.
- The electron next moves by a chemical diffusion gradient into the top TCO. The dye molecule has lost an electron and will decompose if another electron is not provided.
- The dye strips an electron from the electrolyte, creating a redox reaction. This reaction occurs quickly compared to the time that it takes for the injected electron to recombine with the oxidized dye molecule, preventing this recombination reaction that would effectively short-circuit the solar cell.
- The electrolyte then recovers its missing electron by mechanically diffusing to the Pt counterelectrode at the bottom of the cell, where the cathode injects electrons after flowing through the external circuit.

Thus, the surface that holds the dye should be highly porous (or have high surface area) and be able to transport electrons and separate electrons and holes efficiently. TiO_2 nanotubes are well suited for use in dye sensitized solar cells (DSSC). Figures 7.47 and 7.48 show pictures of TiO_2 nanotubes and Figure 7.49 shows a schematic of how

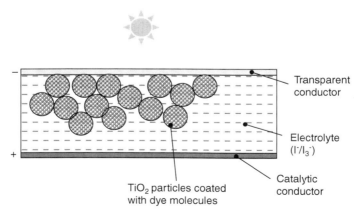

Figure 7.46 Structure of TiO_2 solar cell [48, 49].

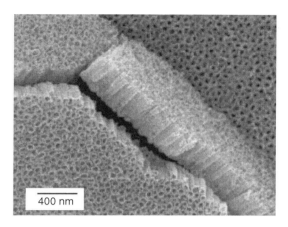

Figure 7.47 Picture of TiO$_2$ nanotube array [50].

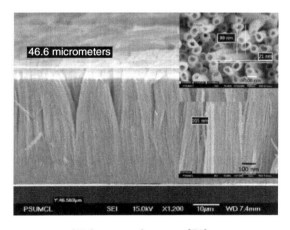

Figure 7.48 SEM picture of TiO$_2$ nanotube array [51].

they are used in DSSCs [50, 51, 52]. Nanotubes are synthesized by anodizing sputtered Ti thin films by placing them in an acidic bath with a mild electric current. Titanium dioxide nanotube arrays grow to about 360 nm in length [51]. The tubes are then crystallized by heating them in oxygen. This process turns the opaque coating of titanium into a transparent coating of nanotubes. The axes of the nanotubes are oriented perpendicular to the substrate, as shown in Figure 7.49. This nanotube array is then coated in a commercially available dye. Dye coated nanotubes make up the negative electrode and a positive electrode seals the cell which contains an iodized electrolyte.

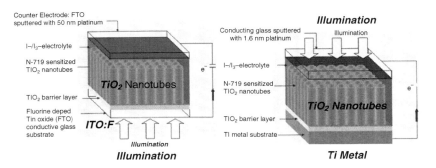

Figure 7.49 Structure of TiO$_2$ nanotube DSSC [52].

TiO$_2$ nanotubes have a well defined and controllable pore size, wall thickness, and tube-length (see Figures 7.47 and 7.48). For a 20 µm nanotube array, the effective surface area is ~3000 times greater than that of a planar unstructured surface. Like most one-dimensional structures, this high-aspect ratio nanotube array architecture promotes efficient harvesting of photons by orthogonalizing the processes of light absorption and charge separation [53]. This tube geometry permits dye to be coated onto the walls or filled into pores of a semiconductor. Both configurations increase exciton production (see above for a discussion of excitons). Two types of nanotubes are being developed for DSSCs: opaque and transparent [54]. Opaque (metallic) nanotubes consist of the acid etched films mentioned above, and transparent NTs are anodized. Metallic NTs are generally 30–40 µm long while transparent NTs are far behind at ~4 µm. The advantage of transparent NTs is that the cell can be illuminated from the front, which reduces absorption by the counter electrode. Metallic NTs must be illuminated from the back of the cell, which causes larger optical losses.

Conversion efficiency generally increases with NT length (more surface area to absorb photons). Conversion efficiency for UV radiation is ~12%, but some think that 15% can be achieved eventually. The main factor that limits higher efficiencies is transport of electrons across the NT network. Recombination at tube interfaces is a problem. While increasing NT length helps, interface states lying below the conduction band edge must also be reduced [54].

Interestingly enough, CNTs are now being developed for use in DSSCs. They act as anchors for highly photoresponsive semiconductors. With recent progress in CNT development and fabrication, there is promise to use some CNT based nanocomposites and nano-structures to direct the flow of photogenerated electrons and assist in charge injection and extraction. The CNT may be overcoated with TiO_2 nanoparticles in some cases [54, 55]. A sol-gel method is used to overcoat multiwall carbon nanotubes (MWCNT) with TiO_2 [54]. This was found to increase the efficiency of the DSSC by ~50% compared to conventional TiO2 DSSCs. CdS quantum dots are also attached to SWCNTs to enhance charge transport and injection [56]. Other semiconductors such as CdSe and CdTe as well as porphyrin and C60 fullerine also show promise [57, 58]. All these nanoparticles must somehow be attached to the SWCNT. Figure 7.50 shows the progression of attaching TiO_2 nanoparticles to a CNT [55].

Nanotube-polymer composites are also being developed for use in DSCCs. These composites combine NTs with high conductivity along their axes in a highly conductive polymer matrix, which disperses the CNTs into the photoactive layer and hopefully increases cell efficiency. The role of CNTs is to increase electron transport. Each CNT-polymer interface essentially acts like a p-n junction. Along this network of CNTs and polymer, electrons and holes can travel toward their respective contacts through the electron acceptor and the polymer hole donor. Increased efficiency is attributed to the introduction of internal polymer/nanotube junctions within the polymer matrix. The high electric field at these junctions can split up excitons, while the SWNT can act as a pathway for the electrons [59]. Poly(3-hexylthiophene)(P3HT) or poly(3-octylthiophene)

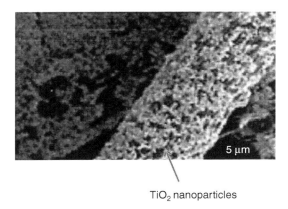

5 µm

TiO₂ nanoparticles

Figure 7.50 CNT before and after attachment of TiO_2 nanoparticles [9].

(P3OT) are often used as the polymer matrix [60]. The structure of the DSCC is then [61]:

- Glass/TCO
- 40 nm thick sublayer of (poly(3,4-ethylenedioxythio-phene)) (PEDOT) and poly(styrenesulfonate) (PSS) to smooth the TCO surface
- 20–70 nm thick Al or LiF4 applied to the photoactive CNT/polymer composite

Using CNTs in the photoactive layer of DSSCs is still in the initial research stages and there is still room for novel methods to better take advantage of the beneficial properties of CNTs. Conversion efficiencies for composite DSSCs are still generally quite low, between 1.3 and 1.5%. However, efficiency ~4.9% has been reported for CNT composites made by sandwiching a SWCNT layer between the ITO and the PEDOT [62].

Another application of CNTs is to build three-dimensional (3D) arrays of CNTs on the surface of Si wafers. CNTs are arranged on the surface in arrays of miniature tower structures, shown in Figure 7.51 [63]. This architecture would capture more photons than conventional Si PV cells and increase conversion efficiency. Photons arriving over a wide range of angles are trapped by the nonreflective CNT array, much like a textured cell structure. PV arrays would eventually be smaller using this concept, and sensitivity to the angle of the sun would be significantly reduced. The CNTs on the surface are ~100 µm tall, and the square arrays are 40 µm by 40 µm with a 10 µm separation.

Figure 7.51 CNT tower structure on Si wafers [63].

Another advantage of this configuration is that because significantly more photons are absorbed, the other active layers in the cell can be made thinner, thus reducing the likelihood of electron hole recombination. Once the CNT towers have been grown, CdTe and CdS p and n layers are deposited on them by MBE. Semiconductor materials are not limited to just CdTe and CdS; other conventional photovoltaic materials can be used. Performance of this cell still must be optimized, but test results show promise as a next generation of solar cell with potentially high conversion efficiencies.

Nanotubes certainly have the potential to improve the efficiency of a broad range of photovoltaic cells. The question remains on their ability for large area coverage and the cost of nanotube structures. They appear to be inexpensive to make, but integration into electro-optic devices may not be economical or straightforward. However, they appear to be a next important step in development of quantum devices.

7.4 Artificially Structured and Sculpted Micro and NanoStructures

We are all aware that microelectronics technology is ever decreasing the size of microcircuit structures (transistors, etc.). Device sizes are approaching nanometer and atomic scales. Progress can be represented by the well-known Moore's law, which predicts the number of transistors on a chip as a function of time. Critical issues here with downsizing to these dimensions are:

- Small size of the features
- Limited number of electrons
- Operation of small features
- Interconnection of small features
- Cost: possible expensive lithography
- Self assembly
- Simpler architecture
- Defect tolerant architecture

Confinement of electrons in dimensions approaching atomic and molecular dimensions requires consideration of quantum behavior. Possible devices include single electron devices, photonic crystals, metamaterials, quantum cellular automata, molecular electronics (OLED, etc.), quantum dot solar cells, and quantum computing.

Artificially structured and sculpted micro and nanoStructures are synthesized by self assembly, nano-patterning (ion beam), plasma etching, and glancing angle of incidence (GLAD) processes. In this section we address photonic band gap (photonic crystal), sculpted thin films, and metamaterials.

7.4.1 Photonic Band Gap (Photonic Crystals) Materials

Photonic band gap (PBG) materials have potential applications in quantum well semiconducting lasers, light emitting diodes, light emitting polymers (OLEDs), waveguides, optical filters, high-Q resonators, frequency selective surfaces (patterned optical microstructures), antennas, and amplifiers. PBG materials, also known as photonic crystals, perform much the same way as an optical high reflector coating, but in three dimensions. Many of their properties correspond to those of an "optical semiconductor" in that photonic crystals have forbidden wavelength regions in which light will not propagate, similar to that of band gaps in optical materials and semiconductors [64, 65, 66]. The PBG photon transport theory also gets as involved as that of semiconductor energy bands [67]. Forbidden optical propagation bands are caused by Bragg reflections off periodic discontinuities in optical densities (optical impedance mismatch), much the same as reflections off Brillouin zone boundaries in k-space for a semiconductor. The Bragg reflections are caused by a periodic dielectric constant (again much like a H/L quarter wave optical stack) and magnetic permeability, shown schematically in 1-dimension in Figure 7.52. This can readily be compared to the "forbidden" wavelength regions of a simple quarterwave high reflector stack tuned in the NIR shown in Figure 7.53 (which has been around for decades and could be considered a predecessor to the photonic band gap material). One can even define

Figure 7.52 Periodic dielectric constant in a photonic crystal.

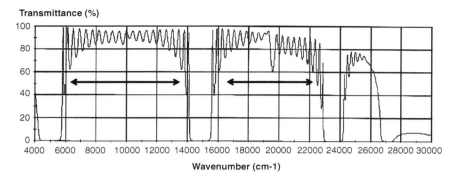

Figure 7.53 "Forbidden" bands in a simple quarterwave optical stack.

a density of states for the various periodic geometries (cubic, FCC, etc.) and the size of the PBG as a function of the index of refraction of the host matrix [67]. It may also be possible to tune the band gap of these structures to some degree [68].

A photonic crystal usually consists of a two or three-dimensional periodic array of optical impedance inhomogeneities, such as rods, spheres, or other shapes in a high index host, much like a crystal lattice on the scale of optical wavelengths. A large refractive index contrast (n_H/n_L) is desirable. One of the simplest examples is an array of rods shown in Figure 7.54 [68]. Here the rods are etched in a silicon crystal and create air gaps with refractive index of 1 (vs 3.6 of Si). The frequency response of photonic band gap materials looks much like that of the energy band diagram of a semiconductor. Figure 7.55a shows the frequency response diagram for the two-dimensional periodic array of cylinders with a hole spacing, or pitch, of about 0.2 µm and a diameter 0.2 µm in a silicon crystal. Figure 7.55b shows the energy band structure of Si for comparison. Also compare this to the transmission of a frequency selective surface with a pitch of 3 µm, which has a resonant wavelength near 5 µm in the infrared [3]. Figure 7.56 shows a three-dimensional array of air holes in a silicon matrix. Compare this to a periodic crystal structure.

Photonic crystals are fabricated by photolithographic patterning and etching, electrochemical etching, and self-assembly. One type of three-dimensional structure is made by introducing SiO_2 spheres in electrodeposited silicon. Self-assembly is also used to construct these structures [67].

(a) Square lattice

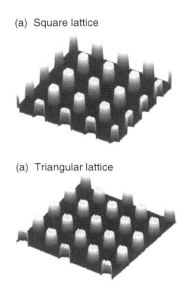

(a) Triangular lattice

Figure 7.54 Two-dimensional photonic crystal etched in silicon.

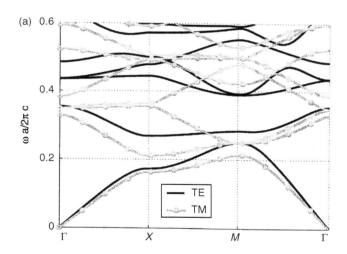

Figure 7.55a Frequency response of a two-dimensional photonic crystal [67].

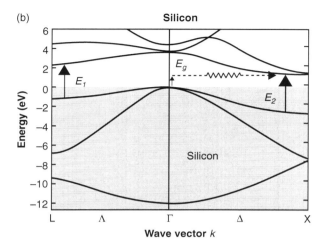

Figure 7.55b Energy band diagram of a Si crystal.

Figure 7.56 Three-dimensional photonic crystal.

7.4.2 Sculpted Thin Film Structures: Glancing Angle Deposition

A wide variety of geometric structures can be engineered by glancing angle deposition (GLAD) [69, 70, 71, 72]. These films can have unique photonic properties. Depending on the structure deposited, these films can polarize light [73], display optical birefringance, create a photonic band gap [74], or be used as a template for LCDs. Chiral structures that polarize light can be made from Al_2O_3, SiO_2, MgF_2, ZnO, ZnS, and other optical materials. The direction that the light is rotated depends on the chirality of the structure (left handed or right handed). For example, Brett reports that MgF_2, SiO_2, and Al_2O_3 structures have rotary powers of 0.16, 0.67, and 1.06 $°/$ µm respectively. Figures 7.57 and 7.58 show Si nanowires and the range of microstructure grown by GLAD.

The optical properties of the chiral structures can be changed by impregnating them with liquids such as water, monomers, and nematic liquid crystals (LC) [73]. If index matching between the chiral structure material and liquid is not achieved, the dielectric constant of the film will be nonuniform and it may have nonlinear or diffuse scattering optical properties. Impregnation by LCs causes such changes in the optical properties.

The chiral and associated structures are deposited by e-beam evaporation, magnetron sputtering, and molecular beam epitaxy (MBE). Geometry of the pillars depends on the angle of incidence, speed of rotation, and direction of rotation. It is also important to reduce the mobility of the deposited atoms/molecules on the substrate to enhance shadowing. Deposition rates for sputtering are quite low. For Si structures we see deposition rates as slow as 0.5 µm in 4 hr.

7.4.3   Artificially Structured Surfaces: Metamaterials

The need for increased functionality of thin films and surfaces has led to the development of new families of artificially

Si Nanowires

Figure 7.57 Si Nanowires grown by GLAD.

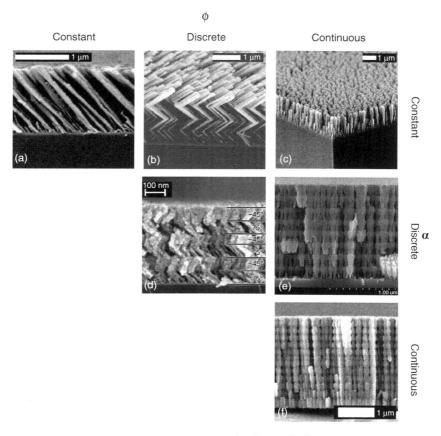

Figure 7.58 Typical microstructures grown by GLAD [71].

structured materials, including metamaterials, low dimensional materials, nanotubes, frequency selective surfaces, photonic band gap structures, and nanocomposites. Metamaterials, are loosely defined are macroscopic composites having a manmade, three-dimensional, periodic cellular architecture designed to produce an optimized combination, not available in nature, of two or more responses to specific excitation [75]. Examples of metamaterials are:

- Negative refractive index materials
- Cloaking device
- Artificial magnetism

- Artificial dielectrics
- Photonic crystals
- Plasmonics

Metamaterials can be tailored in shape and size, lattice constant, and interatomic spacing, and can be artificially tuned and defects can be designed and placed in desired locations. Much like low dimensional materials, metamaterials can have physical properties not possible with conventional materials, literally without violating any law of physics.

We will focus on optical and electromagnetic metamaterials here. They can be designed to guide light and longer wavelength electromagnetic waves around an object, rather than reflect or refract the light. They can also be used in lenses that image below the diffraction limit. Cloaking devices have been demonstrated in the microwave region that can literally guide waves around an object, making it invisible to radar, etc. The same is possible for light, but at a much smaller (and difficult) scale. Hence the object will not be detectable in the wavelength region of interest.

Negative refractive index (NRI) and plasmonics are two of the most interesting examples of metamaterials [76–79]. Subdiffraction super lenses can be fabricated using plasmonic metamaterials. We will deal primarily in phenomenological descriptions of this technology; the mathematics behind this technology is formidable and I refer to you references 2–4 for details. Figure 7.59 conceptually shows how a cloaking device would work [79]. Figure 7.59a shows that the structure to be protected would normally reflect and absorb electromagnetic radiation (light, microwaves, radar, etc.), and surely be visible to an observer. Figure 7.59b shows the situation with a cloaking device. Light and electromagnetic waves are bent around the object and none are reflected or absorbed, thus making the object "invisible". What makes light and electromagnetic waves bend around an object? It takes a *negative refractive index* (NRI) material to bend light in the opposite direction of conventional materials. The result of a negative refractive index, discussed below, is that light bends outward in a more optically dense medium, thought to be impossible in conventional optical materials a decade ago [80].

Figure 7.60 demonstrates the concept of NRI. For a positive refractive index material, a ray of light incident upon a medium

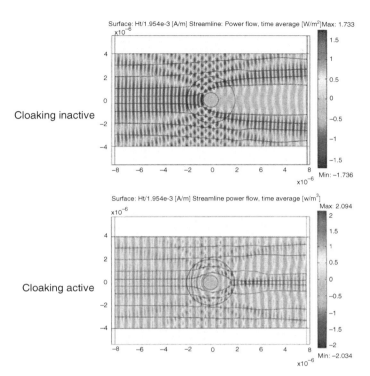

Figure 7.59a Light and electromagnetic waves reflecting from and being absorbed by an object [69].

Figure 7.59b Light and electromagnetic waves being bent around an object by NRI materials [69].

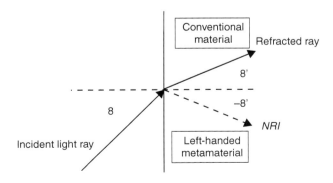

Figure 7.60 Bending of light rays at normal and NRI interfaces.

will be refracted, or bent, to the *right*. The ray will be bent to the *left* in an NIR medium. The required condition for an NRI is [76]

$$\varepsilon'\mu''+\mu'\varepsilon''<0 \qquad (7.9)$$

where the permeability and permittivity of the material are both negative with

$$\mu = \mu'+i\mu'' \text{ and} \qquad (7.10)$$

$$\varepsilon = \varepsilon'+i\varepsilon'' \qquad (7.11)$$

Note that even though both μ and ε are <0, $\mu\varepsilon$ > 0 and recall that n = $(\mu\varepsilon)^{1/2}$, and thus n is real, but mathematically can be <0 (recall that there are positive and negative square roots). If both components are <0, it makes sense that n would be <0. Metals are an exception due to their high electron density. ε < 0 and μ > 0 for metals, making n imaginary, and electromagnetic waves decay in the form of an evanescent wave instead of propagating. *n < 0 is a considerable technical accomplishment (probably one of the best in the last decade) and requires sophisticated quantum and electromagnetic theory and microfabrication techniques.*

As an example of a metamaterial structure, Figure 7.61 shows SEM images of a microfabricated NRI material [81]. The 3D

Figure 7.61 SEM of NRI material [81].

fishnet metamaterial is fabricated on a multilayer metal dielectric stack by using focused ion-beam milling, which is capable of cutting nanometer-sized features with a high aspect ratio. The figure shows the 3D fishnet pattern, which was milled on 21 alternating films of Ag and MgF_2, resulting in ten functional layers. Figure 7.62 plots the refractive index as a function of wavelength, being negative for wavelengths >1500 nm [81]. NRI at visible wavelengths is still a formidable challenge; however, Ames Laboratory has created a metamaterial of index of −0.6 for red light (780 nm) [80].

Metamaterials with n < 0 possess some very interesting properties:

- $\varepsilon'\mu'' + \mu'\varepsilon'' < 0$
- Wavefronts move in opposite direction as energy (Poynting vector). This means that unlike a normal right-handed material, the wavefronts are moving in the opposite direction to the flow of energy, as diagrammed in Figure 7.63.
- Phase velocity moves in opposite direction as group velocity
- Doppler shift is reversed
- Cherekov radiation points the opposite direction

Consider the wave packet shown in Figure 7.63. For propagation of the wave packet in a conventional optical medium, the phase

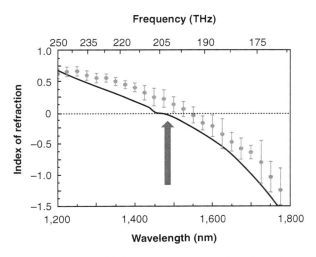

Figure 7.62 Refractive index of fishnet structure as a function of wavelength [81].

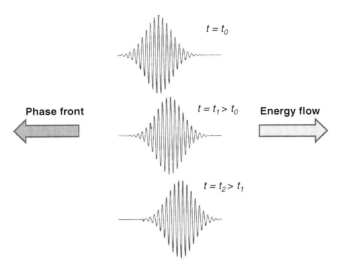

Figure 7.63 Directions of the velocity and phase velocity of a wave packet in a NIR material [2].

and group velocity are in the same direction, the direction of energy flow. For an NRI medium, the direction of the phase front of the wave packet shown in the figure is opposite that of the energy flow (Poynting vector). If the direction of energy flow is used as a reference direction of propagation, and since the n is <0, the wave vector k must be in the opposite direction. The rate of energy flow is described by the group velocity v_g of the wave pack, which carries all information, while the wave vector is related to the phase velocity (velocity at which wavefronts move):

$$k = \frac{\omega}{v_p} = \frac{n\omega}{c} \text{ (must be } < 0 \text{ since } n < 0)$$

thus, in a NRI material, the wavefront moves in the opposite direction as the wave packet, known as a "negative wave".

This has several ramifications:

- For plane waves propagating in optical metamaterials, the electric field (E), magnetic field (H), and wave vector (k) follow a left handed rule, thus giving rise to the name left-handed metamaterials.

- The effect of negative refraction is analogous to wave propagation in a left-handed transmission line.

Note that NRI can also be achieved without this left handed behavior.

Although there are a number of interesting applications to address, superlenses are one of most intriguing since they seem to disobey the laws of physical optics. A conventional lens cannot reconstruct images with size less than that of a wavelength of light, known as the diffraction limit, and this has been the fundamental limit of conventional optics (much to the distress of large telescopes, microscopes, etc.). A super lens, however, is a lens that can image light below the diffraction limit. An electromagnetic wave has both a propagating (far field) component and an evanescent (near field) component. Not going into any detail, NRI materials can focus both propagating and evanescent waves while ordinary optical materials use only the propagating wave (the evanescent wave decays exponentially at the lens interface). Evanescent waves in a negative-refractive medium actually increase in amplitude as they move away from their origin. Figure 7.64 shows ray diagrams for ordinary and plasmonic super lenses [76]. Here we see the far field and near field features.

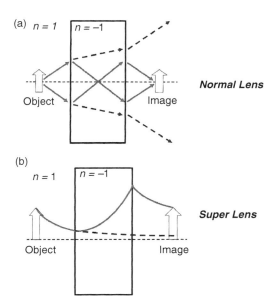

Figure 7.64 Ray diagrams for ordinary and super lens imaging [76].

Images much smaller than a wavelength can be reconstructed using the near field component, which allows for a new range of imaging applications. Optical images are limited by diffraction of the light wave (called diffraction limit). The first NRI super lens was demonstrated at microwave frequencies and provided resolution three times better than the diffraction limit [82]. Actually, the first super lens was constructed of a thin Ag film and used surface plasmon coupling to enhance the evanescent wave [83]. The first optical NRI super lenses consisted of nanoscale grids cut into Ag layers, which resulted in a 3-dimensional composite structure similar to that shown in Figure 7.63 [81]. Ag nanowires were used to construct a 3D structure which exhibited NRI down to 660 nm [84].

7.4.4 Artificially Structured Materials: Metamaterial Designs

Metamaterial structures require extremely precise sophisticated micro-nano-scale geometries. The building blocks of metamaterials are artificially designed and fabricated structural units. The individual units function like atoms or molecules in a material and can be constructed into an artificial crystal lattice, much like layers in a superlattice. Unlike naturally occurring materials, the size, shape, geometry, and spacing can be tailored to achieve specific properties. Defects can even be engineered into the material. Dimensional scales required for metamaterials depend on the wavelength region of interest and are summarized in Figure 7.65 [85]. The feature size indicates the unit cell size. Individual units can be wires, split-ring resonators (SRR) with various shapes, Swiss rolls, fish nets, metal segments, and posts (or cylinders) [85, 86, 87]. Multilayer structures can be formed to simulate multilayer optical coatings, nanolaminates, and superlattices.

All the above structures are designed to create a magnetic or electric response to achieve a negative electric permittivity, negative magnetic permeability, or both. Let's look at some design principles and designs (unfortunately this brief overview cannot address all structures) and their properties. Obviously the goal is to achieve the above properties in the wavelength range of interest. One general rule of thumb is to obtain under-damped and over-screened resonances for corresponding electric (E) and magnetic (H) fields [86]. We will see this for some split resonator designs. While one structure is preferable, two unit structures are often combined to achieve

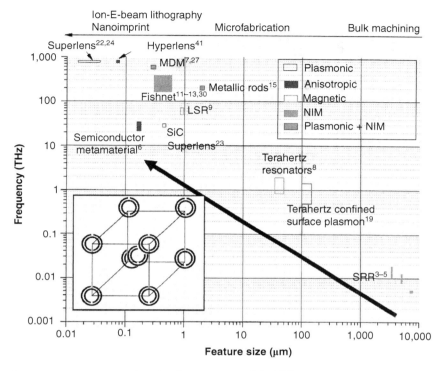

Figure 7.65 Dimensional scales for metamaterials [85].

$\varepsilon < 0$ and $\mu < 0$. Cross talk, however, can be a problem. It is even possible to tune response to a certain degree.

Many design methods employ metallic (conductive) materials and develop an equivalent circuit to balance electromotive forces, and generally involve complex mathematical analysis using Faraday's law and Stokes theorem [86]:

$$\int E.dl = -\frac{\partial}{\partial t}\int\int ds.B$$

where the first integral is over the current loop and the second is over the surface. The electromotive force = emf = $-\partial/\partial t\phi(t)$, where $\phi(t)$ is the magnetic flux.

One design methodology engineers structure uses metals and plasmons with negative permittivity. Models assume that the mass of negative charge (electrons) in the plasma is much less than the mass of positive charge (ions), and only the motion of the electrons contributes to polarization of the material. Consequently, this model includes many metals, whose permittivity is negative at frequencies below the plasma frequency [77]. For metals, we are concerned mainly with optical frequencies. Figure 7.66 shows the real and imaginary parts of the permittivity of a plasma. The wave propagates at frequencies higher than ω_p and is evanescent at lower

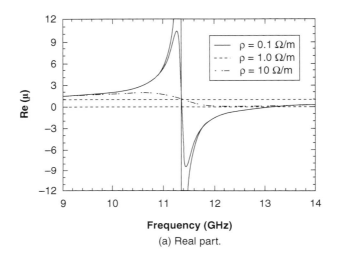

(a) Real part.

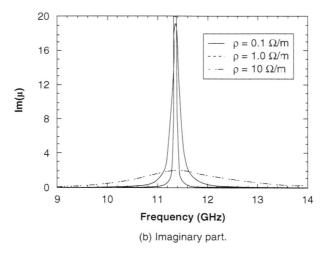

(b) Imaginary part.

Figure 7.66 Real and imaginary parts of dielectric permittivity of a metal [77].

frequencies. The reader is referred to Kittel for a full treatment of the dependence of the dielectric response of metals [77]. Waves with $\omega < \omega_p$ correspond to negative energy solutions and are evanescent. Plasmon modes are bulk plasma oscillations and have the dispersion shown by the dashed line in Figure 7.66. Surface plasmons are confined to the metal/dielectric interface. Metals best suited for generation of plasmons are Al, Cs, Au, Li, Ni, Pd, K, and Ag.

Wire meshes are used to generate low frequency plasmons. An array of thin metallic wires, such as that shown in Figure 7.67, with periodicity a will behave as a low frequency plasma and can be modeled with a negative dielectric permittivity [88, 89, 90]. Assuming a is greater than the radius of the wires (r), an electric field applied parallel to the wires, and a longitudinal plasmonic mode, the plasma frequency can be expressed as

$$\omega_p^2 = \frac{2\pi c^2}{a^2 \ln\left(\dfrac{a}{r}\right)} \tag{7.12}$$

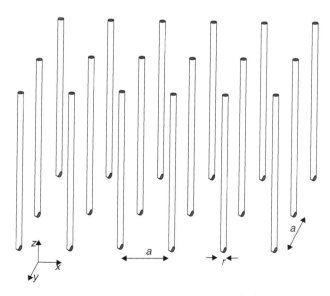

Figure 7.67 Array of metallic wires with periodicity a [90].

[85] and the permittivity is

$$\varepsilon = \frac{1 - \omega_p^2}{\omega \left(\omega + \dfrac{i\varepsilon_o a^2 \omega_p^2}{\pi r^2 \sigma} \right)} \qquad (7.13)$$

Here σ is the conductivity of the metallic wire.

An array of discontinuous wire segments, shown in Figure 7.67, behaves like an array of electric dipoles which act like two capacitors in series per unit cell. The permittivity of this type of array can be large and negative. The capacitance across the wire segment is, therefore, negative due to this large negative permittivity. The expression for the permittivity is extremely complex but shows a negative resonance at near to mid infrared wavelengths.

As discussed above, the magnetic permeability should also be <0. As we have seen, the electric response is achieved over a wide range of wavelengths and frequencies (UV – radio). This is not the case for magnetic resonances, which tend to occur at lower frequencies and die out at high frequencies. Resonance usually occurs in the microwave region. Thus, the goal of metamaterials designers is to move this resonance to higher frequencies (shorter wavelengths). Negative permeability is a significantly harder technical challenge than negative permittivity. The challenge is to design and fabricate subwavelength size structures whose resonance is driven by the magnetic component of an electromagnetic wave. Designs can be highly anisotropic or isotropic, displaying magnetic activity for magnetic fields oriented in one specific direction or uniformly in all directions.

We consider two examples, stacked metallic cylinders and split ring resonators, shown in Figure 7.68 [86]. The cylinders have radius r and lattice period a. Circumferential surface currents are induced by an oscillating magnetic field. These currents generate a magnetization that opposes the applied field. The effective permeability can be expressed as

$$\mu_{eff} = \frac{1 - \dfrac{\pi r^2}{a^2}}{\left(1 + \dfrac{i^2 \rho}{\mu_o \omega r} \right)}, \qquad (7.14)$$

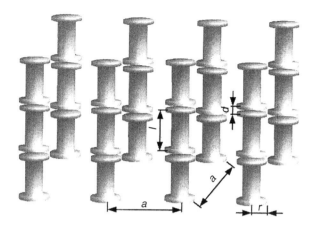

Figure 7.68 Array of stacked cylinders and design of split ring resonator [86].

where ρ is the resistance/unit length of the cylinder.

The split ring resonator (SRR) is probably the most widely used design type. Along with Figure 7.69, Figure 7.70 shows examples of SRRs [85, 86, 87, 88]. SRRs introduce a capacitive response in addition to an inductive response, which significantly enhances the resonance. The split rings introduce capacitive gaps spaced symmetrically around a cylinder. The capacitance of this SRR can be tuned by adjusting the length/size of the arms and introducing a dielectric in the gaps between the arms.

The SRR is essentially an LC circuit, and the electromagnetic field of the incident radiation drives the resonance of this circuit. The SRR thus has a "Q" like any LC circuit. Induced currents flow in the direction shown in Figure 7.64 with charges accumulating at the gaps. The resonant frequency ω_o is give by [86]

$$\omega_o = (LC)^{\frac{-1}{2}} = c\left(\frac{2d_c}{\{(1-f)\varepsilon\, l_c \pi r^2\}^{\frac{1}{2}}}\right) \quad (7.15)$$

Here d_c and l_c are capacitive gaps and $f = \pi r^2/a^2$.
The effective permeability is

$$\mu_{eff} = \frac{1+f\omega^2}{(\omega_o^2 - \omega^2 - i\Gamma\omega)} \text{ and can be } < 0. \quad (7.16)$$

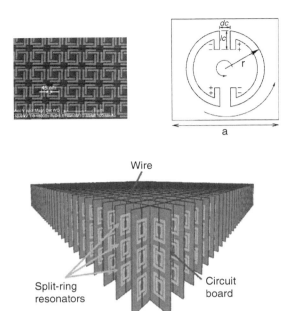

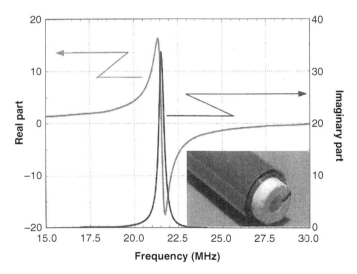

Figure 7.69 Examples of split ring resonators [85]

Figure 7.70 Effective permeability of a SRR with r = 1.5 mm, dc = 0.2 mm and magnetic field along the axis [86].

Note that $\mu_{eff} \sim 1 - f$ for $w >> \omega_0$.

Figure 7.71 shows the real part of μ_{eff} [86] for an SRR tuned in the GHz region. We see that $\mu_{eff} < 0$ for $\omega > \omega_0$ and depends critically on the resistivity of the resonator material.

As another example, Figure 7.72 shows the response of an SRR tuned in the terahertz region [91]. Here we see that the design is more complex than that shown in Figure 7.71, consisting of four metallic L-shaped legs. The permeability response was negative beginning at 60 THz.

The Swiss roll resonator, shown in Figure 7.73, was developed for low frequency applications [92]. This type of resonator consists of a coiled cylindrical structure with each roll separated by an insulator. This structure has a very large self inductance. The inductive response is shown in Figure 7.74 [93]. The structure consists of thin Cu sheet separated by polyimide, with an overall diameter ~1 cm. Permeability is <0 at frequencies above 22 MHz.

A word on scaling and size reduction of these structures to access higher frequencies and visible radiation. It seems obvious from Maxwell's equations that simply reducing the size of the structures will get us down to near infrared and visible wavelength ranges. Nature, however, has not made this straightforward; the materials parameters in these designs are dispersive at optical

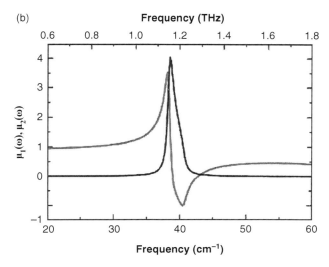

Figure 7.71 SRR tuned at terahertz frequencies [91].

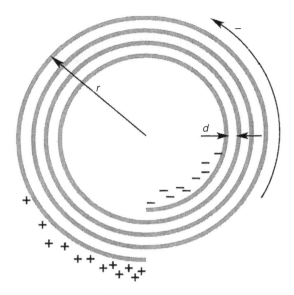

Figure 7.72 Swiss roll resonator [86]

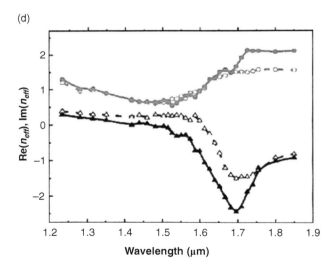

Figure 7.73 Refractive index of a multilayer fish net structure [95].

frequencies. Most metals are no longer ohmic and skin depths (defined $\delta = (2/\mu_{o}\sigma\omega)^{1/2}$) of electromagnetic radiation are large compared to structure size. Also, fabrication of arrays of nanoscale structures becomes more challenging. The Si/Ag/Si fishnet structure shown in Figure 7.63 comes very close to the desired response [94]. Refractive index is shown in Figure 7.74 [95]. Note that it is negative for wavelengths >1.4 µm.

References

1. Joel I. Gersten and Frederick W. Smith, *The Physics and Chemistry of Materials*, Wiley.
2. See Hyperphysics.com
3. Manijeh Razeghi, *Fundamentals of Solid State Engineering*, Kluwer Academic Publishers (2002).
4. S. M. Sze, *Physics of Semiconductors*, 2nd Ed., Wiley Interscience (1981).
5. Charles Kittel, *Introduction to Solid State Physics* (Eighth Ed.), Wiley (2005).
6. Kevin F. Brennan, *The Physics of Semiconductors with Applications to Optoelectronic Devices*, Cambridge (1999) 80.
7. C. Morant, Prieto, P. Forn, A. Picas, J. A. Elizalde, E. Sanz, J. M., *Surface and Coatings Technology* (2004), 180–181(Complete), 512–518.
8. Hoagland, R.G., R.J. Kurtz, and C.H. Henager, Slip resistance of interfaces and the strength of metallic multilayer composites. *Scripta Mater.*, 2004. 50(6) 775–779.
9. Eli Yablonovich, *Phy. Rev. Lett.*, 58(20) (1987) 2059–62.
10. Sajeev John, *Phy. Rev. Lett.*, 58(23) (1987) 2487–89.
11. Manijeh Razeghi, *Fundamentals of Solid State Engineering*, Kluwer Academic Publishers (2002).
12. Joel I. Gersten and Frederick W. Smith, *The Physics and Chemistry of Materials*, Wiley (2001).
13. Charles Kittel, *Introduction to Solid State Physics* (Eighth Ed.), Wiley (2005).
14. Uichiro Mizutani, *Introduction to the Electron Theory of Metals*, Cambridge (2001).
15. T. J. Coutts, *Electrical Conduction in Thin Metal Films*, Elsevier (1974).
16. Joseph Callaway, *Energy Band Theory*, Academic Press (1964).
17. Hyperphysics.com
18. *CRC Handbook of Chemistry and Physics*, 73rd Ed., CRC Press (1991).
19. S. M. Sze, *Physics of Semiconductors*, 2nd Ed., Wiley Interscience (1981).
20. J. W. *et al.*, *Phy Rev Lett* 68(5) (1992) 631.
21. W. G. Jeroen Wilder *et al.*, *Nature*, 391, 6662 (1998), 59–62.
22. C. D. Spataru *et al.*, *AIP Conf. Proc.*, 772 (2005) 1061.
23. C. D. Spataru *et al.*, *Phys. Rev. Lett.*, 92 (2004) 077402.
24. M. P. Anatram and F. Leonard, *Rep. Prog. Phys.*, 69 (2006) 507.
25. Andreas Thess *et al.*, *Science* 273 (1996) 483.
26. Stefan Frank *et al.*, *Science* 280 (1998) 1744.
27. Phaedon Avouris (IBM Watson Research Center), Lecture given at Michigan State University (2000).

28. Savas Berber *et al.*, *Phys. Rev. Lett.*, 84, 20 (2000) 4613.
29. Jianwei Che *et al.*, *Nanotechnology*, 11 (2000) 65.
30. Francois Leonard, *The Physics of Carbon Nanotube Devices*, William Andrew (2009).
31. Y. Miyauchi *et al.*, *Phys Rev B*, 74 (2006) 205440.
32. K. Iakoubovskii *et al.*, *Appl Phys Lett*, 89 (2006) 173108.
33. H. Kataura *et al.*, *Synthetic Metals*, 103 (1999) 2555.
34. S. B. Sinnott and R. Andreys (2001). *Critical Reviews in Solid State and Materials Sciences* 36 (3): 145–249.
35. K. Iakoubovskii *et al.*, *J Phys Chem B* 110 (35) (2006) 17420.
36. K. Iakoubovskii *et al.*, *J. Phys. Chem. C* 112 (30) (2008) 11194.
37. C. D. Spataru *et al.*, *Phys. Rev. Lett.*, 92 (2004) 196401.
38. F. Wang *et al.*, *Phys. Rev. Lett.* 92 (2004) 177401.
39. F. Wang *et al.*, *Science*, 308 (2005) 838.
40. Jared Crochet *et al.*, *J. Am. Chem. Soc* 129 (2007) 8058.
41. N. Ishigami *et al.*, *J. Am. Chem. Soc.* 130 (30) (2008) 9918.
42. J. A. Misewich *et al.*, *Science* 300 (5620): (2003).
43. C. Fantini *et al.*, *Phys. Rev. Lett.* 93 (14) (2004) 147406.
44. H. Kataura *et al.*, *AIP Conference Proceedings*, 544 (2000) 262.
45. D. A. Stewart and F. Leonard, *Phys. Rev. Lett.*, 93 (2004) 107401.
46. D. A. Stewart and F. Leonard, *Nano Lett.*, 3 (2003) 1067.
47. Shlomit Chappel *et al.*, *Langmuir,* 18 (8) (2002) 3336.
48. Artificial Photosynthesis, Anthony F. Collings and Christa Critchley, Ed, Wiley-VCH (2005).
49. A. Hagsfelt and M. Gratzel, *Chem. Rev.*, 95 (1995) 49.
50. K. Shankar *et al.*, *Nanotechnology*, 18 (2007).
51. D. Gong, *J. Mater. Res.*, 16 (2001) 3331.
52. spie/org/x25474.xml.
53. N. S. Lewis, *Science*, 315 (2007) 798.
54. Tae Young Lee *et al.*, *Thin Solid Films*, 515 (12) (2007) 5135.
55. Michael Berger, *Nanowerk LLC* (2007).
56. Istvan Robel *et al.*, *Advanced Materials* 17 (20) (2005) 2458.
57. M. Olek *et al.*, *Journal of Physical Chemistry B*, 110 (26) (2006) 12901.
58. Taku Hasobe *et al.*, *Journal of Physical Chemistry B*, 110 (50) (2006) 25477.
59. E. Kymakis *et al.*, *Progress in Photovoltaics: Research and Applications*, 93 (3) (2003)1764.
60. Hiroki Ago *et al.*, *Advanced Materials*, 11 (15) (1999) 1281.
61. A. J. Miller *et al.*, *Applied Physics Letters*, 89 (12) (2006) 123115.
62. S. Chaudhary *et al.*, *Nano Letters*, 7 (7) (2007) 1973.
63. Electro-Optical Systems Laboratory at the Georgia Tech Research Institute (GTRI).
64. Eli Yablonovich, *Phy. Rev. Lett.*, 58 (20) (1987) 2059–62.
65. Sajeev John, *Phy. Rev. Lett.*, 58 (23) (1987) 2487–89.
66. S. John and R. Rangarajan, *Phys. Rev. B*, 38 (1988) 10101.

67. Kurt Busch and S. John, *Phys. Rev.* E 58 (1998) 3896.

68. M. S. Dresselhaus *et al, Proceedings of the Conference "Chemistry, Physics, and Materials Science of Thermoelectric Materials: Beyond Bismuth Telluride"* (2002).

69. R. Messier, *et al., J. Vac. Sci. Technol.* A 15(4) (1997) 2148–51.

70. M. J. Brett *et al., Proceedings of the 45th Annual Technical Conference of the Society of Vacuum Coaters* (2002) 11–17.

71. Matthew M. Hawkeye and Michael J. Brett, *J Vac Sci Technol* A25(5) 1317.

72. Michael T. Taschuk, Matthew M. Hawkeye and Michael Brett, in *Handbook of Deposition Technologies for Films and Coatings*, Third Edition, Peter M. Martin, Ed, Elsevier (2009).

73. K. Robbie, D. J. Broer and M. J. Brett, *Nature* 399 (1999) 764–66.

74. G. A. Kimmel *et al., J. Chem. Phys.* 114 (2001) 5295.

75. *R. M. Walser, W. S. Weiglhofer and A. Lakhtakia. Ed., Introduction to Complex Mediums for Electromagnetics and Optics, SPIE Press, Bellingham, WA* (2003).

76. MRS Bulletin, 33 (10) (October 2008).

77. Clifford M. Krowne and Yong Zhang, Ed., *Physics of Negative Refractive and Negative Index Materials*, Springer (2007).

78. Clifford M. Krowne, *Phys. Lett.* A, 372 (2008) 3926.

79. Duke Engineering Department news release "First Demonstration of a Working Invisibility Cloak".

80. *Michio Kaku, Physics of the Impossible: A Scientific Exploration Into the World of Phasers, Force Fields, Teleportation, and Time Travel, Doubleday (2008) 22–23.*

81. Jason Valentine *et al., Nature Letters,* 07247 (2008).

82. B. A. Munk, *Metamaterials: Critique and Alternatives* (2009).

83. N. Fang *et al., Science,* 308 (2005) 53.

84. J. Yao et *al., Science,* 321 (2008) 930.

85. Pratik Chaturvedi *et al., MRS Bulletin,* 33 (10) (2009) 915.

86. S. Anantha Ramakrishan and Tomasz M. Grzegorczyk, *Physics and Applications of Negative Refractive Index Materials*, SPIE Press (2009).

87. Jason Valentine *et al., Nature Letters,* 10.1038 (2008) 1.

88. J. B. Pendry, *Phys. Rev. Lett.,* 85 (2000) 3966.

89. J. B. Pendry, *Science,* 306 (2004) 1353.

90. S. Ramakrishan, *Rep. Prog. Phys.,* 68 (2005) 449.

91. W. Wu *et al., Appl. Phys. Lett.,* 90 (2007) 063107.

92. J. B. Pendry *et al., IEEE Trans. Microwave Theory Tech.,* 47(11) (1999) 2075.

93. M. C. K. Wiltsire *et al., IEEE Seminar on Metamaterials for Microwave and (sub)Millimeter Wave Applications*, London, November 2003, paper 13.

94. Jason Valentine *et al., Nature Letters,* (2008) 07247.

95. E. Kim *et al., App. Phys Lett,* 91 (2007) 173105.

8

Multifunctional Surface Engineering Applications

A number of surface engineering technologies can fit into two or more technologies or involve two or more deposition processes covered in previous chapters. For example, these technologies can combine optical coatings with wear resistance, conductive optical coatings (electrical + optical), electro-optical coatings, and photovoltaic thin films. These multifunctional hybrid surface engineering technologies include:

- Thin film transparent conductive oxides
- Thin film solar cells
- Electrochromic and thermochromic films
- Photorefractive thin films
- Photocatalytic and electrochemical thin films
- Gas and water permeation barriers
- Thermal radiation emitters
- Frequency selective surfaces and structures

8.1 Thin Film Photovoltaics

Solar or photovoltaic (PV) cells are applied to solid surfaces to generate electrical energy from electromagnetic energy (solar radiation, infrared radiation). The first generation PV cells were based on crystalline semiconductors, such as Si, GaAs, Ge, and InP. Thin film PV cells constitute the second generation. The entire field of thin film photovoltaic and solar cell technology is too broad to cover here in detail. Many of the technologies used to fabricate thin film cells are presented in Chapters 2, 6, and 7. Deposition processes are discussed in Chapter 2. Thin film semiconductors and metallization are discussed in Chapter 6, and TCO technology is reviewed in Section 8.2. Thin film solar cells are applied to solid surfaces such as glass, stainless steel, aluminum, titanium, and polyimide. The most important thin film materials now used or being developed for photovoltaic cells are (see Figure 8.1) a-Si:H, nanocrystalline Si, polycrystalline Si, CdTe, CdS, GaAs, AlGaAs, $CuInSe_2$ (CIS), $CuIn_xGa_{1-x}Se_2$ (CIGS), and organics. Referring to these materials, we will address only the basics of thin film solar cells.

Basic solar cell structure for a-Si:H, CdTe and CIGS thin film cells is shown in Figure 8.1) and consists of a substrate, conductive bottom layer, p-layer, intrinsic layer, n-layer, and transparent conductive top contact. An antireflection coating is usually applied over the top to reduce surface reflections and inject as much light into the cell as possible. Note that not all cells will have this design,

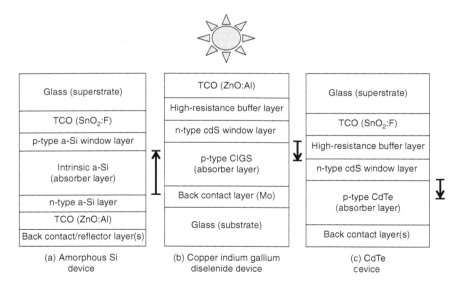

Figure 8.1 Basic layer structure of a-Si:H, CdTe and CIGS thin film solar cells (courtesy National Renewable Energy Laboratory (NREL)).

with some being simpler and some more complex. As shown in Figure 8.2, junctions of p-n or p-i-n layers form energy band structures that are used to separate electron and holes generated in the absorbing layer (usually the p-type layer). The p-n or p-i-n structure generates an internal (or built in) electric field at the layer interfaces (junctions) [2, 3]. Band structure of the semiconductor layers (see Chapter 7) is all important for cell operation. Optimum band gap for the absorber layer is ~ 1.5 eV. Photons with energy greater than the band gap energy, incident on the absorbing layer, can excite an electron from the valence band into the conduction band, as shown in Figure 8.2, forming an electron-hole pair. The hole is created in the valence band. As a result of illumination, addition of electrons into the valence band and holes into the conduction band creates a further imbalance in charge neutrality and if we provide an external current path, electrons will flow through the path to the p side to unite with holes that the electric field sent there, doing work along

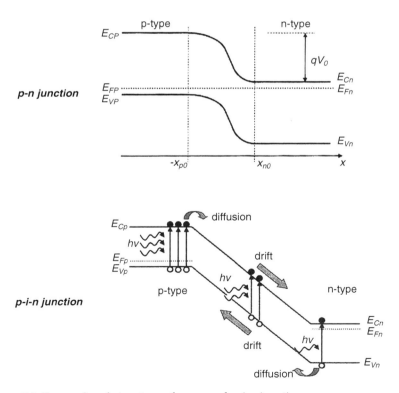

Figure 8.2 Energy band structure of a p-n and p-i-n junctions.

the way [3]. The electron flow provides the current, and the cell's electric field causes a voltage. With both current and voltage, we have power. Recall that power = voltage × current (P = IV).

The efficiency of the PV cell is simply

$$Eff = \eta = \frac{P_{out}}{P_{in}} \qquad (8.1)$$

8.1.1   Amorphous Silicon Solar Cells

The structure of a thin film a-Si PV cell is shown in Figure 8.1. This technology was developed as a lower cost alternative to crystalline Si solar cells, and is a viable candidate for large scale production [4]. A major impediment to full scale production is low deposition rates (~3 Å/s). As stated in Chapter 6, the most widely used deposition process for a-Si:H is PECVD using SiH_4 as a precursor. The homojunction (same materials) cell structure is deposited onto a glass superstrate. The glass also protects the thin film structure from the environment and abrasion. The absorber layer for this cell is the 0.5 μm thick intrinsic a-Si:H layer. This layer is sandwiched between a top p-type layer and a bottom n-type layer. Transparent electrical contacts, fluorine doped tin oxide (SnO_2:F) and aluminum doped zinc oxide (ZnO:Al - AZO), allow solar radiation into the cell. The back contact can be a metal reflector. This cell structure is also deposited in roll-to-roll vacuum systems onto stainless steel and polyimide webs for roofing and building applications. Best reported PV cell efficiencies are near 12.5% with average efficiencies typically.

8.1.2   CdTe Solar Cells

The structure of the CdTe solar cell is shown in Figure 8.3. As stated in Chapter 6, CdTe used in solar cells is deposited by closed space sublimation, a process that has very high deposition rates, but often requires high substrate or post deposition heat treating temperatures. The cell structure is deposited on a glass superstrate. As with all modern thin film PV cells, the top electrical contact is a TCO; in this case SnO_2:F, although $CdSnO_4$ is also used. The high resistance buffer layer consists of Zn_2SnO_4 and $ZnxCd_{1-x}S$ films. The absorber layer for this cell is p-type CdTe and the n-type layer is CdS, forming a heterojunction (different materials). CdTe layers are doped with Cu, Cl, or F and typically have thicknesses near 1.6 μm [5, 6],

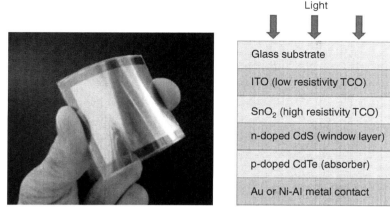

Light

Glass substrate

ITO (low resistivity TCO)

SnO$_2$ (high resistivity TCO)

n-doped CdS (window layer)

p-doped CdTe (absorber)

Au or Ni-Al metal contact

Figure 8.3 CdTe module on PET.

and CdS layers are 0.05 µm thick [6]. High efficiency cells have the CdTe layer doped with Cu from the back electrical contact or from chemical treatment [3, 7]. The bottom electrical contact is either Ti, Au or a Ni-Al alloy. A ZnTe buffer layer is placed between the CdTe and metal contact. Cells can be formed using glass as the top surface or as the bottom surface. In either case, the cell must be encapsulated to prevent environmental degradation (see Section 8.5 on permeation barriers).

Typical PV cell efficiencies range between 14%–18 % and depend to a large extent on substrate temperature [3]. Although most CdTe cells are produced on glass, web processes have been developed for large area applications [8]. Figure 8.3 shows a picture of a CdTe PV module on PET.

8.1.3   CIGS Solar Cells

A Mo back contact is deposited onto a glass substrate or a metal foil (usually stainless steel), as shown in Figure 8.4. The heterojunction is formed between the CIGS and ZnO, buffered by a thin window layer of CdS and a layer of intrinsic ZnO. The CIGS absorber layer is not doped but is p-type due to intrinsic defects. Transparent conductive aluminum doped ZnO (ZnO:Al – AZO) forms the top contact. The CIGS layer is between 1.5 mm and 2.5 µm thick, while the CdS layer is only 0.05 µm thick. Thickness of the Mo layer is ~ 0.5 µm and intrinsic ZnO is ~ 0.0 µm thick.

CIGS PV cells have routine efficiencies between 15% and 20%.

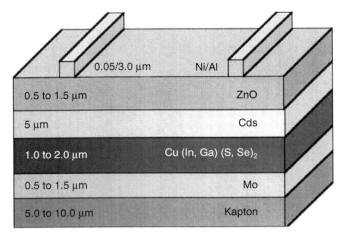

Figure 8.4 Layer structure of CIGS solar cell.

8.2 Transparent Conductive Oxide Thin Films

Although transparent conductive oxides (TCO) have been intensely developed since the late 1970s, they have been around for a century. Cadmium oxide (CdO) was the first transparent conductive coating and was used in solar cells in the early 1900s. Tin oxide (SnO_2) was deposited on glass by pyrolysis and CVD in the 1940s for electroluminescent panels. Since then, applications and deposition processes have mushroomed. Transparent thin films and materials are some of the most commonly used materials that we depend on for a wide range of applications. Of these, transparent conductive oxides (TCOs) head the list. Table 8.1 lists electrical properties of selected n-type transparent oxides. A number of ternary compounds have been developed over the last ten years, including Zn_2SnO_4, $ZnSnO_3$, $MgIn_2O_4$, $(GaIn)_2O_3$, $Zn_2In_2O_5$, and $In_4Sn_3O_{12}$ [1]. They are critical for energy efficiency applications such as low-e windows, solar cells, and electrochromic windows. Transparent conductive thin films are used as the transparent electrical contacts in flat panel displays, sensors, and optical limiters. They also have applications wherever a transparent conductive electrical contact is needed, such as in the artificial lung being developed by Battelle [2]. A new developing application is the charge carrier layer in a transparent transistor and solar cell. Clark Bright has written an excellent review of transparent conductive oxides in SVCs 50th Anniversary book [1].

Table 8.1 Resistivities of selected transparent conductive oxides.

TCO	~Lowest Reported Resistivity ($\mu\Omega$.cm)*
ITO	114 [11]
In_2O_3	100 [12]
SnO_2	400 [12]
ZnO	20 [12]
ZnO:Al	1300 [12]
$CdSnO_2$	130 [13]
CdO:In	60 [12]

*Best available literature.

If we confine ourselves to only thinking about metals, the question is "how can a material be both transparent and conductive?" First consider semiconductors. These materials are transparent in NIR and IR wavelength ranges and can be reasonably conductive due to their band structure. Doping also improves the conductivity while the transmittance is not significantly degraded. Let's review the band structures of metals and semiconductors. Recall that electrical (and heat) conduction in metals is due to free electrons in the conduction band, as shown in Figure 8.5 [1]. Because the occupied energy levels fall right below the Fermi Energy (E_F), electrons can easily be excited into the empty or partially filled conduction band. From Drude theory, the electrical conductivity is simply related to the free electron density by

$$\sigma = \frac{ne^2\tau(E_F)}{m}, \qquad (8.2)$$

where n is the electron density (in the conduction band), e is the charge on the electron, τ is the relaxation time between electron-electron collisions and other scattering events, and m is the mass of the electron.

The situation changes considerably for semiconductors, which have a well defined gap between the valence and conduction

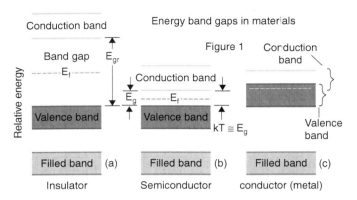

Figure 8.5 Metal energy band structure [1].

bands. Here the conductivity results from electrons and holes and is given by

$$\sigma = e^2 \left(\frac{n_h \tau_h}{m_h^*} + \frac{n_e \tau_e}{m_e^*} \right), \qquad (8.3)$$

where the hole and electron contributions have been separated out and m_h^* and m_e^* are the effective mass of the hole and electron ($\sim (d^2E/d^2k)^{-1}$).

Semiconductors are doped to improve their conductivity. Doping introduces states in the band gap just above the valence band or just below the conduction band, as shown in Figure 8.6 [2]. Defects also create states in the band that can increase or decrease the conductivity. Doping and defect sites have the same effect on transparent insulators, such as ZnO, In_2O_3, CdO, and SnO_2. Both doping and oxygen vacancies are used to increase the conductivity of these materials. Figure 8.7 shows Sn doping sites in an indium oxide (In_2O_3) lattice [14]. The Sn atom occupies an interstitial site and contributes an electron (i.e., a donor), making the doped indium oxide into indium tin oxide (ITO). Figure 8.8 shows the assumed band structure of undoped and Sn doped indium oxide [14]. Note that the occupied states are just below the conduction band and also refer to Figure 8.6b for an n-type semiconductor. Oxygen vacancies are mainly responsible for electrons in the conduction band [15]. Figure 8.9 shows that ITO and other TCOs are highly degenerate n-type semiconductors. Degeneracy simply implies that there are a number of energy bands that overlap and the Fermi energy is close to the edge of the conduction band.

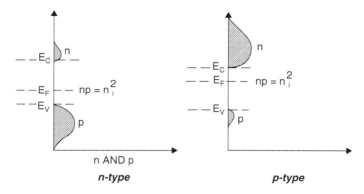

Figure 8.6 Density of states for p and n-type dopants in a semiconductor [2].

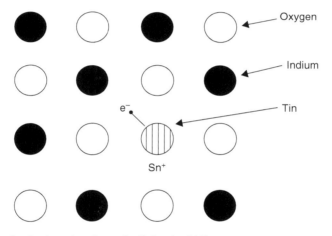

Figure 8.7 Sn doping sites in an In_2O_3 lattice [14].

Recall that for defect sites in solids can also scatter electrons, so a balance must be reached of increased conduction by electrons and scattering by defect sites. Because of this, there is a conductivity maximum or resistivity minimum for TCO when the contributions from doping and scattering are optimized (recall that the actual relaxation time can be expressed as

$$\frac{1}{\tau_F} = \frac{1}{\tau_{lat}} + \frac{1}{\tau_{imp}} + \frac{1}{\tau_{def}} + \frac{1}{\tau_{sscat}};$$

here all except surface scattering are important). Resistivity is high for low doping concentrations and high again for high doping due to scattering from oxygen vacancies [14]. Figure 8.10 shows this

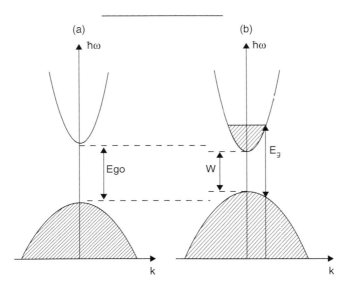

Figure 8.8 Assumed band structure of In_2O_3 and ITO [14].

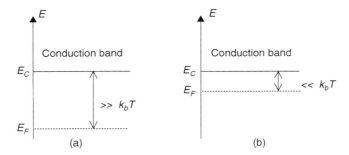

Figure 8.9 Energy band structure of a degenerate semiconductor [14].

"resistivity well" [15] and Figure 8.11 shows this behavior for Sn doping in magnetron sputtered ITO.

As one would expect, the optical properties of TCOs also depend on doping and oxygen vacancies. Low-e and EMI shielding windows take advantage of their high NIR –IR reflectance and high visible transmittance. Free electrons will absorb EMI radiation, which includes photons. Drude theory can also be used to model the optical properties of metals and transparent conducting materials. Recall from Chapter 1 that the frequency-dependent conductivity can be expressed as

$$\sigma(\omega) = \frac{ne^2\tau(1 + i\omega\tau)}{m(1 + \omega^2\tau^2)} \ [16].$$ (8.4)

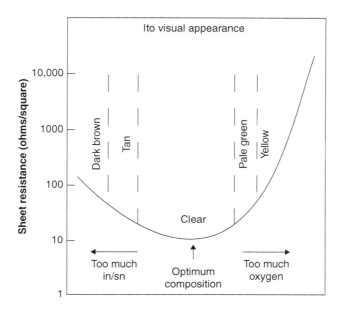

Figure 8.10 Resistivity well in a TCO [9].

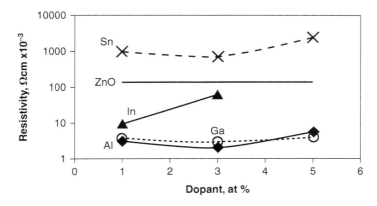

Figure 8.11 Resistivity vs Sn content for magnetron sputtered ITO.

The frequency dependent permittivity is then

$$\varepsilon(\omega) = \varepsilon_0 + \frac{i\sigma(\omega)}{\omega}, \qquad (8.5)$$

and from this we obtain the plasma frequency $\omega_p^2 = ne^2/m\varepsilon_0$. The plasma frequency is the natural frequency of oscillation, or "tuned frequency" of the electron gas. Expressions for the reflectivity of metals and transparent conducting materials are derived from these relations. The plasma wavelength ($\sim 1/\omega_p$) for most n-type TCOs is in the NIR. The dielectric constant for a free electron gas can be expressed as

$$\varepsilon = 1 - \frac{\omega_p^2 \tau^2}{(1 + \omega_p^2 \tau^2)} + \frac{i\omega_p^2 \tau^2}{\omega \, (1 + \omega_p^2 \tau^2)}, \tag{8.6}$$

and Figure 8.12 shows the behavior of the reflectance ($= (\sqrt{\varepsilon} - 1)^2/(\sqrt{\varepsilon} + 1)^2$) near the plasma frequency [16]. The TCO is highly reflective at frequencies below the plasma frequency and transmitting at higher frequencies. The free carriers also absorb electromagnetic radiation near the plasma frequency, which is called the Drude tail.

Thus far I have only covered n-type TCOs, and presented a very simplified picture of the conduction mechanisms. P-type TCOs are now being developed and have very good transmission in the NIR and IR, but higher resistivitiesThe conduction mechanism for p-type TCOs is very different than n-type materials. Their conductivity is not due to electrons but to small polarons [17].

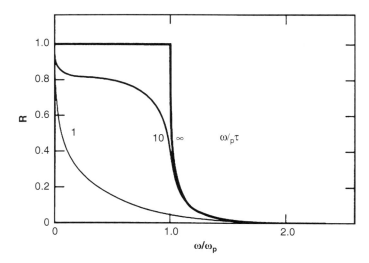

Figure 8.12 Reflectivity as a function of frequency normalized to the plasma frequency ω_p [16].

8.2.1   Indium Tin Oxide

Indium tin oxide (ITO) is the most widely used and developed transparent conductive oxide material. Applications for thin film ITO include transparent electrodes for a range of display, photovoltaic and sensor applications, EMI shielding, low-e windows, transparent heaters, and transparent electronics. The properties of ITO will first be reviewed and then we will examine several applications. As discussed last month, TCOs have a combination of good visible transmittance and electrical conduction, and ITO generally leads the way. High quality ITO has a resistivity in the range of 2×10^{-4} Ω.cm, compared to 1.8×10^{-6} Ω.cm for silver. We showed above that TCOs were highly degenerate semiconductors and that the electrical properties depended on carrier density.

Recall that ITO is actually Sn-doped indium oxide (In_2O_3). Referring to Figure 8.13 shows the general dependence of resistivity (curve A) on Sn content. Here the thickness of the magnetron sputtered ITO film is ~ 140 nm. Curves B and C show how the visible transmittance and electron mobility also depend on Sn content. From this figure, it is evident that these three important properties are interrelated. There is an Sn content ~ 20 wt. % at which the resistivity, visible transmittance, and electron mobility are optimized. At Sn levels below this value, electrical properties are determined primarily by doping concentration, and by scattering off oxygen vacancies for higher Sn levels. At this doping level, the internal transmittance (does not include substrate transmittance) is 91% and the mobility is 40 cm²/Vs. We will see that transmittance is actually optimized by increasing μ. The dielectric constant can be expressed as

$$\varepsilon(\omega) = \varepsilon_0 + \frac{i\sigma(\omega)}{\omega} = \varepsilon_0 + \frac{ine\mu(\omega)}{\omega}, \tag{8.7}$$

where ω is the frequency $\left(\sim \dfrac{1}{\lambda} \right)$.

Since we know that $\sigma = ne\mu$, $\varepsilon(\omega)$ and resulting optical properties also depend on μ, according to the relations

$$1 = T + R + A, \tag{8.8}$$

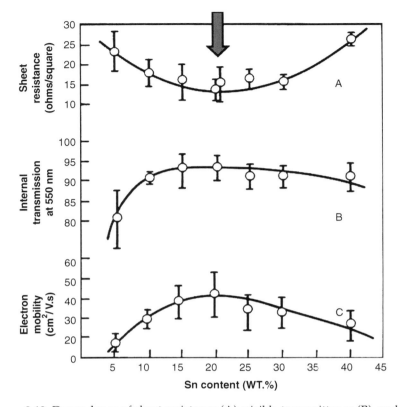

Figure 8.13 Dependence of sheet resistance (A), visible transmittance (B), and electron mobility (C) on Sn content for magnetron sputtered ITO.

$$R = \frac{(n_o - n_{ITO})^2}{(n_o + n_{ITO})^2} \text{ and} \qquad (8.9)$$

$$n_{ITO} = \sqrt{\varepsilon(\omega)} \qquad (8.10)$$

Figure 8.14 shows the spectral transmittance, reflectance, and absorption of a sputtered film, and Figure 8.15 shows the spectral transmittance of a sputtered film on a plastic window. From Figure 8.14 we see that the plasma wavelength ($\sim 1/\omega_p$) is in the NIR and, as expected, the reflectance increases at longer wavelengths (or lower frequencies). The transmittance of the film is high at visible wavelengths and the reflectance increases in the NIR - IR.

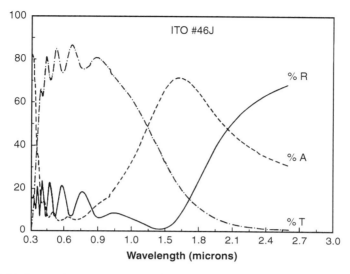

Figure 8.14 Spectral transmittance, reflectance, and absorption of a magnetron sputtered ITO film.

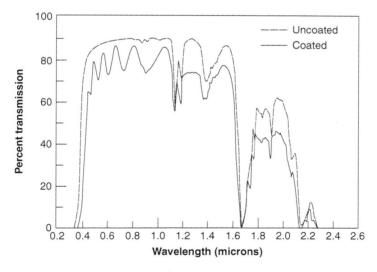

Figure 8.15 Spectral transmittance of a magnetron sputtered ITO film on plastic.

The structure of the transmittance in the NIR in Figure 8.15 is due to the transmittance of the plastic window.

The optical constants of ITO depend to a large degree on the deposition process, Sn content, and oxygen vacancies, and no one set of n and k can represent all films. In general, the refractive index is in the range 1.8–1.9 for most films and the extinction coefficient for transmissive ITO is < 0.01 at visible wavelengths. Figure 8.16 shows the dependence of transmittance on resistivity, again showing the onset of metal-like behavior with decreased resistivity. These spectra demonstrate that conductive films with excellent visible transmittance can be deposited, but visible transmittance generally decreases and NIR – IR reflectance increases with decreasing resistivity, as predicted by Drude. Finally, good ITO is completely clear. If too much Sn is added the films become tan and dark, and the films turn pale green and yellow as scattering from oxygen vacancies increases.

As with all films that depend on electron conduction, defects and grain boundaries also affect the resistivity. The goal of many deposition processes is to obtain films with the largest possible grain size, which also increases the Hall mobility. This is accomplished by heating the substrate during deposition [18]. If possible, films are often heat treated after deposition. Post deposition heat treatment is generally not an option for films on plastic substrates.

One of the more interesting optical phenomena associated with ITO and TCOs in general is the Moss-Burstein shift, which is defined by the red shift in the optical absorption edge (optical band gap) with increased resistivity and decreased carrier density [19]. The Moss-Burstein shift occurs in heavily doped semiconductors, and is due to the filling of states at the bottom of the band minima

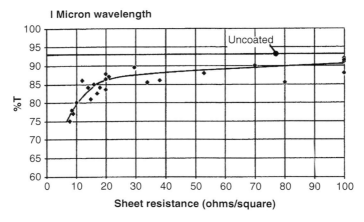

Figure 8.16 Dependence of the transmittance of magnetron sputtered ITO at 1000 nm wavelength on sheet resistance.

as the doping level is increased. This filling of states in the valence band (see Figure 8.6) leads to electron transitions from the valence to conduction band occurring at higher energies (larger band gaps) which increases the onset of absorption by an amount approximately equal to the Fermi energy.

High quality ITO films are deposited on glass, semiconductors, and plastics by planar magnetron sputtering [20], closed field magnetron sputtering [21], ion beam sputtering [22], reactive thermal and electron beam evaporation [23], and CVD and PECVD processes [24]. The sputtering process for ITO films can be tricky, particularly if a metal InSn target is used. The density and optical constants of sputtered films are near the values of the corresponding bulk materials, and deposition rates are high using a metal target (but not as high as evaporation processes). The trick with magnetron sputtering using an alloy target is to avoid poisoning the surface of the target. This is accomplished by mapping out the dependence of the target voltage on oxygen partial pressure and working just at the metal-oxide knee of the hysteresis curve. The two main techniques of process control involve keeping the target voltage constant and monitoring the change in transmittance of an optical monitor trace. The process can also be controlled by keeping the target voltage constant by adjusting the oxygen flow. Both these processes depend on the composition of the target and must be very finely calibrated and monitored to achieve the desired combination of transmittance and electrical resistance.

The easiest and best way to control the magnetron sputtering process is to use a ceramic target with the desired composition and Sn doping. The target consists of a mixture of SnO_2 and In_2O_3 powders. Since the target is conductive, DC magnetron sputtering can still be used with just a very small partial pressure of oxygen to compensate for small changes in stoichiometry due to preferential sputtering. Deposition rates for this method are lower than for the metal target case. RF sputtering can also be used to deposit from a nonconductive ceramic target. As always, not to reinvent the wheel, the recent literature should be checked before any deposition processes are developed. Because ITO applications are so extensive, the specific application or coating properties have already been described in the literature.

Because ITO films can be deposited at low substrate temperature by magnetron sputtering, it is very amenable to vacuum web coating of ITO onto flexible plastics and to plastics in general [24]. In-line processes have also been developed to coat large glass sheets [25]. Figure 8.17 shows the placement of the magnetron cathodes in the in-line coater, but the geometry is much the same in a web coater. Additionally, ITO deposition can be combined with other web coating processes to construct an entire thin film structure, such as thin film solar cells [26].

ITO films are deposited by reactive thermal and electron beam evaporation by introducing oxygen into the vacuum chamber.

In addition to low-e windows, one of the most important applications for ITO films is for transparent electrodes, used in electronic display applications. Figure 8.18 shows the layer structure of an organic light emitting device (OLED) with an ITO layer. The ITO layer is needed to transmit light out of the device. ITO is also used in thin film heterojunction solar cells [25]. Figure 8.19 shows the layer structure of a Si-ITO heterojunction cell. Other thin film cells consist of n-CdTe/ITO layers [26] and organic cells have the layer structure shown in Figure 8.20 [27]. Many of the organic layers were deposited using ink-jet technology.

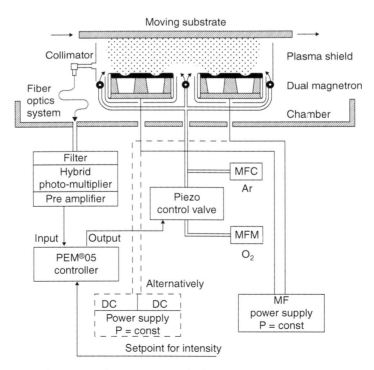

Figure 8.17 Placement of magnetron cathodes in an in-line ITO process [8].

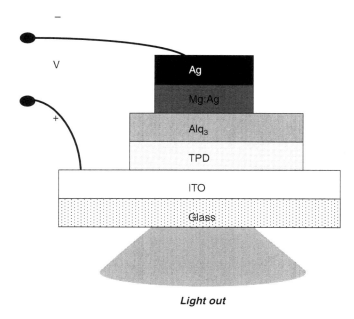

Figure 8.18 Layer structure of an OLED showing placement of the ITO film.

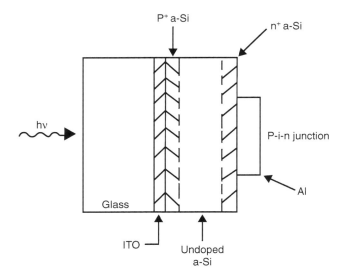

Figure 8.19 ITO heterojunction solar cell [25].

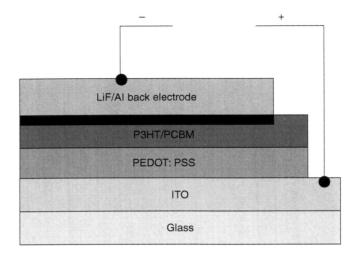

Figure 8.20 Organic solar cell containing ITO layer [27].

ITO is the most important and widely used thin film TCO. However, dwindling world indium supplies is a worry, and as a result, there is an effort to replace ITO due to the indium used in this material.

8.2.2   ZnO and Related Materials

While ITO is the most widely used transparent conductive oxide (TCO), zinc oxide (ZnO) and related compositions are very competitive, and preferred, for many applications. It is widely used as transparent electrical contacts for solar cells, laser diodes, and LEDs. As we have discussed earlier for TCOs, ZnO is an n-type semiconductor with a direct bandgap of 3.37 eV at room temperature. ZnO thin films are becoming more widely used in thin film piezoelectric devices and medical applications [28]. This high conductivity is believed to be due to oxygen vacancies or zinc interstitials, but the subject remains a bit controversial [29]. An alternative explanation has been proposed, based on theoretical calculations, that unintentional substitutional hydrogen impurities are the primary conduction mechanism [30].

The most common applications are in laser diodes and light emitting diodes, since they have exciton and biexciton energies of 60 meV and 15 meV, respectively. It is expected that the exciton properties of ZnO will be improved further by epitaxy.

ZnO films are usually doped with Al (ATO), Gd, excess Zn, or In to achieve high conductivity combined with transparency [31, 32]. This material is also often alloyed with other TCO materials. Table 8.2 lists ZnO and numerous related compositions.

ZnO thin films are mainly deposited by magnetron sputtering [33] and chemical vapor deposition (CVD) [33]. Planar and cylindrical magnetrons with metal and ceramic targets have been used [34]. The physical properties (electrical, optical, piezoelectric) of ZnO films prepared by dc reactive magnetron sputtering mainly depend on the sputtering parameters such as substrate temperature, oxygen partial pressure and sputtering pressure apart from the target-substrate distance, sputtering power, and deposition rate. Resistivity and optical transmittance were found also to depend on the crystallinity of the sputtering target [33]. Crystalline ZnO has a hexagonal structure. Films deposited using a c-axis oriented target have decidedly lower resistivity and higher visible transmittance than those deposited with a noncrystalline target. As with other TCOs reported, electrical and optical properties also depended on placement over the sputtering target. Figure 8.21 compares the dependence of the resistivity of these two types of films on substrate position.

Table 8.2 ZnO family of thin film compositions.

Composition	Application	Typical Resistivity (10^{-4} Ω.cm)
ZnO	Low-e windows, thin film solar cells	1
ZnO:Al (AZO)	Thin film solar cells	1–3
ZnO:Gd (GZO)	Thin film solar cells	< 10
Zn_2SnO_2	Thin film transistors	< 50
$ZnSnO_3$	Thin film solar cells	< 50
$Zn_2In_2O_3$	Thin film solar cells	3
ZnO:As		

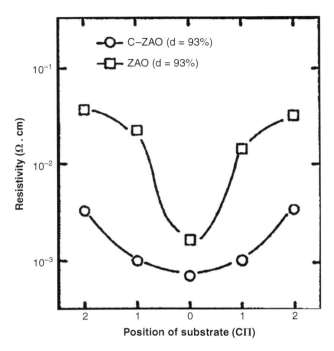

Figure 8.21 Comparison of resistivity of ZnO films deposited using crystalline and noncrystalline sputtering targets [33].

8.2.3   P-type Transparent Conductors, The Quest for High Infrared Transmission

Until the last decade, there was very little progress in the development of p-type transparent conductors (PTCO). Until then, they had very poor conductivity and low transparency. If PTCOs could be a reality, transparent p-n devices could be developed with a wide range of applications. Additionally, PTCOs could have improved NIR and IR transparency, which would open up another new range of applications. Let's begin by reviewing p-type semiconductors. Figure 8.22 compares the energy band structure of p-type and n-type semiconductors [35]. The p-type semiconductor has an acceptor energy level just above the valence band, while the n-type semiconductor has a donor band just below the conduction band. The acceptor and donor levels in the band gap result from perturbation of the periodicity of the crystal lattice, and because donor and acceptor concentrations are so small, no energy band is

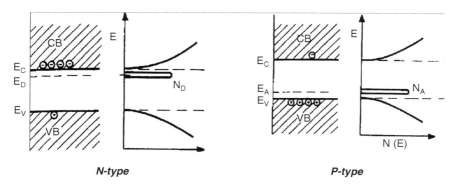

Figure 8.22 Comparison of the energy band structure of p-type and n-type semiconductors [3].

formed. The Fermi energy is skewed toward the valence band in p-type semiconductors. Holes are the majority charge carriers and responsible for conduction in p-type materials. The conductivity of a semiconductor can be expressed as:

$\sigma = q(n\mu_e + p\mu_h)$, where q is the charge, n is the electron concentration, p is the hole concentration, μ_e is the electron mobility and μ_h is the hole mobility. Thus, $p\mu_h$ is greater than $n\mu_e$ in p-type materials.

Recall that free carriers in n-type TCOs induce NIR absorption and high reflectance at IR wavelengths. Free holes do not behave like free electrons in the conduction band. The optical properties of p-type semiconductors can be very different than n-type materials due to the dependence of the plasma frequency ω_p and free carrier absorption α on effective mass m* and relaxation time τ. A few relations should be sufficient to demonstrate the optical behavior of PTCOs. The dependence of free carrier absorption on m* and τ can be expressed as:

$\alpha = Ne^2/m^*\varepsilon_o nc\tau\omega^2$ [36], where N is the carrier density, n is the dielectric constant and ω is the frequency. Also,

$$\omega_p^2 = \frac{Ne^2}{\varepsilon_{opt}\varepsilon_o m^*},\qquad(8.11)$$

where ε_{opt} is the dielectric constant.

The effective mass depends on the energy band structure of the semiconductor and is beyond the scope of this Chapter. The jury is still out on the exact conduction mechanisms for PTCOs; there appears to be several. In addition to free hole conduction, it has been hypothesized that plasmons and small polarons [17] also determine the electrical and optical properties of PTCOs.

Several types of PTCO's are being developed:

- Delafossites: $CuFeO_2$ [37], $CuAlO_2$, $AgCoO_2$ [38]
- Spinels: $NiCo_2O_4$, [39] $ZnIr_2O_4$ [40]
- N and Ga doped ZnO [41, 42], Ga-doped SnO_2 [43]
- Cu_2O structure [44]: $CuAlO_2$, $CuGaO_2$, $CuInO_2$, $SrCu_2O_2$, and LaCuOCh (Ch = chalcogen)

Because free electrons are not involved in the conduction process, the electrical and optical properties of PTCOs are very different from their n-type cousins and conduction mechanism and optical properties of each of these groups is very different. Although conductivity is not as high as n-type TCOs, their transparency range stretches far into the infrared, and it is now possible to build transparent p-n, p-i-n devices and solar cells [46]. Typical resistivities are in the 10^{-2} Ω·cm range.

Films with the spinel structure are deposited by reactive magnetron sputtering, spray pyrolysis, and spin casting. For $NiCo_2O_4$ spinel films . there appears to be a tradeoff between IR transmission and conductivity. The performance of reactively sputtered films was superior to that of the spin cast films.

Transparent thin film p-n junctions usually involve n-ZnO or n-ZAO, and include p-$AgCoO_2$/n-ZnO [37], p-$CuYO_2$/n-ZnO:Al [45], p-$Li_{0.15}Ni_{0.85}$/n-ZnO [46], n-ZnO/p-$SrCu_2O_2$ [47] and p-Cu-CoO_x/i-Cu-CoO_x/i-ZnO/n-AZO [48, 49]. Diodes often use ITO as a back electrical contact.

8.3 Electrochromic and Thermochromic Coatings

A number of types of coatings and coating structures are used for solar control, including low-e coatings, solar selective coatings, and switchable coatings. Electrochromic and thermochromic coatings are the most widely used thin film switchable coating designs, which means that transmission can be varied by applying an external "switch" such as a voltage or temperature. Liquid crystal and electrophoretic devices also employ thin films [50]. Electrochromic (EC) devices are the most complex of these systems, and are used in applications ranging from rear view mirrors, sun roofs, sun glasses and windows in buildings [50, 51, 52].

Figure 8.23 shows the layer design of a simple EC coating. The basic EC coating consists of a transparent top electrode (usually ITO), which also injects electrons into the EC layer beneath. By far the most studied and most promising EC material is WO_3, which can have either an amorphous or crystalline microstructure. WO_3 has a high coloration efficiency and capacity [53, 54]. Other inorganic materials that are presently being developed for solar control applications are various forms of WO_3, NiO, $WMoO_3$, MoO_3 and IrO_X [53]. Many of these are doped with a conduction ion such as Li^+ or H^+. The ion conductor is used to transport protons, which are supplied by the ion storage or counterelectrode layer, into the EC material. The most promising ion conductors are certain immobile solvent polymer systems, ionic glasses, and open-channel metal oxide structures such as perovskites. Oxides of Ta and Nb also show promise. The bottom layer is another transparent conducting coating. It is possible to inject both electrons and protons into the EC material, thereby inducing a strong absorption band in a given region, e.g., the visible spectrum, that results in a color change. In an EC device, an externally applied electric field is used to control the injection process (voltage is applied across the two transparent electrodes). The coloration remains for some time even after the external field is removed. The system returns

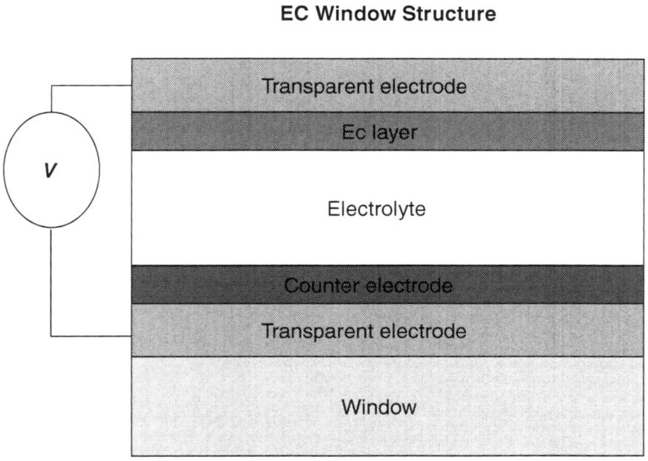

EC Window Structure

Figure 8.23 Layer design of EC coating.

to the initial state upon reversing the polarity of the external field [54]. It is interesting to note that EC windows being developed by the National Renewable Energy Laboratory (NREL), in addition to several independent facilities, now employ solar cells to facilitate switching [55, 56].

There are two major categories of EC materials: transition metal oxides (including intercalated compounds) and organic compounds (including polymers). Table 8.3 shows a few of the most common EC materials. Organic EC materials are based on the viologens, anthraquinones, diphthalocyanines, and tetrathiafulva-lenues. With

Table 8.3 Electrochromic materials and their dominant ion conductor and ion storage materials.

Electrochromic Layer	Ion Conductor or Electrolyte	Ion Storage
a-LiXWO$_3$	LiClO$_4$+PC	Redox Cple/NiO
a WO$_3$	Ta$_2$O$_5$	Ir$_x$Sn$_y$O$_2$:F
a-LiXWO$_3$	LiClO4+PC	Prussian Blue
a-H$_x$WO$_3$	Polymer	Polyaniline
a-HXWO$_3$	SiO$_2$/metal	WO$_3$
Viologen	PM MA+organic	none
a-WO$_3$	Ta$_2$O$_5$	a-IrO$_2$
c-WO$_3$	Li-B-SiO Glass	IC/LiXV$_2$O$_5$
a-WO$_3$	Li-PEO	CeOX
a-LiXWO$_3$	PPG LiClO$_4$ MMA	LiYV$_2$O$_5$
a-HXWO$_3$	Ta$_2$O$_5$	NiO
a-WO$_3$	Li-PEO	NiO
NiO	a-PEO copolymer	Nb$_2$O$_5$,WO$_3$:Mo
a-WO$_3$	a-PEO copolymer	Polymer
Polyaniline	HCl	–

a, amorphous; c, crystalline; PC, polycarbonate; PMMA, poly (methyl methacrylate); PEO, poly (ethylene oxide); IC, ion conductor; PPG, poly (propylene glycol); MMA, methyl methacrylate.

organic compounds, coloration is achieved by an oxidation-reduction reaction, which may be coupled to a chemical reaction. The violo-gens are the most studied of the organic ECs. Originally, organic ECs tended to suffer from problems with secondary reactions during switching, but recently more stable organic systems have been developed [53]. However, their reliability in harsh environments is questionable.

The EC effect occurs in inorganic compounds by dual injection (cathodic) or ejection (anodic) of ions (M) and electrons (e^-). A typical reaction for a cathodic coloring material is [57]:

$$WO_3(\text{colorless}) + yM^+ + ye^- \leftrightarrow MyWO_3 \text{ (blue) (1)}$$

where M is H^+, Li^+, Na^+, Ag^+, etc. A typical anodic reaction is:

$$Ir(OH) \times (\text{colorless}) \leftrightarrow IrO_x \text{ (black)} + XH^+ + xe^- \text{ (2)}$$

Figure 8.24 shows the spectral data for the switching of a tungsten oxide film over the solar spectrum; the spectral response of the coating in the "bleached" and "switched" states.

EC nickel oxide (NiO) is also being developed. NiO has been combined with manganese oxide, cobalt oxide, and niobium oxide in all alkaline devices [58]. The optical response of an electrochromic $NiO/polymer/Nb_2O_5$ device is shown in Figure 8.25.

The major advantages of EC coatings are that (1) they only require power during switching, (2) they have a long-term memory (12–8 h), (3) they require a small voltage (1–5 V) to switch, (4) they are specular in all states, (5) they have potential for large-area fabrication, (6) they

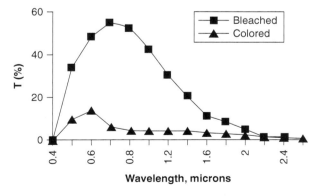

Figure 8.24 Spectral switching of WO_3 EC device.

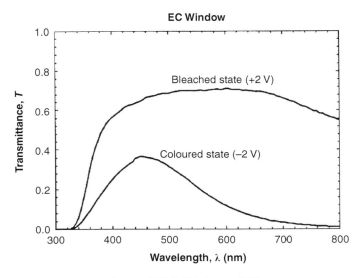

Figure 8.25 Spectral switching of NiO EC device [58].

are lightweight, and (7) EC devices have the advantage of user control [50]. Figure 8.23 shows the spectral response of the coating in the "colored" and "bleached" states.

EC devices are well suited for both solar control and thermal control. A window can be controllably darkened during periods of bright sunlight and switched back to transparency during the evening or during cloudy periods. The EC device can be used to control transmission or reflection of a window or mirror, as shown in Figure 8.26 [56]. Similarly, the reflectance of a mirror can be decreased from that of the reflective layer (usually Al) by coloring the EC layer, thus reducing glare, and increased by bleaching the EC layer.

The most well known thermochromic material is vanadium oxide (VO_2) [59, 60]. This material has promising applications for thermal control of space structures, where surfaces are exposed to intense solar radiation and periods of darkness at very low temperatures. Stoichiometric VO_2 darkens at a temperature near 68 °C due to a semiconductor to metal transition. Deposition of this material must be very precise or less active V-O phases will be formed [59]. The resistivity of VO_2 decreases by three orders of magnitude, and the emittance can be increased up to 90% at the transition temperature. Excellent examples of emittance variations are given in Takahashi's paper [60].

Emissivity control with EC structure

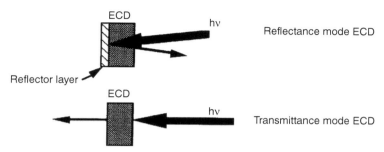

Figure 8.26 Transmittance and reflectance modes of an EC device.

8.4   Thin Film Permeation Barriers

Barrier materials have been used historically to protect items from environmental degradation such as sunlight, air (oxygen), moisture, and dirt. For decades food packaging has been the primary application for these materials; however, recent applications that require significantly higher permeation barriers include:

- Oxygen and water vapor barriers for organic and molecular electronics (organic light emitting devices (OLED)), organic electronics
- Engineered plastic substrates
- Flexible displays
- Encapsulate thin film solar cells
- Encapsulate thin film batteries

Before we address thin film barrier materials it will be instructive to briefly review diffusion mechanisms and permeability. Diffusion is transport of mass in gases, liquids, and solids from regions of higher concentration to regions of lower concentration, typically by atoms or molecules. We will be interested here in diffusion of gas through a solid interface. Diffusion across a barrier is described by Henry's, Graham's and Fick's laws [16]. Graham's law states that:

When gases are dissolved in liquids, the relative rate of diffusion of a given gas is proportional to its solubility in the liquid and inversely proportional to the square root of its molecular mass.

Henry's law states that:

> *When a gas is in contact with the surface of a liquid, the amount of the gas which will go into solution is proportional to the partial pressure of that gas.*

Fick's law states that:

> *The net diffusion rate of a gas across a fluid membrane is proportional to the difference in partial pressure, proportional to the area of the membrane and inversely proportional to the thickness of the membrane. Combined with the diffusion rate determined from Graham's law, this law provides the means for calculating exchange rates of gases across membranes.*

The permeability P of a thin film or membrane is defined as diffusion coefficient × solubility = DS. Solubility is the amount of transferred particles (atoms or molecules) retained or dissolved in the film under equilibrium. Fick's first law, shown schematically in Figure 8.27, relates the diffusive flux to the concentration field, by postulating that the flux goes from regions of high concentration to regions of low concentration, with a magnitude that is proportional to the concentration gradient (spatial derivative). In one (spatial) dimension (x), this can be expressed as:

$$\text{Diffusive flux} = J = -D\frac{dc}{dx} \tag{8.12}$$

where c = concentration, dc/dx is the concentration gradient, and D = diffusion coefficient. We see that very small values of D are required to minimize the diffusive flux. Diffusion coefficients for oxygen in various materials are listed in Table 8.4 [61]. Referring to the table we see that D for O_2 into polymers is orders of magnitude higher that for diffusion of O_2 into quartz and silica.

If diffusion was the only mechanism by which O_2 and water vapor could penetrate a thin film or membrane, there would be no need for permeation barriers. However, permeation can more readily occur through voids, pin holes, grain boundaries, and structural defects, and will increase D by orders of magnitude [61, 62]. It is well known that thin films generally have many of the above defects and as result, have diffusion coefficients significantly higher than comparable bulk materials. It has been determined experimentally

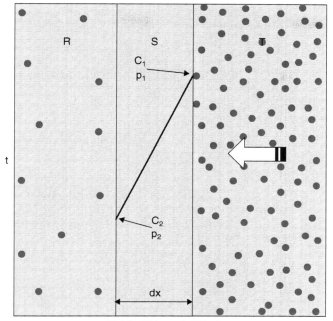

Fick's 1st law of diffusion

Figure 8.27 Fick's first law of diffusion [2].

Table 8.4 Oxygen diffusion coefficients of various solids.

Material	Diffusion Coefficient (cm²/s): D_{O2}
Aluminum oxide (Al_2O_3)	~ 10^{-30}
Silica (SiO_2)	10^{-7}
Quartz	10^{-23}
Polypropylene	10^{-7}
Polyester	10^{-9}
PET	4×10

that, above a certain thickness, permeation though a thin film will not decrease and will generally increase, as shown in Figure 8.28 [62, 63, 64].

In order to account for this result and permeation mechanisms, a number of models have been developed to predict and describe

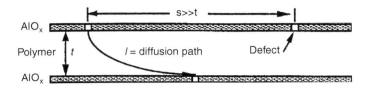

Figure 8.28 WVTR permeation through thin films [63].

permeation of gas and water vapor though a thin film, including pin hole model, lag time model [62], tortuous path model [62], and coverage model [62, 63, 64]. It is beyond the scope of this book to describe all models. However, the lag time model put forth by Graff, *et al.*, combines many mechanisms of the aforementioned models by calculating an effective diffusion coefficient for multilayer AlO_x/polymer films on PET. Defect spacing in the AlO_x layer is $2l$. Referring to Figure 8.29, water permeation through PET is described by

$$D_{eff}(AlO_x) = D_{P1}f_D + D_{AlOx}f_b \qquad (8.13)$$

Here D_{P1} is the diffusion coefficient through the polymer layer, D_{AlOx} is the diffusion coefficient through bulk AlO_x, f_D and f_b are the fractional area of defects exposed to water. Based on measurements made on actual barrier films, $D_{eff}(AlO_x)$ was calculated to be 1.4×10^{-14}, compared to $\sim 10^{-30}$ for bulk AlO_x.

Thus, the ideal permeation barrier would be defect free, pin hole free, and structurally perfect and have a diffusion coefficient of bulk. No such membrane or thin film material exists. Permeation, additionally, depends on substrate surface morphology and deposition process. Even the best thin film will not be an effective permeation barrier when deposited on a rough surface.

The following factors, which depend on deposition process and type of substrate, are required to minimize permeation of O_2 and water (WV) through a thin film or membrane:

- Minimize pin holes
- Minimize porosity
- Use materials with inherently low diffusion coefficients
- Minimize structural defects (i.e., cracks, grain boundaries, voids, growth defects)
- Substrate must be atomically smooth

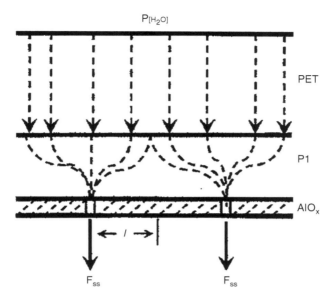

Figure 8.29 Flow paths through polymer layer and defects in thin film layer. Defect spacing is $2l$.

A large number of food products are stored in flexible plastic packaging. Flexible packaging such as PET or mylar, however, is susceptible to oxygen and water permeation. Permeation rates are generally expressed as $cc/m^2/day$ (gas) or $gm/m^2/day$ (water). Typical permeation rates for unprotected plastics are ~ 1 $cc/m^2/$ day [64, 65]. Thin opaque metal coatings, primarily Al, are therefore applied to flexible food packaging to improve barrier performance and protect the contents from oxygen and water vapor. The disadvantage of this barrier coating is that it is not transparent, which is acceptable for many applications. The main advantages of using metal barrier coatings are low cost and low permeation values.

Transparent thin film barrier coatings consist of one to tens of layers and are also used in food packaging. Additional applications that require transparent barriers and encapsulation are OLEDs, solar cells, thin film batteries, and flexible displays. Transparent single layer thin film materials generally used are SiO_2, Si_3N_4, TiO_2, and Ta_2O_5. Ideally, coatings should be amorphous (to reduce grain boundaries) and exceedingly smooth (defect free). Single layer thin films are deposited by magnetron sputtering, PECVD, ALD. Table 8.5 lists several types of thin film barrier coatings, the

Table 8.5 Oxygen and water vapor barrier performance of different barrier materials [65].

Barrier Material	Thickness of Substrate or Coating	Oxygen Transmission [ccm/m²/day/atm]	Water Vapor Transmission [g/m²/day/atm]	Deposition Process	Film Composition	Reference
PET/Blank	12.00 µm	100	64.64			
PVDC	24.00 µm	8	0.3			[68]
EVOH	24.00 µm	0.16–1.86*	N/A			[68]
m-OPA	15.00 µm	30				
Aluminized PET+ (single)	~ 30 nm	0.31–1.55	0.31–1.55	Evaporation	Al	[69, 70]
Aluminized PET++ (double)	~ 30 nm each	0.03	N/A	Evaporation	Al	[69, 70]
Aluminum on PE	7 µm Al	0.001	N/A	Laminated	Al	
SiO on PET	10–80 nm	0.35–10**	0.46–1.24	Evaporation	SiOx	[71–74]
SiO on PET	10–80 nm	0.08–1.55	0.5–5.0	PECVD	SiOx	[75–77]
Al2O3 on PET	20 nm	1.5	5.0	Reactive Evaporation	Al2O3	[78, 79]
Al_2O_3/SiO_x on PET	50 nm	2.0–3.0	1.0	Evaporation	Al_2O_3/SiO_x	[78, 79]
Diamond-like Carbon on PET	20 nm	2	1.50	PECVD	Diamond	[80]

*depending on relative humidity and ethylene content
**depending on used process
+ Al 30 nm/PET 12 µm
++ Al 30 nm/PET 12 µm/AL 30 nm

substrate used, their thicknesses, O_2 and WV transmission rates, deposition process, and composition [65, 66]. Note that improvements in O_2 and WV transmission rates as high as four orders of magnitude have been achieved with these simple treatments.

Due to extreme reactivity with O_2 and water, applications such as OLEDs, however, require O_2 and WV transmission rates ~ 10^{-6} cc(g)/m²/day, which cannot be achieved using single layer treatments. Barrier performance can be significantly improved by building multilayer structures consisting of dielectric and polymer layers [62, 67]. It is imperative that the substrate be as smooth as possible to achieve high levels of barrier performance. Plastic and flexible plastic web substrates are not smooth enough to ensure proper performance of thin film barrier coatings. In fact, features in their surface morphology are often larger than thickness of the thin film barrier. Physical vapor deposition (PVD) and CVD processes grow thin films atom-by-atom and essentially reproduce the morphology of the substrate surface. No surface smoothing occurs. The vacuum polymer deposition (VPD), described in Section 2.4.1, process mitigates this problem by being able to smooth rough surfaces, and can mask defects up to 10 µm in size [81, 82]. Unlike any other vacuum deposition process, VPD films actually smooth the surface of the substrate. A VPD layer, in contrast, does not grow atom-by-atom upward from the substrate; a gas of monomer vapor condenses on the substrate as a full- thickness liquid film that covers the entire substrate surface and its features. The liquid film is then cross-linked into a solid layer by ultraviolet (UV) or electron beam (EB) radiation. The resulting surface is glassy with virtually no defects or pin holes. The VPD layer can be combined with conventional PVD, CVD, PECVD, etc. layers to form low defect, ultra-smooth thin film structures.

VPD technology permits ultra-fast deposition of polymer films in the same vacuum environment as conventional physical vapor deposition (sputtered or evaporated) thin films. With this technology, polymer films can be deposited on moving substrates at speeds up to 1000 feet per minute and thicknesses ranging from a few angstroms to 1.3 mm with excellent adhesion to substrates and thickness uniformity of ± 2%. The VPD process has two forms, evaporative and non-evaporative. Each begins by degassing the working monomer, which is a reactive organic liquid. In the evaporative process, the monomer is metered through an ultrasonic atomizer into a hot tube where it flash evaporates and exits through a nozzle as a monomer gas. The monomer gas then condenses on the substrate as a liquid film that is subsequently cross linked to form a solid polymer by exposure to UV radiation or an electron beam. In the non-evaporative process, the degassed liquid monomer is extruded through a slotted die orifice onto the substrate. It is then cross-linked in the same fashion as in the evaporative process. Salts, graphite or oxide powders, and other nonvolatile materials can be deposited in a homogeneous mixture with the monomer. Such mixtures cannot be flash evaporated, but are required for electrolyte, anode, cathode, and capacitor film layers. The evaporative process can produce thicknesses up to approximately 10 microns at speeds as great as 1000 feet per minute. The non-evaporative process can deposit thicknesses from 10 μm to about 1.3 mm at substrate speeds approaching several hundred feet per minute.

Another advantage of conventional VPD is that it has been found to be quite efficient at smoothing rough substrates. Surfaces with a roughness less than 10 Å RMS are routinely obtained with UV cured PML/Oxide/PML barrier films deposited on both 150 Å RMS PET substrates and 5 μm RMS metal plates.

Up to ten layer dielectric/polymer barrier coatings, known as ultrabarrier coatings, are deposited by the VPD method. The first layer is always a polymer smoothing layer, followed by dielectric/polymer pairs. Figure 8.30 shows a SEM micrograph of a six-layer Al_2O_3/acrylate barrier coating used to encapsulate OLEDs [82]. The Al_2O_3 layer is ~ 100 Å thick and the acrylate layer is ~ 5000 Å thick. The reason why this coating is so effective is that it creates a tortuous path for the O_2 or water molecule to follow. The permeant molecule must first find the defect in the first Al_2O_3 layer, pass through the next acrylate layer and then again find the defect in the next Al_2O_3 layer, and so on. Oxygen and WV permeation rates for this structure are below the measurement capability of conventional OTR and WVTR equipment. However, this structure has protected OLEDs for a minimum of 10,000 hr of operation. A ten-layer structure is deposited onto PET and other flexible plastics to form a product "flexible glass", marketed by Samsung.

Advances in this technology may be realized by improving the quality of the inorganic dielectric layer and reducing the diffusivity and solubility of the polymer layers [62]. This may be accomplished by polymer selection (hydrophobic moieties or organic/inorganic copolymers), physical modifications (such as ion bombardment or crosslinking), or chemical modification (reactive etch

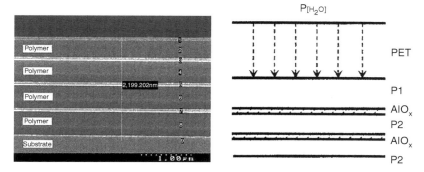

Figure 8.30 SEM micrograph of a six-layer Al_2O_3/acrylate barrier coating [82].

or plasma surface treatment). The range of improvement possible by polymer selection/ modification, however, may be small relative to improvements of the inorganic layer, since the effective D of the inorganic layer is at least four orders of magnitude lower than that of the polymer interlayers. Graff also suggests that permeation can be reduced by thinning the polymer interlayers, but this must be accomplished without compromising the smoothing properties of the polymer.

8.5 Photocatalytic Thin Films and Low Dimensional Structures

8.5.1 Hydrophylic Surfaces

Thin film and low-dimensional structure technologies in conjunction with microtechnology are rapidly replacing many technologies and devices based on conventional bulk materials, particularly in medical and energy-related applications. Thin film photocatalytic and photo-active materials are now used in a number of applications not possible with bulk materials, including oxygen production via photolytically driven electrochemistry (PDEC), energy storage, photovoltaic devices, dye sensitized solar cells (DSSC), self cleaning surfaces, drug delivery, and fuel cell electrodes [83–91, 10]. DSSC is discussed in Section 7.2.2.

Similar to solar cell operation, when a photon with energy of hv (usually ultraviolet) exceeds the energy of the band gap, an electron (e^-) is excited from the valence band into the conduction band leaving a hole (h^+) behind. In electrically conducting materials, i.e., metals, the charge carriers immediately recombine. In semiconductors a fraction of photoexcited electron-hole pairs diffuse to the surface of the catalytic particle (electron hole pairs are trapped at the surface) and can take part in chemical reactions with the adsorbed donor (D) or acceptor (A) molecules. Referring to Figure 8.31, the following reactions can take place:

- Holes can oxidize donor molecules: $D + h^+ \rightarrow D^+$
- Conduction band electrons can reduce appropriate electron acceptor molecules: $A + e^- \rightarrow A^-$

Thus, many metal oxides are strong oxidizers as a result of hole formation. Holes can react in a one electron oxidation step to produce high reactive OH radicals. Both holes and OH radicals are high reactive oxidants:

$$H_2O + h^+ \rightarrow \cdot OH + H^+$$

Atmospheric O_2 acts as an electron acceptor and forms a super oxide ion ($\cdot O_2^-$) according to the reaction

$$O_2 + e^- \rightarrow \cdot O_2^-$$

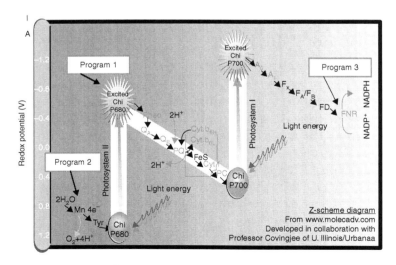

Figure 8.31 Schematic of Photosystem II.

These highly reactive ions can readily oxidize organic material, which is the basis for self- cleaning surfaces. The first self-cleaning glass was based on a TiO_2 thin films. The glass cleans itself in two stages. The photocatalytic reaction above breaks down the organic dirt on the glass using UV light and makes the glass hydrophilic (normally glass is hydrophobic). While in the "hydrophilic" state rain can wash away the dirt, leaving almost no streaks, because hydrophilic glass spreads the water evenly over its surface.

8.5.2   Photolytically Driven ElectroChemistry (PDEC)

PDEC processes are being developed primarily to generate oxygen using photons to "split" water. This and many of the above applications are inspired from Photosystem II which is an integral subprocess of photosynthesis [2, 92]. Photosystem II (or water-plastoquinone oxidoreductase), shown in Figure 8.31, is the first protein complex in photo-activated reactions occurring in photosynthesis, and is located in the thylakoid membrane of plants, algae, and cyanobacteria [84]. Photons absorbed by this enzyme are used to eject electrons that are then transferred through a variety of coenzymes and cofactors and to reduce plastoquinone to plastoquinol. The energized electrons are replaced by oxidizing water to form H ions and molecular oxygen. By obtaining these electrons from water, Photosystem II provides the electrons necessary for photosynthesis to occur. This process has several steps: (1) photons (UVA + visible) are absorbed by chromophores (D1 and D2), (2) charge separation occurs with the ejection of an electron, leaving the hole behind, (3) a chemical change from redox reaction driven by electrons to H^+ flow occurs due to energy capture to produce chemical products ATP and $NADPH^+$, and (4) an O_2 molecule is formed from water supplying the electrons. Referring to Figure 8.31, water is "slit" by the following reaction [84]:

$$2H_2O \rightarrow 4H^+ + O_2 + 4e^-$$

Figure 8.32 depicts the various types of photocatalytic reactions and Figure 8.31 compares these reactions with basic photosynthesis processes. Each uses photocatalytic activity differently to achieve a specific function, as depicted in Figure 8.33. Comparing this to photosynthesis, we see that oxygen is generated using

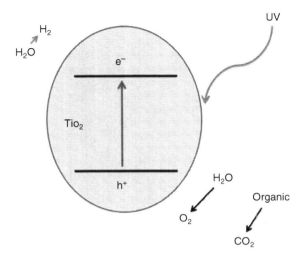

Figure 8.32 Types of photocatalytic reactions.

Photosystem II [84], and the fundamental photosynthesis reactions can be summarized as:

$$6CO_2 + 12H_2O - [h\nu + chlorophyll] \rightarrow C_6H_{12}O_6 + 6O_2 + 6H_2O$$

Thus, all these phenomena rely on the creation of an electron-hole pair (exciton) when the surface is illuminated with ultraviolet radiation, usually in the 350 nm–400 nm spectral range and the absorption of a photon, as described in Section 7.2.2. The anatase phase of TiO_2 has demonstrated the best photocatalytic properties to date.

TiO_2 films with the good reported photocatalytic performance are deposited by magnetron sputtering [89, 90], PECVD [83], laser ablation [92], ALD [93], and sol gel [10] processes. High surface areas and porosities are required in many applications to enhance photocatalysis. This can be accomplished by using nanocrystalline films and low dimensional structures such as nanoparticles, nanocomposites, nanotubes, or sculpted nanowire GLAD films in much the same way as being developed for dye sensitized solar cells (DSSC) [86–89, 94–98]. TiO_2 nanoparticles are even being attached to carbon nanotubes to improve efficiency of DCCSs [93, 94, 99]. It is not out of the question to use quantum dots. Figure 8.34 shows the surface morphology and surface profile of a magnetron sputtered TiO_2 film [9], demonstrating a relatively uniform array

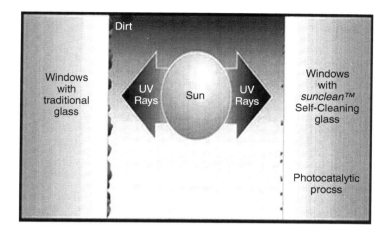

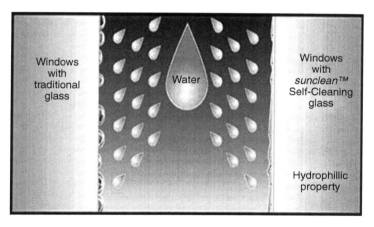

Figure 8.33 Comparison of photocatalytic reaction with photosynthesis.

of submicron (-0.25 -0.5 μm diameter) agglomerates constitut-
ing the photoactive thin film. 2D RMS roughness was 13–15 nm
as shown in Figure 8.35. This figure shows a cross section surface
profile from the image in Figure 8.34, showing a peak-to-valley
height of 70 nm (red triangles) on a vertical scale of ±100 nm. These
results demonstrate that reactive magnetron sputtering results in a
stochastically uniform array of nanocrystallites on the surface.

Other photocatalytic metal oxide candidates include WO_3, Fe_2O_3,
CeO_2, ZnO_2, and ZnS [20]. Additionally, TiO_2 and many of these
materials can be doped to modify the band gap and as a result,

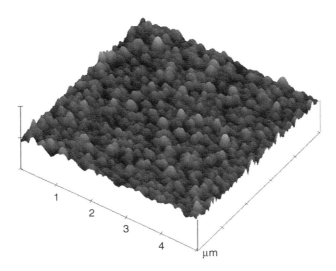

Figure 8.34 Surface morphology of TiO$_2$ film.

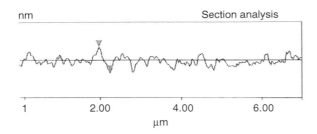

Figure 8.35 Cross section surface profile of film shown in Figure 8.34.

make it possible to use longer wavelength photons. Nitrogen is now being used to dope TiO$_2$ and move the band gap to visible wavelengths [100, 101], which is another example of band gap engineering described in Chapter 7.

8.6 Frequency Selective Surfaces

A frequency selective surface (FSS) is any *surface construction* designed as a filter for electromagnetic plane waves. An FSS can be engineered and constructed to operate at optical wavelengths to mm-waves and consist of metallic geometric structures (called elements) that are typically fabricated using photolithographic patterning techniques discussed earlier in this chapter [102, 103]. Similar to their thin film cousins, they operate as pass bands or stop bands with an angular and frequency (wavelength) dependence. They are periodic, similar to a crystal lattice, but in two dimensions and usually are narrow band. Figure 8.36 provides a survey of some of the shapes used. Elements form dielectric/metal inductive-capacitive (LC) circuits that have a frequency response that depends on shape, size, spacing, and loading. Simple mesh (wire-grid) or square patterns are very effective FSSs. The theory behind FSS is extremely complex and the interested reader is referred to Ben Monk's book on the subject [102]. Much of FSS theory and construction is used in design and fabrication of metamaterials (Section 7.3.3).

FSS elements must be aligned in a periodic array to achieve the required frequency performance. Each element acts like an LC circuit with a resonance at a specific frequency. Transmission

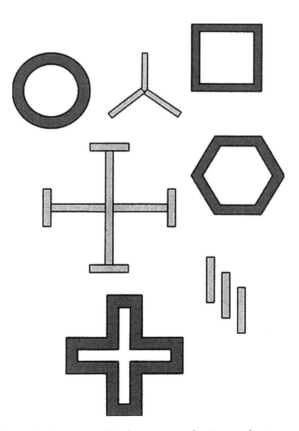

Figure 8.36 Element shapes used in frequency selective surfaces.

and reflection FSS can be made using inverse images of element arrays. Figure 8.37 shows complementary FSS elements. Reflecting structures consist of arrays of isolated metallic elements separated by dielectric or insulating regions. Transmitting structures consist of slots, holes, dipoles , etc. into continuous metal films or foils. Elements in the array are separated by conducting regions. Inductive FSSs have resonant transmission while capacitive FSSs have resonant reflection.

Figure 8.38. shows a transmitting structure consisting of a hexagonal array of micron size holes etched in a silver thin film. Referring to the circular and rectangular meshes shown in Figure 8.39, circular FSS geometry is defined by hole diameter (d), hole spacing (g), and thickness t. Rectangular geometry is defined by line spacing g, line width 2a, and thickness t. The resonant wavelength of the FSS is on the same order as element size and spacing. The refractive index (n)

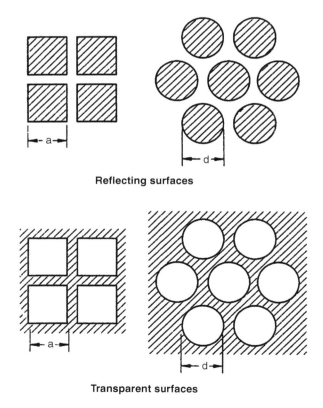

Reflecting surfaces

Transparent surfaces

Figure 8.37 Complementary FSS structures [102].

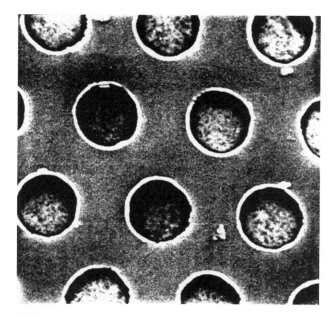

Figure 8.38 FSS with hexagonal array of micron size holes etched in a silver thin film.

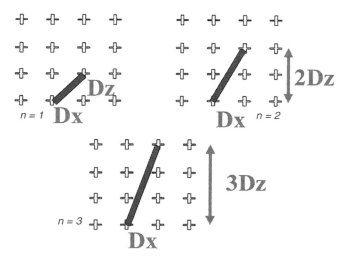

Figure 8.39 Circular and rectangular mesh parameters.

of the substrate is also important. Figure 8.40. shows the relationship between the product ng and resonant, or peak wavelength of the circular mesh for a variety of substrates. Figure 8.41 compares the measured optical transmission of a circular mesh tuned at

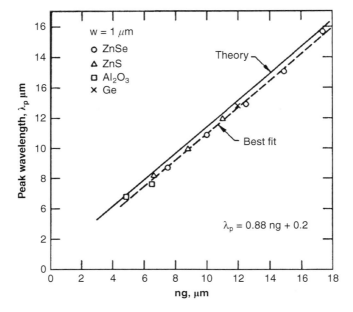

Figure 8.40 Relationship between ng and resonant wavelength of a circular mesh.

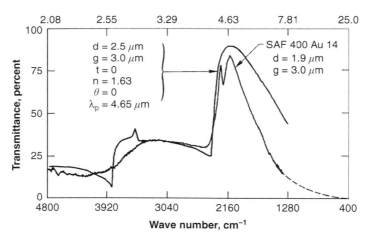

Figure 8.41 Optical transmission of a circular gold mesh with d = 1.9 μm and g = 3.0 μm on a sapphire substrate.

~ 5 μm wavelength with the predicted transmission. The gold mesh was on a sapphire substrate with d = 1.9 μm and g = 3.0 μm.

This has been just a brief introduction to frequency selective surfaces. FSSs are used extensively in microwave to mm-wave shielding applications and can be also used as optical filters. Design of FSS is extremely complex and usually performed on a computer.

References

1. Manijeh Razeghi, *Fundamentals of Solid State Engineering*, Kluwer Academic Publishers (2002).
2. S. M. Sze, *Physics of Semiconductor Devices*, 2nd Ed, Wiley Interscience (1981).
3. Peter Wurfel, *Physics of Solar Cells*, Wiley-VCH (2009).
4. Timothy A. Gessert, *SVC Bulletin* (Fall 2010) 16.
5. Byung Tae Ahn, *et al.*, *Solar Energy Materials & Solar Cells*, 50 (1998) 155.
6. L. Kazmerski, SVC Plenary Lecture, 2008.
7. Nazar Abbas Shah, *et al.*, *J. Non Cryst Solids*, 355 (2009) 1474.
8. M. Raugei, *et al.*, *Thin Solid Films*, 451 (2004) 536.
9. Clark I. Bright, Chapter 7, *50 Years of Vacuum Coating Technology and the Growth of the Society of Vacuum Coaters*, SVC (2007) 42.
10. R. J. Gilbert *et al.*, *J Appl Phys* 102 (2007) 073512.
11. C. May *et al.*, *Proceedings of the 43rd Annual Technical Conference of the Society of Vacuum Coaters* (2000) 137.
12. T. Minami, *MRS Bull*, 25 (2000) 38.
13. X. Wu *et al.*, *Proceedings of the 39th Annual Technical Conference of the Society of Vacuum Coaters* (1996) 217.
14. C. G. Granqvist, *Spectrally Selective Surfaces for Heating and Cooling Applications*, TT 1 SPIE (1989).
15. Clark I. Bright, SVC C-304 class notes.
16. Joel I. Gersten, *The Physics and Chemistry of Materials*, Wiley Interscience (2001).
17. Charles Kittel, *Introduction to Solid State Physics*, Eighth Ed., Wiley (2005).
18. H. Hosano, *et al.*, *Vacuum*, 66 (2002) 419.
19. N. Malkomes, M. Vergöhl, and B. Szyszka, *Journal Vac.Sci. Technol.*, A 19 (2001) 414.
20. T. G. Krug, *et al.*, *Proceedings of the 34th Annual Technical Conference of the Society of Vacuum Coaters* (1991) 183.
21. D. R. Gibson *et al.*, *Proceedings of the 49th Annual Technical Conference of the Society of Vacuum Coaters* (2006) 260.

22. L. J. Meng, *et al.*, *Proceedings of the 49th Annual Technical Conference of the Society of Vacuum Coaters* (2006) 679.
23. C. I. Bright, *Proceedings of the 36th Annual Technical Conference of the Society of Vacuum Coaters* (1993) 63.
24. R. L. Cormia, *et al.*, *Proceedings of the 41st Annual Technical Conference of the Society of Vacuum Coaters* (1998) 452.
25. Y. Zeigler and D. Fischer, *Proceedings of the 49th Annual Technical Conference of the Society of Vacuum Coaters* (2006) 609.
26. C. Menezes, *et al.*, Photovoltaic Solar Energy Conference; *Proceedings of the Fourth International Conference*, Stresa, Italy, May 10–14, (1982)
27. Virang G. Shah and David B. Wallace, Proceedings IMAPS 37th Annual International *Symposium on Microelectronics* (2004) 1.
28. P. M. Martin, *et al., Thin Solid Films*, 379 (2000) 253.
29. D. C Look, *et al., Phys Rev Lett* 82 (12) 2552.
30. A. Janotti and C. G. Van De Walle, *Nat Mater* 6, 1 (2007) 44.
31. W. H. Hirschwald, (1985). *Accounts of Chemical Research* 18, 8 (2007) 228.
32. G. Braunstein, *et al., Appl Phys Lett* 87 (2005) 192103.
33. R. Yoshimura, *et al., Proceedings of the 35th Annual Technical Conference of the Society of Vacuum Coaters* (1992) 362.
34. F. Milde, *et al., Proceedings of the 47th Annual Technical Conference of the Society of Vacuum Coaters* (2004) 255.
35. S. M. Sze, *Physics of Semiconductor Devices*, Second Ed, Wiley Interscience (1981).
36. Mark Fox, *Optical Properties of Solids*, Oxford Press (2001).
37. A. Barbabe, *et al., Mat. Lett.*, 60(29–30) (2006) 3468.
38. K. A. Vanaja, *et al.*, Bull. Mater. Soc., 31(5) (2008) 753.
39. Charles F. Windisch *et al., J Vac. Sci. Technol.* A 19(4) (2001) 1647.
40. Matthijn Dekkers *et al., Appl. Phys. Lett*, (2007) 021903.
41. H. Katayama-Yoshida *et al., Applied Physics* A, 49(1) (2007) 19.
42. X. Li, *et al., J Vac. Sci. Technol.* A 21(4) (2003) 1342.
43. Yixian Huang, *et al., Applied Surface Science*, 253 (11) (2007) 4819.
44. B. Roy, *et al.*, NREL/CP-520–33959 (2003).
45. A. Skumanich, *50th Annual Technical Conference Proceedings of the Society of Vacuum Coaters* (2007) 188.
46. M. K. Jayaraj, *et al.*, Mat. Res. Symp. Proc., 666 (2001) F4.1.1.
47. L. Zhuang and K. H. Wong, *Applied Physics* A, 87(4) (2007) 787.
48. A. Kudo, *et al., Appl. Phys. Let.* 75 (1999) 2851.
49. Shingo Suzuki, *et al., Vac. Sci. Technol.* A 21(4) (2003) 1336.
50. C. M. Lambert, *42nd Annual Technical Conference Proceedings of the Society of Vacuum Coaters* (1999), 197.
51. C. M. Lambert,*Proceedings of SPIE*, Vol. 3788, (1999) 2.
52. S. E. Selkowitz, *Proceedings of SPIE* Vol. 2255 (1994) 226.

53. P. M. Monk, R. J. Mortimer, and D. R. Rosseinsky, *Electrochromism: Fundamentals and Applications*, VCH Publishers, Weinheim, Germany, 1995.

54. C. Granqvist, "Electrochromic Tungsten-Oxide Based Thin Films: Physics, Chemistry, and Technology," *Physics of Thin Films*, Vol. 17 (1993) 301.

55. F. Pichot, *et al.*, "Self-powered electrochromic coatings", *SPIE Proceedings* Vol, 3788, (1999) 59.

56. Courtesy, Sienna Technologies Inc. (Woodinville, WA).

57. C. B. Greenberg, *J. Electrochem. Soc.*, Vol. 141 (1993) 3332.

58. C. M. Lampert, "Smart Window Switch on the Light," *Circuits and Devices*, March 1992, 19.

59. K. Le, R. O. Dillon, and N. Oanno, "Thermochromic VO_2 Deposited by Active Control of Direct Current Magnetron Sputtering", Presented at the 2001 International Conference on Metallurgical Coatings and Thin Films, San Diego, CA, April 30–May 4, 2001.

60. I. Takahashi, M. Hibino, and T. Kudo, *SPIE Proceedings* Vol. 3788, 26.

61. Charles A. Bishop, *Roll-to-Roll Vacuum Deposition of Barrier Coatings*, Wiley-Scrivener (2010).

62. Gordon, L. Graff *et al.*, *J Appl Phys*, 96(4) (2004) 1840.

63. B. M. Henry, *et al.*, *41*st *SVC Annual Technical Conference Proceedings* (1998) 434.

64. A. G. Erlat, *et al.*, *47*th *SVC Annual Technical Conference Proceedings* (2007) 654.

65. W. Decker and B. M. Henry *45*th *SVC Annual Technical Conference Proceedings* (2002) 492.

66. E. G. Erlat, *et al.*, *44*th *SVC Annual Technical Conference Proceedings* (2001) 448.

67. G. L. Graff, *et al.*, *43*rd *SVC Annual Technical Conference Proceedings* (2000) 354.

68. E. Schaper, "The Future of EVOH - The World's Most Popular Barrier Resin," *Proceedings Barrier Pack '89*, Chicago, Illinois, US.

69. D. Chahroudi, "Glassy Barriers From Electron Beam Web Coaters," *Proceedings Barrier Pack '89* Chicago, Illinois, US.

70. AIMCAL Metallizing Committee, Technical Reference Book, Issue 1.

71. C. Sakamaki, "Application Study of High Barrier Ceramics Deposited Film," *Proceedings Barrier Pack '89*, Chicago, Illinois, US.

72. J.W. Jones, U.S. Patent #3, 442, 686, May 6, 1969, Assigned to Du PontdeNemours and Company, Wilmington, Delaware.

73. R. Phillips, *et al.*, U.S. Patent # 4, 702, 963, Oct. 27, 1987, Assigned to Optical Coating Laboratory Inc., Santa Rosa, California.

74. Direct communication, Flex Products Inc., 2793 Northpoint Pkwy, Santa Rosa, CA 95407.
75. R.B. Heil, *et al.*, "Mechanical Properties of PECVD Silicon-Oxide Based Barrier Films on PET," *38th Annual Technical Conference Proceedings of the Society of Vacuum Coaters* (1995).
76. W. Nassel, "Production, properties, processing and application of SiO-coated films," *7th Intern. Conf. Vac. Web Coating*, Miami, FL, USA(1993) 258 .
77. E. Finson, "Transparent Barrier Coatings Update: Flexible Substrates," *7th Intern. Conf. Vac. Web Coating*, Miami, FL, USA (1993) 242.
78. S. Schiller *et al.*, "How to produce Al2O3 coatings on plastic films." *7th Intern. Conf. Vac. Web Coating*, Miami, FL, USA (1993) 194.
79. Y. Yamada *et al.*, *38th Annual Technical Conference Proceedings of the Society of Vacuum Coaters*, (1995) 28.
80. D.S. Finch, *et al.*, "Diamond Like Carbon, a Barrier Coating for Polymers used in Packaging Applications," *Packaging Technology and Science*, Vol. 9 73–85 (1996).
81. J. D. Affinito *et al.*, *39th SVC Annual Technical Conference Proceedings* (1996) 392.
82. M. E. Gross and P. M. Martin, Chapter 11: *Handbook of Deposition Technologies for Films and Coatings*, 3rd Ed., P M Martin, Ed.,Elsevier (2009).
83. P. Hajkova *et al.*, *Society of Vacuum Coaters 50th Annual Technical Conference Proceedings* (2007) 117.
84. *Artificial Photosynthesis*, Anthony F Collings and Christa Critchley, Ed, Wiley-VCH (2005).
85. A. Hagsfelt and M. Gratzel, *Chem Rev* 95 (1995) 49.
86. Suresh Baskaran *et al.*, J. Am. Ceram Soc 81(2) (1998) 401.
87. Jun Liu *et al.*, *ChemSusChem* (2008) 1.
88. Donghal Wang *et al.*, *ACS Nano* 3 (4) (2009) 907.
89. C. Xu, X. Wang, JJ Zhu, *J Phys Chem* C112 (2008) 19841.
90. Chang Yao *et al.*, Mater. *Res. Soc. Symp*. Proc. Vol. 951 (2007)
91. *Artificial Photosynthesis*, Anthony F Collings and Christa Critchley, Ed., Wiley-VCH (2005).
92. S. J. Wang *et al.*, *J Vac Sci Technol* A 26(4) (2008) 898.
93. Rodjanna Phreamhorn et al., *J Vac Sci Technol* A 24(4) (2006) 1535.
94. N. S. Lewis, Science, 315 (2007) 798.
95. Tae Young Lee, *et al.*, *Thin Solid Films*, 515 (12) (2007) 5135.
96. Michael Berger, *Nanowerk LLC* (2007).
97. Jason B. Baxter and Eray S. Aydil, *Sol. Energy Mater. & Sol. Cells*, 90 (2006) 607.
98. A. Kay, M. Gratzel, *Chem Mate*r 14 (2002) 2930.
99. Sung-Hwan Lee, "Photocatalytic Nanocomposites Based on TiO$_2$ and Carbon Nanotubes", Ph.D. Dissertation, University of Florida (2004).

100. K. W. Böer, *Survey of Semiconductor Physics*, Van Nostrand Reinhold: New York, 1990.

101. M. L. Kaariainen *et al.*, *Society of Vacuum Coaters 50th Annual Technical Conference Proceedings* (2007) 335.

102. Ben A. Monk, *Frequency Selective Surfaces, Theory and Design*, Wiley Interscience (2000).

103. Richard Renski, Brian Gay and Lisa Ma, Frequency Selective Surfaces, Design and Analysis Using Ashcroft Product Suite, Presentation #4.

9

Looking into the Future: Bio-Inspired Materials and Surfaces

While several of the engineered structures and their properties presented Chapters 7 and 8 are not yet fully developed and many exciting properties are yet to be discovered, surface engineering technology is also evolving toward bio-inspired materials. Bio-inspired processes include artificial photosynthesis, ferritin regulation, drug delivery, self-healing, self-cleaning, and biological photovoltaics. Self-assembled nanostructures and biophotonic materials are also being actively developed.

9.1   Functional Biomaterials

One of the most active areas of development in this technology is functional organic and biomaterials. Biological and organic applications include:

- Cancer research
- Antimicrobial materials

- Self-cleaning
- Self-healing
- Sterilization (antimicrobial)
- Oxygenation of blood
- Materials from renewable resources (e.g., algae)
- Bio-sensors
- Drug delivery
- Implants
- Dialysis
- Gas permeable membranes

We introduced photocatalytic thin films in the previous two chapters, with emphasis on self-cleaning surfaces and oxygen generating surfaces. Two of the most important applications for photocatalytic materials are decomposition of organic matter and self-cleaning windows and surfaces, which makes them also useful for sterilization and chemical remediation [1, 2]. Organic compounds can be converted to CO_2, H_2O, NO_3^- or other oxides, halide ions, phosphates for environmental remediation [2]. Environmental decontamination by photocatalysis can have advantages over conventional chemical oxidation methods because semiconductors are inexpensive, easily produced, nontoxic, and capable of extended use without substantial loss of photocatalytic activity.

Remediation of biological contaminants can be accomplished using TiO_2-based photocatalytic reactions. Photocatalytic destruction of warfare agents, nerve agent stimulant (organophosphorus compounds), and mustard gas stimulant (organosulfur compounds), has been demonstrated [3–7]. Chemical compounds can be completely mineralized via multiple steps which include several intermediate byproducts. Similar to the degradation of organics, CO_2, H_2O, and inorganic salts are the final products and no hazardous final byproducts are formed. However, accumulation of partially oxidized intermediate species on the catalyst surface can form a "scum" and retard photocatalytic oxidation of TiO_2, thus poisoning the process (much like target poisoning in sputter deposition of oxides) if the photocatalytic removal rate is not high enough [8].

Additionally, photocatalytic nanocomposite systems composed of TiO_2 shells with a carbon nanotube core (see Section 7.2.2) are being developed as antimicrobial treatments to destroy microorganisms and toxins in the environment as well as chemical agents.

Photocatalytic TiO_2 is attractive for controlling environmental pollutants because of the following characteristics:

- TiO_2 is an environmentally benign material
- The same basic technology can be applied to both water and air media
- The use of a catalyst eliminates the need for chemical oxidants
- Microbes are completely mineralized
- Either black lights (UV) or solar radiation can serve as the excitation source
- The scientific basis of the technology has already been established
- High destruction rates enable the system to be compact

Interestingly enough, if photocatalysts can be readily applied to contaminated surfaces such as toilet seats, kitchen countertops, and sinks, only sunlight or black light would be needed to disinfect them, and as a result, they would be self cleaning.

Photocatlaytic processes have been developed to remove organic materials from surfaces. TiO_2 is the most commonly used photocatalytic material. We first review the photocatalytic processes that occur in water and their lifetimes [9]:

- $TiO_2 \to^{(h\nu)} h_{VB}^+ + e_{CB}^-$ (10^{-15} s)
- OH^{*+} radical formation at the TiO_2 surface: $h_{VB}^+ + >Ti^{IV}OH \to \{>Ti^{IV}OH^*\}^+$ (10^{-9} s)
- Electron trapping: $e_{CB}^- + >Ti^{IV}OH \leftrightarrow \{>Ti^{III}OH^*\}^+$ (10^{-10} s) – shallow trap
- Electron trapping: $e_{CB}^- + >Ti^{IV} \leftrightarrow \{>Ti^{III}\}$ – deep trap
- Charge carrier recombination: $e_{CB}^- + >Ti^{IV}OH^*\}^+ \to Ti^{IV}OH$ (10^{-7} s)
- Charge carrier recombination: $h_{VB}^+ + \{>Ti^{III}OH^*\} \to Ti^{IV}OH$ (10^{-9} s)
- Interfacial charge transfer: $\{>Ti^{IV}OH^*\}^+ + Red \to Ti^{IV}OH + Red^{*+}$ (10^{-7} s)
- Interfacial charge transfer: $e_t^- + O_x \to Ti^{IV}OH + O_x^{*-}$ (10^{-3} s)

Referring to Figure 9.1, here VB and CB label the valence and conduction bands, h = hole and e = electron. Thus we see that there

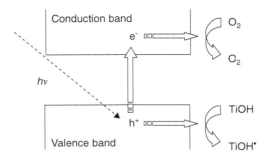

Figure 9.1 Photocatalytic processes occurring with water [9].

are a number of "fast" and "slow" processes that can occur using water.

The widespread use of antibiotics has made many virulent strains of microorganisms resistant to these conventional treatments, and, as a result, TiO$_2$-based photocatalysts are also being developed as alternative sterilization technologies [10]. Upon illumination by near UV radiation (~ 365 nm), microbial cells are killed when contacting TiO$_2$-Pt catalysts [11]. A practical photochemical device in which TiO2 powder was immobilized on an acetyl cellulose membrane was constructed based on this technology [12]. Drinking water has also been disinfected using this technology by introducing TiO$_2$ nanoparticles in direct contact with target microbes [13–16]. OH$^-$ radicals are the active agents here. Fungi, tumor cells, and even cancer cells have also been successfully deactivated by TiO$_2$ photocatalysis [10, 16]. Disinfectant processes of various methods are summarized in Table 9.1 [2]. TiO$_2$-carbon nanotubes systems are being developed to enhance the photocatalytic performance of technologies described above (and below).

Table 9.1 Disinfectant processes involving TiO$_2$ [2].

Method	OH*	O$_2^-$, H$_2$O$_2$	Cl	hv	Adsorption	Capture
Photocatalysis	x	x		x	x	x
UV(254 nm)				x		
Chlorine			x			
HEPA filter						x

A number of interesting bio-solar cell concepts are being developed, some surprising. The dye sensitized solar cell (DSSC), discussed in Section 7.2.2, employs any number of very different materials and phases to convert sunlight into electricity. During photosynthesis the photons in the solar spectrum are absorbed by various types of chlorophyll. Chlorophyll is a solar absorbing "dye" and has the absorption spectrum of chlorophyll as shown in Figure 9.2. Dyes synthesized for DSCCs attempt to simulate the solar absorptive properties of chlorophyll [17].

A surprising example of an artificial bio-solar cell is a cell constructed from spinach, shown in Figure 9.2 [18]. Note how this structure has many of the properties mentioned above. The device is essentially a spinach sandwich. A protein complex (Photosystem 1 – PS1) derived from spinach chloroplasts is assembled on a peptide membrane to form the electric circuit, much like the one shown in Figure 9.3. The cell is constructed from ground up spinach which is purified to isolate a protein complex used in the cell. Similar to Figure 9.4, the chromophores attach to the protein scaffolding. A thin Au film deposited on glass helps the spinach PS1 assemble in layers. A polymeric insulator is then deposited over the spinach to insulate between membranes (see above), and a top semitransparent electrical contact is applied over this assembly. It is estimated that this device is capable of achieving an efficiency of ~ 20%.

Finally, work is proceeding on using bacteria to simulate the photosynthesis process [19]. The structure of an apparatus that absorbs

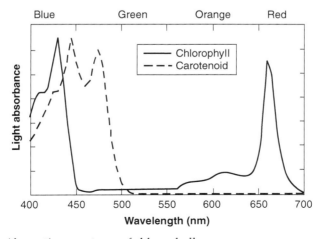

Figure 9.2 Absorption spectrum of chlorophyll.

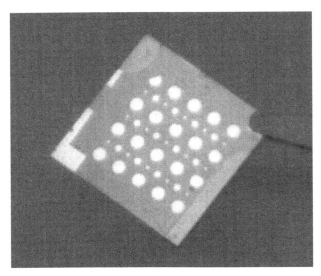

Figure 9.3 Spinach based artificial bio-solar cell [18].

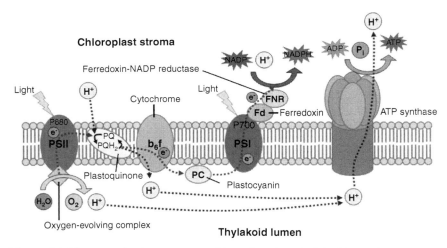

Figure 9.4 Photosynthesis processes taking place in a thylakoid membrane.

sunlight in cyanobacteria (blue-green algae shown in Figure 9.5), has been discovered. The structure has 96 chlorophylls being held at close distances by a protein complex. Each of the chlorophylls absorbs sunlight and delivers its energy to a central chlorophyll pair that utilizes it to electronically charge a cell membrane, creating a highly efficient biological solar cell, as discussed above. This

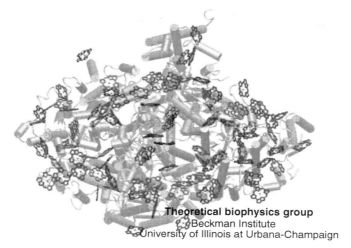

Theoretical biophysics group
Beckman Institute
University of Illinois at Urbana-Champaign

Figure 9.5 Chromophores attached to a protein complex [19].

system is very forgiving in that several of the chlorophylls can be taken off and performance will not be significantly degraded.

To summarize, bio-solar cells are being developed that use electrons generated in the photosynthesis process. These electrons generate an electric current across cell membranes that can be connected to a load to generate power, similar to a semiconductor solar cell. We have seen that natural and artificial photosynthesis can be used to generate electrons within cell membranes and create an electric current and biological solar cell. Work is proceeding on many fronts to develop structures that absorb light over the entire solar spectrum to optimize the efficiency of the bio-solar cell, much like semiconductor photovoltaics. There appears to be a limitless, and low cost, supply of materials for bio-photovoltaics, which is very promising for the economical use of photovoltaics. There are still many problems to overcome, stability being the most important. Although competitive bio- solar cells are at least 20 years in the future, this is one of the most promising areas of green energy development.

9.2   Functional Biomaterials: Self Cleaning Biological Materials

We will see that two very different cleaning phenomena are involved in self-cleaning biological materials compared to inorganic materials (see Section 8.5). An entirely new family of functionality is encompassed by biologically inspired synthesis, hierarchical structures, and stimulus-responsive materials systems [20]. Much like inorganic self-cleaning materials, bioinspsired self-cleaning materials are based on superhydrophobicity [21, 22]. The goal in many cases is to replicate the phenomena at work in the lotus leaf. The leaf's surface morphology consists of micro and nanostructures that suspend water droplets [23, 24, 25]. This is in comparison to photocatalytic processes that reduce surface energy.

As mentioned above, the lotus leaf effect is the basis for this bioinspired technology [26]. Figures 9.6–9.8 display this effect. Dirt particles are picked up by water droplets due to a complex micro- and nanoscopic architecture of the surface, which reduces surface tension. This effect is also found in many other plants (see Figure 9.8), e.g., tropaeolum, cane and columbine, and on wings of certain insects.

Due to their high surface tension, water droplets tend to minimize their surface area by achieving a spherical shape. On contact with a surface, adhesion forces result in wetting of the surface. Either complete or incomplete wetting will occur depending on the structure of the surface (see Figure 9.6) and the fluid tension of the

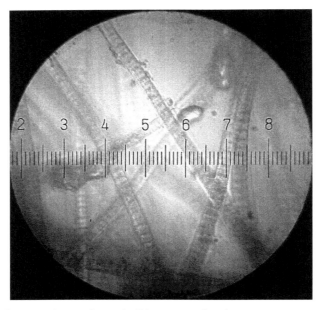

Figure 9.6 Picture of cyanobacteria (blue green algae).

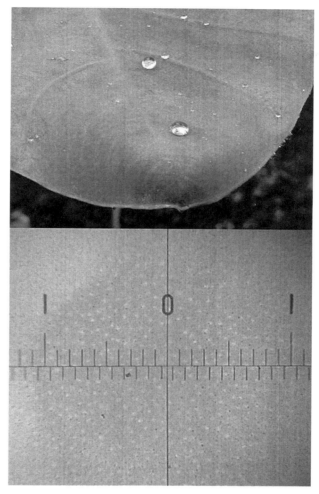

Figure 9.7 Water beading on a lotus leaf.

droplet. Self-cleaning properties occur as a result of the hydrophobic water-repellent double structure of the surface [25], thus enabling the contact area and the adhesion force between surface and droplet to be significantly reduced resulting in a self-cleaning process. This effect results from two structures on the surface: a characteristic epidermis (outermost layer called the cuticle) and covering waxes. The epidermis of the lotus plant possesses papillae (see Figure 9.7) with 10 to 20 μm heights and 10 to 15 μm width on which epicuticular waxes are layered. These layered waxes are hydrophobic and form the second layer of the double structure [27, 28].

Figure 9.8 Microstructure of a lotus leaf.

It will be instructive to briefly review how hydrophobicity is measured. The hydrophobicity of a surface is determined by the contact angle θ_C, shown graphically in Figure 9.9. All surfaces have an energy with the surface. Contact of water with a surface results from thermodynamic equilibrium between three phases:

- Liquid phase of the droplet (L)
- Solid phase of the substrate (S)
- Gas/vapor phase of the ambient (V) (which is a mixture of ambient atmosphere and an equilibrium concentration of the liquid vapor)

The V phase could also be considered another (immiscible) liquid phase. At equilibrium, the chemical potential in the three phases must be equal. Interfacial energies form the basis of contact angle:

- Solid-vapor interfacial energy (see surface energy) - γ_{SV}
- Solid-liquid interfacial energy - γ_{SL}
- Liquid-vapor energy (i.e., the surface tension) - γ

Assuming a perfectly smooth planar surface, the contact angle is then defined by Young's equation [29]

$$0 = \gamma_{SV} - \gamma_{SL} - \gamma\cos\theta_C$$

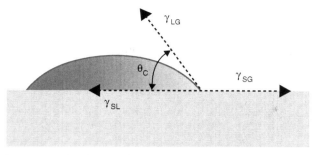

Figure 9.9 Contact angle of a liquid on a solid surface.

The higher the contact angle the higher the hydrophobicity of a surface. Surfaces with a contact angle < 90° are referred to as hydrophilic and those with an angle >90° as hydrophobic. Some plants show contact angles up to 160° and are called super-hydrophobic meaning that only 2–3% of a drop's surface is in contact. Plants with a double structured surface like the lotus can reach a contact angle of 170° whereas a droplet's actual contact area is only 0.6%. All this leads to a self-cleaning effect. Dirt particles with an extremely reduced contact area are picked up by water droplets and are thus easily cleaned off the surface. If a water droplet rolls across such a contaminated surface the adhesion between the dirt particle, irrespective its chemistry, and the droplet is higher than between the particle and the surface.

Any deviation from the above criteria (caused by surface roughness and impurities) will cause a decrease in θ_C. Even in a perfectly smooth surface a drop will assume a wide spectrum of contact angles between the highest (advancing) contact angle, θ_A, and the lowest (receding) contact angle, θ_R. In fact, θ_C can be calculated from θ_A and θ_R using the relation [30]

$$\cos \theta_A = r\cos\theta_C \qquad (9.1)$$

Here r is the roughness ratio.

Thus, structuring the surface has a definite effect on contact angle. A rough surface has more surface area and essentially becomes an air-material composite, and hence surface energy is higher. However, not all fluid can penetrate all the "nooks and crannies" of the surface so some surface area is not active. Thus we have water sitting on the surface and air. The contact angle for this case is given by [31]

$$\cos\theta_C = f_1\cos\theta_1 - f_2, \text{ where} \tag{9.2}$$

θ_1 is the contact angle of the surface with water, f_1 is the fraction area in contact with the surface, and f_2 is the fractional area beneath the drop. The micro and nanostructure of the surface will have considerable effects on the contact angle: θ_C will increase when the drop advances across a surface and will decrease when the drop recedes. Figure 9.10 summarizes the contribution of roughness to contact angle for various models [31, 32, 33], and Figure 9.11 presents various types of naturally occurring composite and engineered surface structures [20].

Referring to Figure 9.10, it is obvious that the contact angle of water and thus the hydrophobicity of a surface can be engineered based on designs found in nature. Consider the force needed to make a drop of water slide off a surface. The contact angle of the drop goes through a hysteresis from "advancing - θ_{adv}" to "receding - θ_{rec}" as discussed above. Hysteresis is not observed for a perfectly smooth surface, but increases with increased surface roughness. It makes sense then that the force needed for the drop to slide is directly proportional to the contact angle hysteresis [34]:

$$F \sim \gamma_{LV}(\cos\theta_{rec} - \cos\theta_{adv}) \tag{9.3}$$

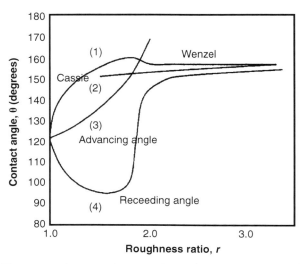

Figure 9.10 Contact angle vs. surface roughness for various models [32].

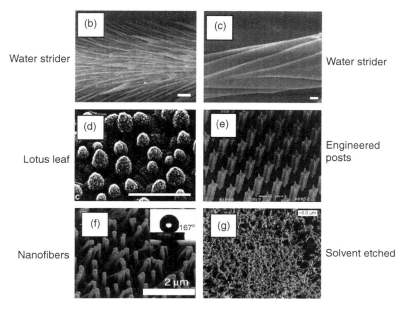

Water strider

Water strider

Lotus leaf

Engineered posts

Nanofibers

167°

Solvent etched

2 μm

Figure 9.11 Naturally occurring composite and engineered surface structures [20].

This force can be minimized by reducing contact angle hysteresis. Surfaces with near zero hysteresis are termed "superhydrophobic" and this is the basis of self-cleaning surfaces. The cleaning process is defined as follows:

- Debris and other particulates are encountered by nearly spherical drops of water as they roll around on the surface
- As a result of roughness and few contact points (see Figures 9.7 and 9.10), debris is only loosely bound to the surface
- Particles are attracted to the drop by surface tension
- Because it is loosely bound to the surface, the drop (with particulates attached) slides off the surface, thus cleaning the surface

Note that this process does not eliminate organic contaminants, as discussed for photocatalytic reactions described below. We see that micro and nanostructure, i.e., simulating structures found in nature, are the most important factors that govern self-cleaning and superhydrophobicity. The structures can be engineered by techniques discussed in Chapters 7 and 8. Self-assembly, colloidal assembly, and other wet methods also show promise [35, 36, 37].

9.3   Functional Biomaterials: Self-Healing Biological Materials

An entirely new family of functionality is encompassed by biologically inspired synthesis, hierarchical structures, and stimulus-responsive materials systems [38]. The previous section addressed self-cleaning biological systems used to generate super-hydrophobic surfaces to reduce surface tension and oxidize organic materials on a surface. The goal in many cases is to replicate the phenomena at work in nature. In this section we address progress in self-healing systems. Self-healing materials are a class of smart materials that have the structurally incorporated ability to repair damage caused by mechanical usage over time. These materials have been of interest for several decades. The motivation comes from biological systems, which have the ability to heal after being wounded. Initiation of cracks and other types of damage on a microscopic level has been shown to change thermal, electrical, and acoustical properties, and eventually lead to whole scale failure of the material. In biological systems, chemical signals released at the site of the fracture initiate a systematic response that transports repair agents to the location of the injury and initiates healing, as shown in Figure 9.12 [38]. We don't have these types of processes in

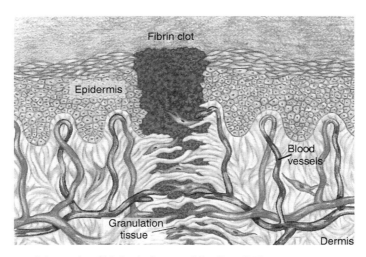

Figure 9.12 Schematic of biological wound healing [20].

non-biological materials. The self- healing process is complex and comprised of three distinct steps [39]:

1. Inflammatory phase
2. Fibroblastic phase (wound closure)
3. Scar maturation phase (matrix remodeling)

The inflammatory phase occurs immediately following the injury and lasts approximately 6 days. Coagulation begins immediately. After ~ 24 hr the fibroblastic phase occurs at the termination of the inflammatory phase, and the wound starts to close. This phase can last as long as 4 weeks. Scar maturation begins at the fourth week and can last for years. Here extracellular matrix is synthesized and remodeled as tissue grains strength and function. The culmination of these biological processes results in replacement of normal skin structures with fibroblastic mediated scar tissue.

To date, no engineered materials system can fully simulate the biological self-healing process. This problem has been addressed by the following strategies:

- Reduce the likelihood of a crack forming through the development of tribologial materials (nanolaminates, etc.)
- Healing cracks that have formed
- Thermal and radiational healing of stress related bond breakage and cracking

Important applications include healing of materials used in space. Industrially, cracks are usually mended by hand, which is difficult because they are often hard to detect. A material (polymers, ceramics, etc.) that can intrinsically correct damage caused by normal usage could lower production costs of a number of different industrial processes through longer part lifetime, reduction of inefficiency over time caused by degradation, as well as prevent costs incurred by material failure [40].

Virtually every engineered self-healing process is polymer based [41]. The most widely used self-healing strategies are

- Microcapsule polymer composites
- Reversible polymers and chain breaking
- Thermal healing
- Healing by radiation-induced effects

The self-healing polymer composite (CSHC) consists of a polymer that contains a catalyst and micron-size bubbles filled with monomer. A typical catalyst contains 2.5% Ru, and the micron-size spheres are filled with dicyclopentadiene (DCP) [42]. The DCP was encapsulated in poly-urea formaldehyde in an oil-water emulsion. Diameter of the spheres ranged from 10 to 10,000 µm. Self-healing is based on the monomer breaking out of the spheres as a result of fracture and stress and filling microcracks. When a crack reaches the microcapsule, shown in Figure 9.13, the capsule breaks and the monomer bleeds into the crack, where it can polymerize and mend the crack

Self-healing structures require the following salient features: [41]

- Microcapsules must survive polymer processing.
- Localized stress needed to break open the monomer microcapsule should be as low as possible.
- Catalysts dispersed in the polymer should not degrade the polymer.
- The microcapsule system should not affect the performance or lifetime of the polymer.
- The self-healing process should occur rapidly, i.e., the DCP should diffuse rapidly into the cracks in the polymer.
- The microcapsules should be firmly bound in the polymer matrix to facilitate rapid response to applied stress and minimize stress required to rupture them.

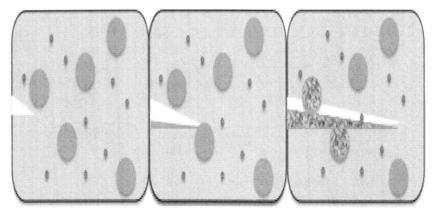

Figure 9.13 Depiction of crack propagation through microcapsule-imbedded material. Monomer microcapsules are represented by pink circles and catalyst is shown by purple dots.

Several factors determine how the microcapsules rupture, including wall thickness and mechanical properties, and the usual polymer thermal properties: glass transition temperature (T_G) and melting temperature (T_M). Wall thickness usually ranges between 160 nm and 220 nm.

This technology is limited by [41]:

1. Only mechanically-induced micron-size defects, such as microcracks, delamination, and chain slippage can be healed.
2. There is no healing resulting from damage due to large localized stresses.
3. There is a low temperature limit (T_M) at which the microcapsules will not release monomer.
4. The monomer inside the microcapsules is not stable in an oxidizing environment, such as low earth orbit.
5. Catalyst residue may degrade and destabilize the polymer.
6. The catalyst will decompose when heated and may degrade the polymer.
7. Diffusion of monomer in the polymer matrix is temperature dependent.
8. Radiation (in space) may adversely affect the microcapsules
9. Lifetimes that are not yet known.

A word on how the polymeric microcapsules fracture. Cracks in polymers are caused by breaking of chemical bonds in the polymer. Organic polymers consist of long chains of molecules, which can be broken by an applied mechanical stress though cleavage of sigma bonds [43]. While newer polymers can break in other ways, traditional polymers typically break through homolytic or heterolytic bond cleavage. Figure 9.14 illustrates these types of bond breaking for PMMA. The factors that determine how a polymer will break include: type of stress, chemical properties, level and solvents used, and temperature [44].

After a bond is broken homolytically, two radicals are formed which can recombine to repair damage or can initiate other homolytic breaks which can in turn lead to more damage. Regarding heterolytic bond breakage, cationic and anionic species are formed which can in turn recombine to repair damage, can be passivated by a solvent, or can react destructively with nearby polymers [41].

Some polymeric systems are reversible, meaning that they can revert to the initial state, which can be a monomer, oligimer, or not cross linked. Since a polymer is stable under normal conditions, the reversible process usually requires an external stimulus. For a reversible self-healing polymer, if the material is damaged by heating, for example, it can be reverted back to its original constituents, and thus repaired or "healed" to its polymer form by applying the original condition used to polymerize it.

In addition to the microcapsule approach, there are a number of polymer systems that self heal using the various phenomena described above:

- Diels-Alder [45]
- Retro Diels-Alder [45]
- Thiol-based polymers
- Autonomic polymer healing

Diels-Alder (DA) polymers respond to a mechanical stress very differently than most common polymers [46]. DA response to stress is reversible; they undergo a reversible cycloaddition, in which stress breaks two sigma bonds in a retro Diels-Alder reaction [45]. The result is the addition of pi-bonded electrons as opposed to radicals or charged chain sections [46]. A reversible DA and RDA reaction using cross-linked furan-meleimides based polymers is shown in Figure 9.14 [47]. When heated, DA polymers (poly(ethylene terephthalate-co-2,6-anthracenedicarboxylate, for example) break down to their original monomeric units according to the RDA reaction (shown in Figure 9.15) and the polymer reforms upon cooling

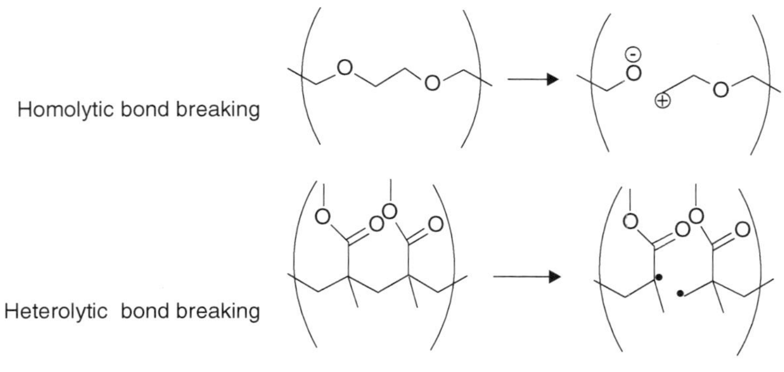

Figure 9.14 Homolytic and heterolytic bond breaking in PMMA [43].

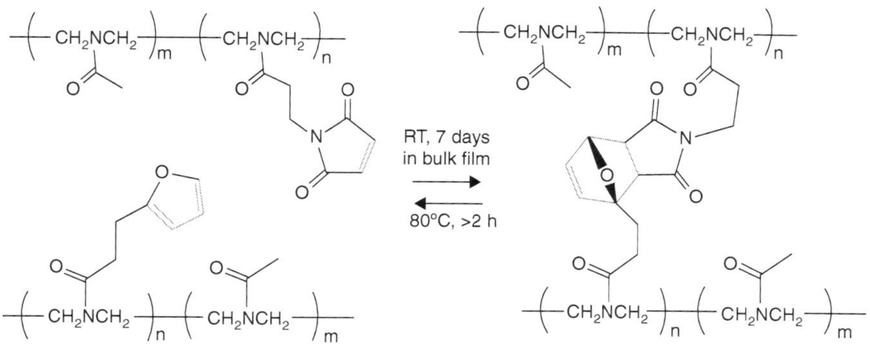

Figure 9.15 DA reaction for cross-linked furan-meleimides based polymer [47].

or through any other conditions that were initially used to make the polymer. During the last few decades, two types of reversible polymers have been studied [46]:

- Polymers where the pendant groups, such as furan or maleimide groups, cross-link through successive DA coupling reactions
- Polymers where the multifunctional monomers link to each other through successive DA coupling reactions

Figure 9.16 depicts the reversible cross-linking in the highly cross-linked furan-maleimide based polymer network (compare to Figure 9.15) [48]. The DA reaction for multifunctional monomer occurs in the backbone of the polymer to construct or reconstruct the polymer and not as a linking process. During heating and cooling cycles (120 C - ~ room temperature) a furan-maleimide based polymer (Tris-maleimide (3M) and tetra-furan (4F)) was de-polymerized via the RDA reaction into its starting materials and then restored (upon cooling).

Thiol based polymers round out our discussion on reversible self-healing reactions. These polymers have disulfide bonds that can be reversibly cross-linked through oxidation and reduction. Under reducing conditions, disulfide (SS) bridges, shown in Figure 9.17, in the polymer break and a monomer results. But when oxidized, thiols (SH) of each monomer form the disulfide bond, cross-linking the starting materials to form the polymer [49].

We end this brief overview with "living polymers" [50]. A living polymer is defined as a polymerization process without chain

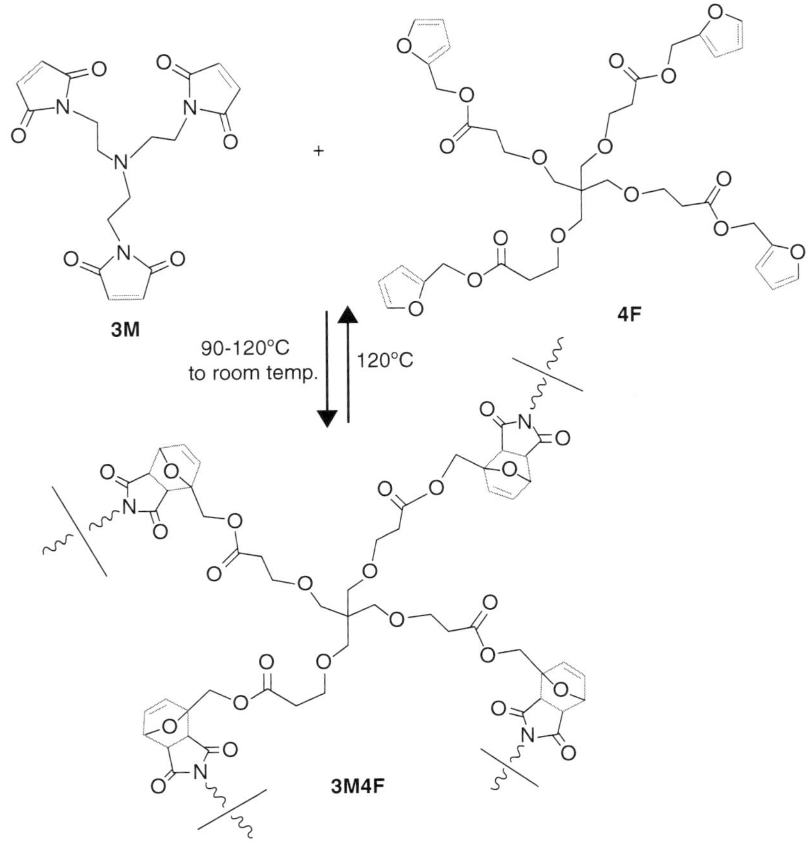

Figure 9.16 Reversible cross linking in furan-maleimide based polymer network [48].

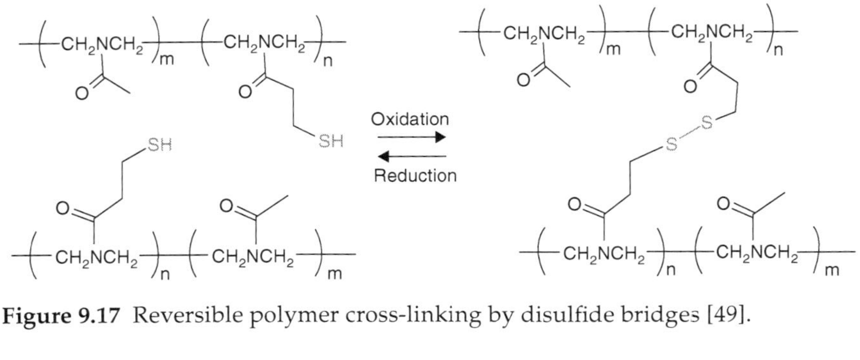

Figure 9.17 Reversible polymer cross-linking by disulfide bridges [49].

breaking (transfer and termination) reactions [50, 51, 52]. The ends of the polymer chains remain active and can grow if more monomer is added. The equilibrium state of this system consists of polymers and monomers. This type of polymer has the following salient features:

- Polymerization proceeds until all monomer has been consumed. If additional monomer is added, the polymer continues to grow
- The average molecular weight is a linear function of monomer converted (obvious)
- The number of polymer chains remains constant; only their length changes
- The molecular weight of the polymer can be controlled by the stoichiometry of the reaction
- The distribution of molecular weights is relatively narrow
- Block polymers (ABA, AB, ABC) can be synthesized by adding more monomer sequentially
- Polymerization depends linearly on reaction time

It should be noted that tight control of growth conditions is often required to synthesize this type of polymer.

One method of synthesis, shown in Figure 9.18, is known as reversible addition fragmentation chain transfer (RAFT). RAFT technology is capable of synthesizing polymers with predetermined molecular weight and narrow molecular weight distributions over a wide range of monomers with reactive terminal groups that can be controllably modified. Further polymerization is possible [53]. RAFT can be used for virtually all methods of free radical polymerization, including solution, emulsion, and suspension polymerizations. This technique introduces a chain transfer agent (CTA) into a conventional free radical polymerization reaction (must be devoid of oxygen, which terminates propagation). The CTA enables additional growth of polymer chains. A CTA is usually a di- or tri-thiocarbonylthio compound, which produces the dormant form of the radical chains. This type of polymerization is much more difficult to control that homolytic bond formation and breaking (see above). The CTA for RAFT polymerization affects polymer length, chemical composition, rate of the reaction, and the number of side reactions that may occur and thus must be judiciously chosen.

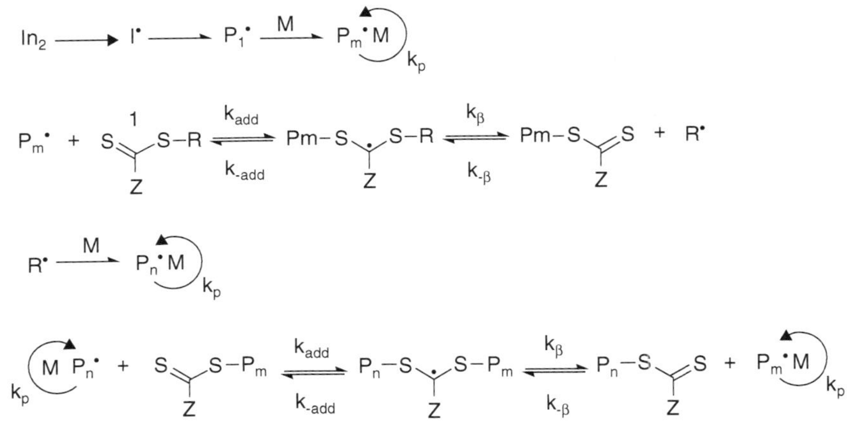

Figure 9.18 RAFT reactions for living polymers [53].

Ionizing and UV radiation can damage polymers by breaking chains and forming radicals. Living polymers are one method used for self-healing of radiation damaged polymers [39]. Generally, radiation damage will consist of active and inactive polymer chains. The equilibrium condition for this system will consist of a constant number of active (also known as macroradicals) chains. Under the influence of ionizing radiation, new free radicals are produced, which means that if equilibrium is to be preserved, some chains must be deactivated. Deactivation occurs via recombination (combination of free radicals), and thus the polymer is healed.

An excellent review of other technologies is given by Moshe Levy [54].

To date, manmade self healing polymer technology does not even come close to that of biological systems. Polymers can be damaged by heat, ionizing and UV radiation, and mechanical stress. One important application is healing polymers located in low earth orbit. Self-healing polymers have been developed over the last four decades using a variety of processes which heal broken chains, replace broken chains, and add to broken chains. One of the most promising self-healing techniques appears to be the microcapsule approach which introduces monomer in response to an external stimulus (usually a stress). Future self- healing systems could also incorporate fully autonomic circulatory networks capable of healing large damage volumes [39].

9.4 Self-Assembled and Composite Nanostructures

We are all aware that microelectronics technology is ever decreasing the size of microcircuit structures (transistors, etc.). Device sizes are approaching nanometer and atomic scales. Progress can be represented by the well-known Moore's law, shown in Figure 9.19, which predicts the number of transistors on a chip as a function of time. Critical issues here with downsizing to these dimensions are:

- Small size of the features
- Limited the number of electrons
- Operation of small features
- Interconnection of small features
- Cost: possible expensive lithography
- Self-assembly
- Simpler architecture
- Defect tolerant architecture

Confinement of electrons in dimensions approaching atomic and molecular dimensions requires consideration of quantum behavior [55]. Possible devices include single electron devices, quantum cellular automata, molecular electronics (OLED, etc.), and quantum computing.

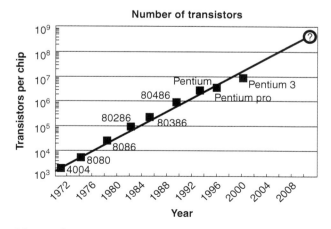

Figure 9.19 Moore's law.

Self-assembled nanostructures, essentially quantum dots, are being developed for a number of applications, including microelectronics, thermoelectric power generation, memory storage, bio-sensors [56, 57, 58, 59, 60, 61]. Examples of self-assembled nanostructures are:

- Ge nanostructures on Si
- PbTeSe quantum dots
- Au quantum dots on Si for memory storage
- Au quantum dots on Si for bio-sensors
- Si quantum dot luminescence [62]
- InGaAs quantum dot/GaAs solar cells [62]
- CaF$_2$ nanostructures on Si [63]
- Quantum platelets [64]

Ge nanostructures are being formed on Si substrates to reduce feature dimensions and extend Moore's law to higher densities [58]. Self-assembly, not lithography, is used to form the low dimensional structures. Strains from a lattice mismatch between two materials are used to form nanostructures and quantum dots [55, 56, 57, 58, 63, 64]. Referring to Figure 9.20, the SiGe lattice spacing is larger than the Si spacing and is compressed when SiGe is deposited over a Si layer [55]. This compression forms strains in the SiGe lattice, and with assistance from surface diffusion, Ge nanostructures and quantum dots are formed. See Figure 9.21. Typical layer thickness are 3 nm for SiGe and 3.5 nm for Si, deposited by PECVD. Shape of the nanostructure depends on how strain is obtained in the layered

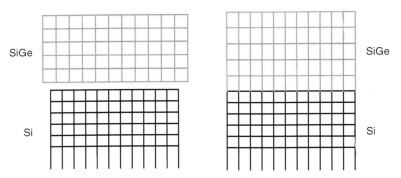

Figure 9.20 Compression of SiGe lattice when deposited over Si [58].

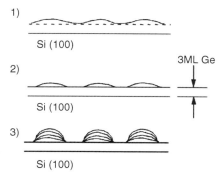

Figure 9.21 Formation of Ge quantum dots [1].

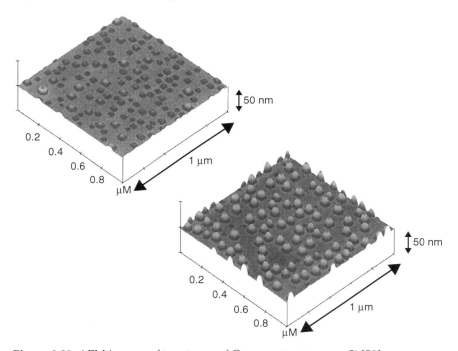

Figure 9.22 AFM images of two types of Ge nanostructures on Si [58].

structure and is directly related to deposition conditions (gas flow, pressure, and temperature). Figure 9.22 shows AFM images of two configurations of Ge nanostructures on Si: quantum dots and "domes". The height of dome-shaped structures levels off at ~ 6 nm with diameter ~ 70 nm.

To be useful for nanoelectronic applications, the size and spatial distributions have to be uniform. This is accomplished by control of strain energy, reaction kinetics, and thermodynamic assembly. Island shape is determined by:

- Volume energy
- Surface energy
- Interface energy
- Edge and corner energies

Adding HCl during deposition significantly affects island size; island size decreases and density increases with increased HCl partial pressure. Post deposition etching with HCl also sculpts the shape, resulting in shorter and narrower structures.

As with most electronic grade semiconductors, the nanostructures can also be doped. Phosphorous is introduced using PH_3 and also can be used to control island size, shape, and uniformity. As with other self-assembly processes, islands can be aligned and ordered by patterning the Si substrate prior to deposition. Recall that this is also true for GLAD films [65]. The Ge islands become more aligned with addition of subsequent SiGe/Si layers. As shown in Figure 9.23.

Ge island formation is also determined by the degree of stress anisotropy [66]. A lattice mismatch in "x" and "y" directions results in equi-axed islands. Anisotropic islands (wires) form when the lattice mismatch is in different directions. Here growth is constrained in one direction, and hence the formation of wires instead of domes.

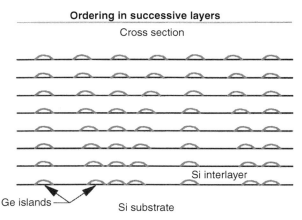

Figure 9.23 Ordering of Ge islands with addition of SiGe/Si layers [58].

Growth, however, is by no means perfect. Best efforts have resulted in several percent defects. Strategies to live with this level of defects include using them in a structurally simple architecture, such as a crossbar array or parallel straight wires. Additionally, defects could be mapped and architecture configured around them. Using self-assembled, self-ordered nanostructures may require a defect tolerant computer that can accept and design around a limited number of defects.

We have seen in the last chapter that the band structure of low dimensional structures can be very different from that of bulk semiconductors. Figures 9.24, 9.25, and 9.26 show arrays of quantum

Figure 9.24 Array of PbTeSe/PbTe quantum dots for thermoelectric power generating devices [56, 57].

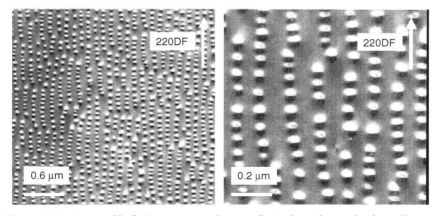

Figure 9.25 Array of InGaAs quantum dots on GaAs for enhanced solar cell conversion [68].

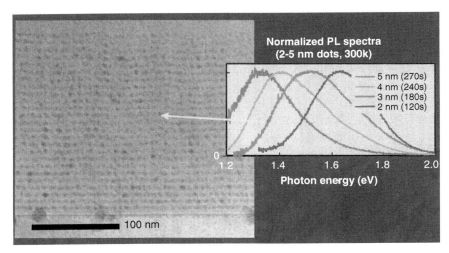

Figure 9.26 Array of Si quantum dots with enhanced photoluminescence.

dots used to enhance electrical and thermoelectric properties of semiconductors [56, 67, 68]. PbTeSe/PbTe quantum dot thermoelectric devices show promise of very high figures of merit [69]. SiGe/Si quantum dots (described above) are also being developed for thermoelectric power generation [69]. Intermediate band solar cells, shown in Figure 9.27, integrate quantum dots into their structure and have a theoretical efficiency near 63%.

Si quantum dot arrays show enhanced photoluminescence as shown in Figure 9.28. Peak luminescence shifts to higher energies with decreased dot size. The bandgap of Si quantum dots depends on their size, as shown in Figure 9.28. Again we see a dependence on size of the dot. The bandgap begins to increase from that of crystalline Si (~ 1.2 eV) for particle sizes < 3 nm. The shift is dramatic, with an increase to ~ 2.5 eV for 0.5 nm particle sizes.

9.5   Introduction to Biophotonics

We all know that electromagnetic radiation interacts with biological materials and systems. We have all experienced a sunburn. Photosynthesis keeps vegetation alive and provides much needed oxygen to sustain life. Biophotonics is the study of the interaction of light with surface of biological materials and systems. Photonics uses light much the same way electric circuits use charge carriers

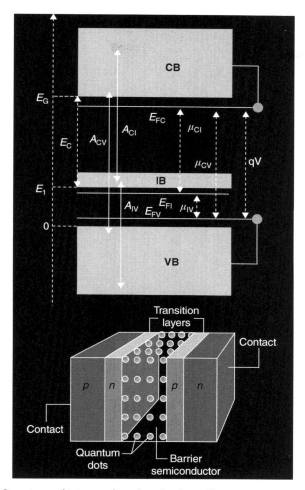

Figure 9.27 Structure of intermediate band solar cell.

(electrons and holes). This technology encompasses much of optical technology, including lasers, optical materials, control of light, sensors, waveguides, and nonlinear optical materials. Figure 9.29 shows the general relationship between photonics and biotechnology [70]. We are exploring new ground here folks. Photonics is used for optical diagnostics, light activated and light guided therapy, early detection of diseases, bioimaging, biosensing, tissue engineering, and manipulation of cells. Common medical applications are wrinkle and hair removal by lasers, plastic surgery, cancer treatment, eye surgery, prostate surgery, and general surgery. The

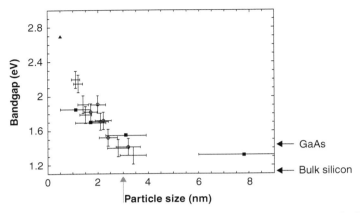

Figure 9.28 Modification of Si/GaAs band gap using Si quantum dots [15].

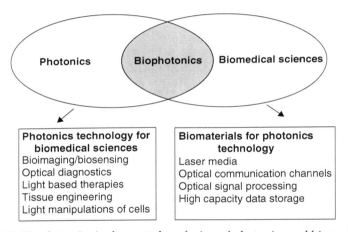

Figure 9.29 Biophotonics is shown to be a fusion of photonics and biomedical sciences.

artificial lung being developed by PDEC technology, for example, is a new and exciting biophotonic application [71].

Essentially biophotonics involves a basic understanding of how light interacts with bio-molecules, and integrates chemistry, physics, engineering, biological sciences, and clinicians. This interaction can be used to perform a function or to extract information about a biological system. Due to the complexity of this technology, only the basics will be presented here.

Photosynthesis is a good place to start this discussion; essentially that's where it all began. This process occurs in plants, algae, and many types of bacteria. This reaction uses light (photons) to produce oxygen (among other chemicals that sustain the plant). The process steps are shown in Figure 9.30. Light is absorbed by chromophores (chlorophyll for example), which act much like an antenna to collect the photons. Figure 9.31 shows the absorption spectrum of chlorophyll a [72]. Note the strong absorption in the UV and red wavelengths. Other types of chlorophylls and carotenoids absorb at different wavelengths to cover the visible spectrum. The chromophores act as a reaction center to transform light energy into chemical energy. Light is absorbed and causes a reaction that generates compounds such as ATP (adenosine triphosphate) that the plant turns into energy. The basic reactions are:

$$\text{PSI: } 2H_2O + 2NADP^+ + \chi ADP + \chi P_i + \chi h\nu \rightarrow O_2 + 2NADPH + 2H^+ + \chi ATP + \chi H_2O$$

$$\text{PSII: } 12\,NADPH^+ + 12H^+ + 18\,ATP + 6CO_2 + 12H_2O \rightarrow C_6H_{12}O_6 + 12\,NADP^+ + 18ATP + 18P_i.$$

Here, NADP is nicotinamide-adenine dinucleotide phosphate, ADP is adenosine diphosphate, P_i is phosphorus, χ is a concentration, and the star of the show is the photon $h\nu$. Note in the second

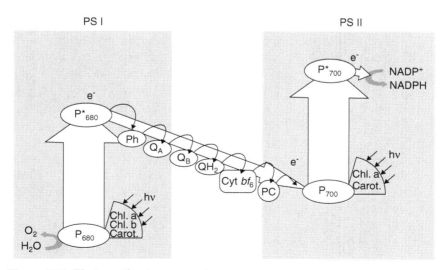

Figure 9.30 Photosynthesis process steps.

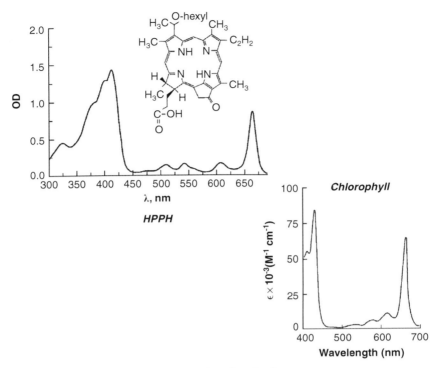

Figure 9.31 PSI and PSII systems used to absorb photons.

equation that carbon dioxide (CO_2) is converted to glucose ($C_6H_{12}O_6$). The ($^+$) indicates a reduction (loss of an electron). Referring to Figure 9.31, PSI and PSII are photon absorbing systems, essentially composed of chlorophyll and carotenoids (absorption spectrum of chlorophyll is shown in Figure 9.2). PSI converts photons to oxygen and energy, and PSII converts CO_2 to sugar and energy. The absorbed photon energy is transferred via an electron transfer chain to reaction centers, P_{680} and P_{700}, which indicates the absorption wavelength maximum. Water is used to supply electrons to reduce the NADP and NAPDH, which reduce carbon dioxide and eventually produce sugars. To summarize this process: the leaves of the plant interact with the environment, harvest light energy, utilize carbon dioxide to form sugar, and convert water into oxygen.

Biological systems are basically an assembly of cells, which are composed primarily of water, amino acids, carbohydrates (sugar), fatty acids, and ions. Macromolecules (proteins, peptides, polysaccharides (complex sugars), DNA, RNA, and phospholipids) make up the remainder of the cell mass. All these materials are organized in a semipermeable membrane of the cell wall. There are two types of cells; prokaryotic cells (bacteria for example) have little internal structure and eukaryotic cells have much more complex inerds, including a membrane bound nucleus. Figure 9.32 shows the hierarchy of cells and their sizes.

The chemical building blocks of life exist in cells, nucleic acids (RNA, DNA), proteins, saccharides, and lipids. DNA and RNA consist of three building blocks: nitrogen containing ring compounds that are either purine or pyrimidine bases, sugar and phosphate. The four bases that constitute DNA have the familiar labels A (adenine), G (guanine), T (thymine), and C (cytosine). T is replaced by U (uracil) for RNA. Proteins are formed during a polymerization process that links amino acids. Only 20 amino acids are used to form the vast array of proteins that are used in biological systems. We are all familiar with sugars (saccharides). They are the primary

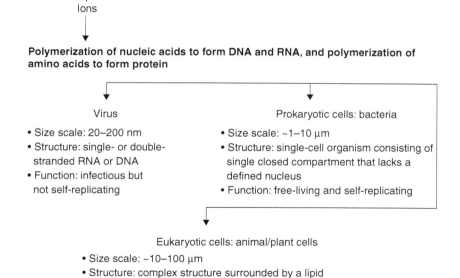

Small molecules:
Amino acids
Nucleic acids
Water
Lipids
Ions

Polymerization of nucleic acids to form DNA and RNA, and polymerization of amino acids to form protein

Virus

- Size scale: 20–200 nm
- Structure: single- or double-stranded RNA or DNA
- Function: infectious but not self-replicating

Prokaryotic cells: bacteria

- Size scale: ~1–10 μm
- Structure: single-cell organism consisting of single closed compartment that lacks a defined nucleus
- Function: free-living and self-replicating

Eukaryotic cells: animal/plant cells

- Size scale: ~10–100 μm
- Structure: complex structure surrounded by a lipid membrane, contains an organized nuclear structure
- Function: self-replicating and able to assemble to form tissues

Figure 9.32 Hierarchy of cell structure.

source of energy for biological systems. Lipids form a group of compounds in living systems that are soluble in nonpolar solvents. Most lipids are derived from fatty acids and represent "fat" in the biological system.

So, how do we integrate the chemistry and structure of cells and photonics? Tissues contain cells and are self-supporting media, which means that they can have optical properties in a macro or bulk scale. Therefore, light can interact with tissue to give us some information on the structure and the chemical makeup of the tissue or be used to modify the structure and chemistry of the tissue. As shown in Figure 9.33, light can interact with tissues and cells by absorption, reflection, refraction, and scattering. Absorption is due to various intercellular as well as extracellular constituents of the tissue. Scattering effects are very pronounced in tissues. The turbidity or haziness of a tissue results from multiple scattering events from its heterogeneous structure, which consists of macromolecules, cell organelles, and, of course, water. Scattering from tissue involves several mechanisms.

Inelastic scattering from biological tissue is relatively weak. Raman scattering, however, is significant due to molecular vibrations. Raman scattering can provide valuable information on the chemistry and structure of the cells and tissue and detect small changes in these properties. Changes in tissues as a result of diseases and mechanical stress can be detected by Raman techniques [73, 74]. Carden and Morris [73, 74] measured spectra of minerals and proteins in prosthetic implants. They observed changes in the

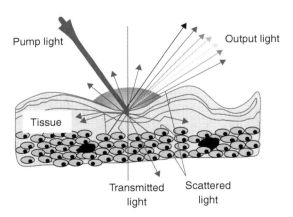

Figure 9.33 Optical interactions of light and tissue.

Raman spectra of the inorganic and organic compounds due to mechanical stress on the implants. These spectra revealed ruptures of collagen cross links due to shear stress exerted by the implant passing across the bone. More recent work [73] has used Raman spectroscopy to examine premature fusion of the skull bones at the sutures, which is a birth defect in the face and skull.

We all know that tissue can absorb light. As with all materials, this absorption is wavelength dependent. Figure 9.34 shows the penetration of various types of laser radiation into tissue. It's interesting to note that the penetration depth decreases with increased blood content in the tissue.

Absorption of light also induces the following phenomena in tissues [70]:

- Radiative
  - ꙩ Tissue autofluorescence
  - ꙩ Fluorescence from various parts of the tissue
- Nonradiative
  - ꙩ Photochemical (excited state reaction)
  - ꙩ Photoablation (breaking of cellular structure)
  - ꙩ Photodisruption (fragmentation and cutting of tissue)
  - ꙩ Thermal (light -> heat)
  - ꙩ Plasma-induced ablation (dielectric breakdown -> ablation)

All the above processes can be used to probe the structure and composition of the tissue and cells, and also to perform operations

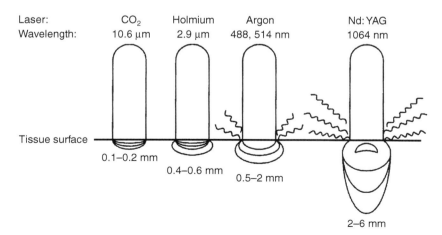

Figure 9.34 Penetration depths of laser light into tissue.

on them. Several spectroscopic techniques (in vivo spectroscopy in particular) can take advantage of these optical processes to give us information on cellular and tissue structure, abnormal tissue structure, and biological functions. *In vivo* spectroscopic techniques include absorption, Raman scattering, back-scattering, and fluorescence. For example, Figure 9.35 shows Raman spectra for benign and malignant breast tissue [75]. The spectra for the benign tissue, benign tumor, and malignant tumor are significantly different. This method is also being pursued for brain tumors [76].

Bioimaging is an important biomedical tool for diagnosis and treatment of human diseases. While present imaging techniques (MRI, CAT scan, ultrasound, radioisotope, x-ray) examine gross

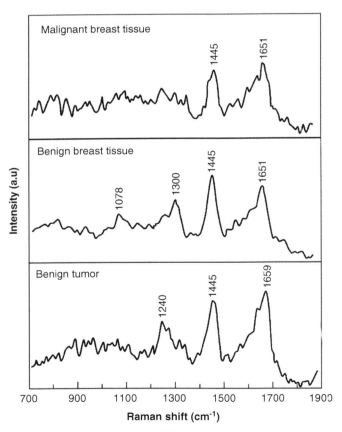

Figure 9.35 Raman spectra from normal breast tissue, benign breast tumor, and malignant breast tumor.

or bulk features, biophotonic technology is developing methods to image cells and tissues for early detection, screening, and diagnosis of life threatening diseases. Optical imaging also overcomes the following limitations of other imaging techniques:

- Harmful ionizing radiation
- X-ray imaging on young patients
- Harmful radioactivity
- Inadequate resolution

Optical imaging utilizes the spatial variation of the optical properties (transmittance, reflectance, scattering, absorption, fluorescence) of a tissue sample (biopsy for example) to obtain information on the sample. Lasers are commonly used as the light source. Imaging to the scale of 100 nm has been accomplished in in vitro, in vivo, and ex vivo specimens. Imaging methods include optical and electron microscopy, fluorescence microscopy, multiphoton microscopy (two-photon laser scanning microscopy), optical coherence tomography, total internal reflection fluorescence microscopy, near field optical microscopy, spectral and time resolved imaging, fluorescence resonance energy transfer, fluorescence lifetime imaging microscopy, and nonlinear optical imaging.

The future of optical imaging lies in multifunctional imaging, high resolution imaging using two wavefronts, combining various types of microscopy, and further miniaturization of microscopes. Ultrasonic imaging techniques are also being developed (these are not optical, but use optical materials).

9.6   Advanced Biophotonics Applications

We complete this chapter with a brief review of microarray technology for genomics and proteomics, light activated therapy, tissue engineering with light, laser tweezers and scissors, and bionanophotonics.

Microarrays are used for rapid analysis of large numbers of DNA, protein, cellular, and tissue samples. Optical methods are used for detection and readout of microarrays. Analytical processes involve immobilization of capturing biorecognition elements, scanning and readout of large amounts of micro-size samples. While the most widely developed microarray devices analyze DNA, proteins, cells and tissues, the best developed are those that evaluate DNA. These are commonly called "a laboratory on a chip" and "biochips". Using these microarrays, thousands of DNA and RNA species can be analyzed simultaneously with impressive precision and sensitivity to assess cellular phenotyping and genotyping. This technology has significantly accelerated drug development and disease research. The microarray used a micropatterned array of biosensors. Figure 9.36 shows the analysis scheme and Figure 9.37 shows some types of biological microarrays. Much of this scheme should look familiar as discussed in the previous section. New additions are patterning of microarrays and scanning the optical signal readouts. These arrays are useless without high rate scanning and high throughput; it would be just another large set of samples.

The DNA microarray is a good example of this device. This device consists of an array that uses 5 μm to 150 μm sized spots supported on a glass substrate, and is used to identify DNA sequences of genes by fluorescence imaging. Fragments of single-stranded DNA ranging from 20 to over 1000 bases are attached to these spots. These fragments are used to immobilize DNA fragments to be studied and identified. The array is scanned using a high speed fluorescence detector.

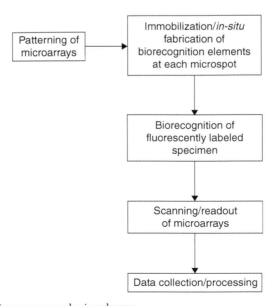

Figure 9.36 Microarray analysis scheme.

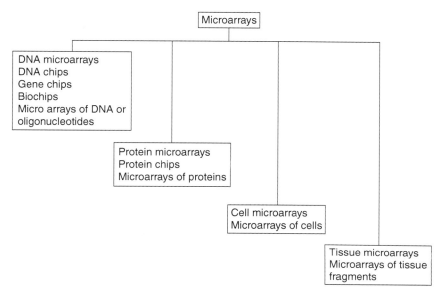

Figure 9.37 Examples of biological microarrays.

The microarray device consists of the microarray slide spotter and the scanner [70]. The spotter is shown in Figure 9.38. The important steps in assembling this device are the printing of the microarray and sample preparation. The steps involved in printing are:

- Coat the glass slide with polylysine
- Robotically print probe materials (c-DNA or oligonucleotides) onto the glass plate (much like an ink jet printer)
- Blocking of remaining exposed amines of polylysine with succinic anhydride
- Denaturing of double stranded DNA to form single stranded DNA, if necessary

The diameter of the DNA spots ranges from 50 μm to 150 μm, which can provide as many as 10,000 spots on a device. The spots are "printed" onto the glass slide by a number of techniques, including noncontact (micro-droplets ejected from a dispenser-similar to ink jet printing) and contact printing, with direct contact between

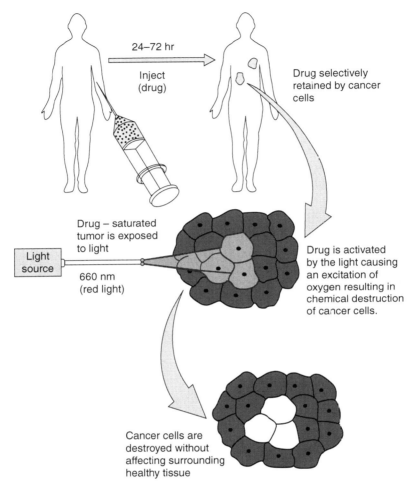

Figure 9.38 Steps in photodynamic therapy (PDT).

the print head and the slide. Samples are prepared by the following steps [70]:

- Isolate RNA and produce multiple copies of the required DNA fragments
- Convert m-RNA to c-DNA by reverse transcription
- Attach labels to the c-DNA with red fluorescent dye and the control material with green dyes
- Hybridize with the microarray probes
- Wash away the unhybridized material

- Optically scan the microarray with a laser scanner or confocal microscope
- Analyze the data

The fluorescence intensities of the DNA and reference are ratioed by computer and the level of gene expression is obtained. Current technology can scan a maximum of 31,000 elements.

This has been just one example of how microarray technology is used. Protein or tissue microarrays are beyond the scope of this section, but we will touch on how this technology is used to molecularly profile tumors and identify the molecular signatures of cancer and other diseases. It provides a method of rapidly screening large amounts of data to track the progression of cancer, in particular breast cancer, diffuse large B-cell lymphoma, leukemia, colon adenocarcinoma, and ovarian cancer. DNA array technology provides a rapid method to genotype which facilitates diagnosis and treatment of disease. Each of these diseases has a signature gene that can be expressed and compared to genes of normal tissue. Once the diseased gene is identified, a treatment that targets this specific type of cancer can be initiated.

Light-activated, or photodynamic therapy, uses light for therapy as a medical procedure of treatment of disease. This type of therapy is a promising treatment for cancers and diseases that utilize activation of external chemical agents (photosensitizers) by light. The photosensitizer is administered intravenously or topically to the affected site and this site is then irradiated with the specific wavelength of light that is absorbed by the photosensitizer. The drug absorbs the light, producing activated oxygen species that destroys the tumor. Figure 9.38 summarizes this type of therapy. Key steps in this process are:

- Introduction of the photosensitizer drug (PTD drug)
- Selective prolonged retention time of the PTD drug in the targeted tissue/tumor
- Delivery of the light (usually supplied by a laser) to the affected region
- Light absorption by the photosensitizer, producing highly reactive oxygen that destroys the cancer
- Removal of the drug to stop the reaction

The success of photodynamic therapy relies on the ability of the PDT drug to have a greater affinity to cancer cells than to normal tissue. Both types of tissue absorb the drug, only the abnormal tissue retains it longer. The cancerous tissue is irradiated after it desorbs from the normal tissue. The mechanism is called singlet oxygen quencher. There are a number of photosensitizers used in PDT, but they all must have the following basic properties:

- Must be able to target specific cancer cells
- Hydrophobicity for good penetration into the cancer cells
- The photosensitizer should absorb strongly in a narrow wavelength band but in a transparency region for normal tissue
- The photosensitizer should not activate in the dark
- The photosensitizer should not aggregate
- There must be a method to rapidly excrete the photosensitizer from the body

Photosensitizers include porphyrin derivatives, chlorines and bacteriochlorins, benzoporphyrin derivatives, 5-aminolaevulinic acid (ALA), texaphyrins, phthalocyanines, and cationic and dentritic photosensitizers.

Applications for PDT include cancer therapy for non-small-cell lung cancer, endobronchial lung tumors, cancer of the esophagus, other lung tumors, skin cancers, breast cancer, brain cancer, colorectal tumors and gynecologic malignancies [70]. This therapy also has applications to cardiovascular diseases, chronic skin disease, rheumatoid arthritis, macular degeneration, wound healing, endometriosis, and precancerous conditions. Figure 9.39 shows fluorescence images of a tumor before and after PDT treatment [77]. Note that the tumor is virtually eliminated.

The primary types of tissue engineering, shown in Figure 9.40, are tissue welding, tissue contouring and reconstruction, and tissue regeneration. Dermatology and ophthalmology primarily employ tissue reconstruction and contouring. Dermatology applications include treatment of vascular malformations, removal of tissue lesions and tattoos, wrinkle removal and tissue sculpting, and hair removal. Ophthalmologic applications of lasers are used extensively in refractory surgery (photorefractive keratectomy), to repair blockages and leaky blood vessels in the retina, in situ keratomileusis, and cataract surgery. Tissue welding applications include direct welding, laser soldering, and dye-enhanced laser soldering.

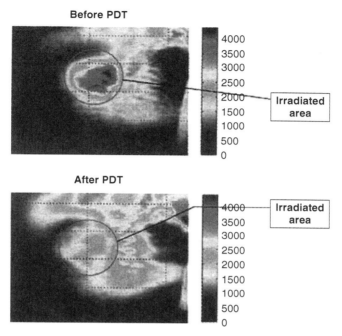

Figure 9.39 Fluorescence images of a tumor before and after PDT treatment [77].

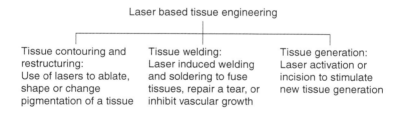

Figure 9.40 Summary of applications of tissue engineering.

Laser treatment has been shown to accelerate tissue regeneration and repair after injuries.

Lasers play a very important role in this biophotonic application. Lasers are now being used in angioplasty to remove plaque and open up blockages in blood vessels. Laser ablation is also being applied to soft tissue extraction and extraction of hard and soft tissues in dentistry.

Figure 9.41 summarizes dermatological applications of lasers, and Figure 9.42 shows ophthalmic applications of lasers. The basic principle involved is that certain wavelengths of light are absorbed

Procedure	Skin resurfacing		Hair removal				Tattoo Removal	
	CO_2 laser	Er:YAG laser	Alexandrite laser	Diode laser	Nd:YAG laser	Ruby laser	Q-switched frequency-doubled Nd:YAG laser	Q-switched alexandrite laser
Commonly used lasers								
Wavelength pulse duration fluence (energy)	10.6 μm 800 μsec 3.5–6.5 J/cm² (0.250–0.4J)	2.94 μm 0.3–10 msec 5–8 J/cm² (1–1.5J)	0.755 μm 2–20 msec 25–40 J/cm²	0.81 μm 0.2–1 sec 23–115 J/cm²	1.064 μm 10–50 msec 90–187 J/cm²	0.694 μm 3 msec 10–60 J/cm²	0.532 μm 10–80 nsec 6–10 J/cm²	0.752 μm 50 nsec 2.5–6 J/cm²
General references and websites	1,4		2 – 4				4 – 6	

Figure 9.41 Dermatology applications of lasers.

Procedure	Laser photocoagulation			Laser thermal keratoplasty (LTK)	Laser-Assisted *In Situ* keratomileusis (LASIK)
Commonly used lasers	Argon ion laser	Krypton ion laser	Laser diode	Ho:YAG laser	ArF excimer laser
Wavelength operation regime (pulse duration)	514.5 nm CW (0.1–1.0sec)	647 nm CW (up to 10sec)	810 nm CW (up to 2sec)	2.1 μm pulse (0.25–1sec)	193 nm pulse (15–25 nsec)
Power (energy)	0.05–0.2 W	0.3–0.5 W	2 W	20 mJ	50–250 mJ
General references and websites	1,2			3–5	6–8

Figure 9.42 Ophthalmic applications of lasers.

preferentially (targets) by damaged tissue, compared to surrounding undamaged tissue. The tissue heats up by a process called photothermolysis and is destroyed. Figure 9.43 shows the removal of a tattoo by laser technology. Laser light causes very rapid heating of the tattoo pigment granules, which kills the cells that contain them. The pigment escapes from the fractured cells.

Lasers are now routinely used to correct vision problems by photorefractive keratectomy and laser in situ keratomileusis. In both these techniques a pulsed laser beam flattens the cornea by selectively removing tissue from the center rather than from the middle. This changes the focal length of the cornea farther back toward the desired spot on the retina and corrects the vision for distance.

Low levels of light have been shown to help heal wounds and repair tissue after injury. In fact, my chiropractor routinely uses a laser to speed healing of damaged tissue and reduce inflammation. He has an interesting portfolio of actual cases of wounds and skin disease being healed by laser therapy. The skeptics are always there, but this appears to be a viable process to aid healing. The tissue regeneration process can be summarized with the following steps:

- Healing begins with blood formation
- Blood clot formation leads to formation of scar tissue, which retards healing of normal tissue
- There is very little blood clot formation with laser ablation
- The absence of blood clots accelerates the regeneration of native tissue

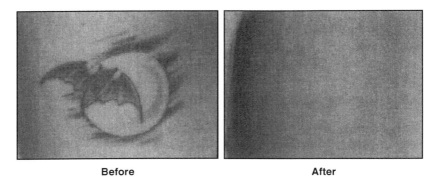

Before   **After**

Figure 9.43 Removal of a tattoo by laser therapy.

Thus, laser treatment produces virtually no scar tissue that inhibits normal healing. There is strong evidence that recovery after laser surgery is much faster than after conventional (scalpels, etc) surgery. Figure 9.43 shows muscle regeneration of muscle tissue after laser surgery compared to scalpel incision. Note the formation of normal, non-fibrous tissue with the laser incision.

New applications in this technology are emerging at a rapid pace, and the use of light to enhance and simulate biological mechanisms and functions is improving medical technology and patient care.

References

1. P. Hajkova *et al.*, *Proceedings of the SVC 50th Technical Conference* (2007) 117.
2. Sung Hwan Lee, Photocalalytic Nanocomposites Based on TiO_2 and Carbon Nanotubes, Dissertation, University of Florida (2004).
3. K. E. O'Shea *et al*, *J of Photochem and Photobio A: Chemistry*, 107 (1997) 221.
4. A. V. Vorontsov *et al*, *Applied Catalysis B: Environmental*, 32 (2001) 11.
5. A. V. Vorontsov *et al*, *Environ Sci Technol*, 36(2002) 5261.
6. Y -C Chen *et al*, *Photochem Photobio Sci*, 2 (2003) 694.
7. I. Martyanov, K. J. Klabunde, *Environ Sci Technol*, 37 (2003) 3448.
8. D. Panayotov *et al.*, P. Kondratyuk, J. T. Yates, Jr., *Langmuir*, 20 (2004) 3684.
9. Sung Hwan Lee, Photocalalytic Nanocomposties Based on TiO_2 and Carbon Nanotubes, Dissertation, University of Florida (2004).
10. P. C. Maness *et al.*, S. Smolinski, *Appl and Environ Microbiol*, 65 (1999) 4094.
11. T. Matsunaga *et al.*, *FEMS Microbiol Lett*, 29 (1985) 211.
12. T. Matsunaga *et al.*, *Appl and Environ Microbiol*, 54 (1988) 1330.
13. R. J. Watts *et al.*, *Water Research*, 29 (1995) 95.
14. W. A. Jacoby *et al.*, *Environ Sci Technol*, 32 (1998) 2650.
15. P. C. Maness, *et al. Appl Environ Microbiol*, 65 (1999) 4094.
16. *Artificial Photosynthesis*, Anthony F. Collings and Christa Critchley, ed., Willey-VCH (2005).
17. Peter Wurfel, *Physics of Solar Cells*, 2nd Ed., Willey-VCH (2009).
18. *MIT Tech Talk*, 49 (2) (2004) 1.
19. Melih K. Sener *et al.*, *J Phy Chem B*, 106 (2002) 7948-7960.
20. Jeffrey P. Youngblood and Nancy R. Sottos, *MRS Bulletin*, 33 (2008) 732.
21. C. W. Extrand, *Langmuir*, 22 (2006) 1711.
22. M. Ma and R. M. Hill, *Curr Opin Colloid Interface Sci*, 11 (2006) 193.

23. W. Barthlott and C. Neihhuis, *Planta*, 202 (2005) 1.
24. R. Furstner *et al*, *Langmuir*, 21 (2005) 956.
25. C. Neihuis and W. Barthlott, *Ann Bot*, 79 (1997) 667.
26. *A. Lafuma and D. Quere, Nature Materials, 2 (2003). 457.*
27. P. Forbes, London: *Fourth Estate* (2005) 272.
28. P. Forbes, *Scientific American*, 299 (2) (2008) 67.
29. J. E. Greene, Thin Film Nucleation, Growth and Microstructural Evolution: An Atomic Scale View, in *Handbook of Deposition Processes for Films and Coatings*, 3rd Ed, P M Martin, Ed. (2009).
30. Rafael Tadmor, *Langmuir*, 20 (2004) 7659.
31. A. B. D. Cassie, *Trans Faraday Soc*, 75 (1952) 5041.
32. R. N. Wenzel, *Ind Eng Chem*, 28 (1936) 988.
33. R. E. Johnson and R. H. Dettre, Surface and Colloids, E Matijevic, Ed., Wiley Interscience (1969).
34. C. W. Extrand and Y. Kumagai, *J Colloid Interface Sci*, 170 (1995) 515.
35. R. M. Jisr *et al.*, *Angew Chem Int* Ed, 44 (2005) 782.
36. G. Zhang *et al.*, *Langmuir*, 21 (2005) 4713.
37. S. Agarwal *et al.*, *Macromol Mater Eng*, 291 (2006) 592.
38. Jeffrey P. Youngblood and Nancy R. Sottos, *MRS Bulletin*, 33 (2008) 732.
39. A. J. Singer and R. A. F. Clark, *N Engl J Med*, 341 (1999) 738.
40. M. Q. Zang, *Polymer Letters*, 2 (2008) 238.
41. M. Chipara and K. Wooley, *Mater Res Soc Symp* Vol. 851, NN4.3.1.
42. S. R. White *et al.*, *Nature*, 409 (2001) 794.
43. S. R. White *et al*, *Chem Rev*, 109 (2009) 5755.
44. F. R. Jones *et al.*, *J Royal Soc*, 4 (2007) 381.
45. S. D. Bergman and F. Wudl, *J Matls Chem* 18 (2008) 41.
46. M. Q. Zang, *Polymer Letters*, 2: (2008) 238.
47. D. A. Schiraldi *et al.*, *Macromolecules*, 32 (1999) 5786.
48. F. Wudl, *Science*, 295 (5560) (2002) 1698.
49. T. Saegusa *et al*, *Macromolecules*, 26 (1997) 883.
50. K. Matyjaszewski, *Cationic Polymerizations: Mechanisms, Synthesis and Applications*, M Decker (1996).
51. M. Szwarc, *Nature*, 176 (1956) 1168.
52. G. W. Coates *et al*, *J Polym Sci A: Polym Chem*, 40 (2002) 2736.
53. Graeme Moad and David H. Solomon, *The Chemistry of Radical Polymerization*–2nd Ed., Elsevier (2006).
54. Moshe Levy, "Living Polymers" – 50 Years of Evolution: weizmann. ac.il/ICS/booklet/18/pdf/levy.
55. Francesco Ruffino *et al.*, *Toward Functional Nanomaterials*, Zhiming Wang, ed., Springer (2009).
56. T. C. Harman *et al*, *Electron. Mater.* Vol. 29, pp. L1-L4.
57. T. C. Harman *et al*, *Science*, Vol. 297, pp. 2229–2232.
58. Ted Kamins, "Self Assembled Nanostructures: Ge on Si", Applied Materials Epitaxy Symposium, Santa Clara, CA, September 19, 2002.

59. P. M. Martin, *Vacuum Technology & Coating*, February 2007 8.
60. P. M. Martin, *Vacuum Technology & Coating*, February 2006, 6.
61. A. Kirakosian *et al.*, *J. Appl. Phys.*, 90, No. 7 (2001) 3286.
62. Lawrence Kazmerski, "Solar Photovoltaics Technology: The Beginning of the Revolution", Presented at the 51st Annual SVC Technical Conference, Chicago, IL (2008).
63. J. Viernow *et al*, *Appl. Phys. Lett.*, 74, No. 15 (1999) 2125.
64. Adam Li *et al.*, *Phys. Rev. Lett.*, 85. No. 85 (2000) 5380.
65. P. M. Martin, Ed., *Handbook of Deposition Technologies for Films and Coatings*, 3rd Ed., Elsevier (2009).
66. Y. Chen, *Appl. Phys. Lett.* 76 (2000 4004.
67. P. M. Martin, SVC Fall 2008 Bulletin.
68. M. L. Lee and R. Venkatasubramanian, *Appl. Phys. Lett.*, 92 (2008) 053112.
69. G. Chen *et al.*, *International Materials Reviews*, Vol. 48 (2003).
70. Paras N. Prasad, *Introduction to Biophotonics*, Wiley-Interscience, Hoboken (2003).
71. P. M. Martin, B. F. Monzyk, E. C. Burckle, J. R. Busch, R. J. Gilbert, K. A. Dasse, "Thin Films are Helping in the Fight Against Pulmonary Disease", *Vacuum Technology & Coating*, August 2004, 42–49.
72. Pan *et al*, *J. Am. Chem. Soc.*, 124 (2002) 4857–4864.
73. A. Carden and M. D. Morris, *J. Biomed. Opt.*, 5, (2000) 259–268.
74. A. Carden *et al*, *Calcif. Tissue Int.* (2003).
75. R. R. Alfano *et al*, *Laser Life Sci*, 4, (1991) 23–28.
76. W. C. Lin *et al*, *Photochem Photobiol.*, 73 (2001) 396–402.
77. A. Pifferi *et al*, *Photochem. Photobiol.*, 72 (2000) 690–695.

Index

Also of Interest

Forthcoming and published related titles from Scrivener Publishing

Atomic Layer Deposition: Operation and Practical Applications
By **Tommi Kääriäinen, David Cameron, Marja-Leena Kääriäinen,** *Lappeenranta University of Technology,* and **Arthur Sherman**
Forthcoming spring 2012
The first part of the book will detail the fundamentals of ALD including films, materials and oxides. The second part will be devoted to applications and will cover Roll-to-Roll and Continuous ALD, diffusion barriers, coatings, low temperature processing, rigid and flexible substrates, etc.

Introduction to Plasmas for Materials Processing
By **Scott Walton** and **Richard Fernsler,** *Plasma Physics Division, Naval Research Laboratory, Washington, D.C.*
Forthcoming fall 2012
The aim of this work is to instruct engineers and product developers that use plasma-based processing equipment and provide them with a skill set that empowers them to take a more proactive approach by fully understanding the plasma-surface interface.

Handbook of Biomedical Coatings and Deposition Technologies
Edited by **Arpana Bhave,** *Boston Scientific Corporation* and **David A. Glocker,** *Isoflux Incorporated*
Forthcoming fall 2012
The Handbook will serve as the key resource for those wanting to learn how to develop and/or produce functional coatings for biomedical applications. It will be unique by focusing specifically on those coating technologies important in the medical industry.

Atmospheric Pressure Plasma for Surface Modification
By **Rory A. Wolf**, *Enercon Industries*
Forthcoming spring 2012
The book's purpose is to impart an understanding of the practical application of atmospheric plasma for the advancements of a wide range of current and emerging technologies. Specifically, the reader will learn the mechanisms which control and operate atmospheric plasma technologies and how these technologies can be leveraged to develop in-line continuous processing of a wide variety of substrates.

Roll-to-Roll-Vacuum Deposition of Barrier Coatings
By **Charles A. Bishop**, *C.A. Bishop Consulting, Ltd*
Published 2010   ISBN 978-0470-60956-9
This is a practical guide to provide the reader with the basic information to help them understand what is necessary in order to produce a good barrier coated web or to improve the quality of any existing barrier product.